Vanadium in Soils and Plants

Vanadium is an essential element for humans and animals. The toxicity of vanadium at higher concentrations could be a global environmental concern and a significant issue for both environmental protection and economic benefits. The relevance of anthropogenic vanadium in the environment has increased significantly in recent years due to an increased demand for vanadium in high-temperature industrial activities. This book summarizes vanadium's current research and explains its behavior and mobilization in the environment, especially in soils, sediments, water and plants. Through case studies from various countries, it discusses critical limits set and risk assessment approaches and remediation approaches of vanadium-contaminated soils.

FEATURES

- Provides a comprehensive overview of vanadium in the total environment
- Covers the role of vanadium in various environments such as soils, sediments, water and plants
- Includes bioavailability studies and further case studies from various countries around the world
- Focuses on a better understanding of biogeochemical processes of vanadium
- Is written by international experts who present the current stage of the knowledge including innovative remediation and management approaches of vanadium-contaminated sites

This book will be of use to upper-level undergraduate and graduate students taking courses in soil science, environmental science, soil ecology, water science, plant science, ecotoxicology, geology and geography as well as scientists, lecturers, environmental and technical engineers, ecologists, applied ecological scientists and managers.

Advances in Trace Elements in the Environment

Series Editor: H. Magdi Selim, Louisiana State University, Baton Rouge, USA

The nature of trace elements in the environment depends largely on their interactions in the soil, water and air. Understanding the fate and transport of trace elements in the environment is essential for humans and animals, plants, water bodies, and soils. This series provides timely information on the biological, chemical, and physical processes governing the interactions of trace elements in the environment. Such knowledge is needed for minimizing the negative impact of trace elements on soil and water quality, plants, and human health. Books from this series are of value to industrial and mining engineers, government regulators, scientists, and consultants in the environmental arena.

Dynamics and Bioavailability of Heavy Metals in the Rootzone
Edited by H. Magdi Selim

Trace Elements in Waterlogged Soils and Sediments
Edited by Jörg Rinklebe, Anna Sophia Knox, and Michael Paller

Competitive Sorption and Transport of Heavy Metals in Soils and Geological Media
Edited by H. Magdi Selim

Nickel in Soils and Plants
Edited by Christos Tsadilas, Jörg Rinklebe, and Magdi Selim

Phosphate in Soils
Interaction with Micronutrients, Radionuclides and Heavy Metals
Edited by H. Magdi Selim

Permeable Reactive Barrier
Sustainable Groundwater Remediation
Edited by Ravi Naidu and Volker Birke

Nitrate Handbook
Environmental, Agricultural, and Health Effects
Edited by Christos Tsadilas

Vanadium in Soils and Plants
Edited by Jörg Rinklebe

For more information on this series, please visit: www.routledge.com/Advances-in-Trace-Elements-in-the-Environment/book-series/CRCADVTRAELE

Vanadium in Soils and Plants

Edited by

Jörg Rinklebe

CRC Press is an imprint of the
Taylor & Francis Group, an **informa** business

First edition published 2023
by CRC Press
6000 Broken Sound Parkway NW, Suite 300, Boca Raton, FL 33487–2742

and by CRC Press
4 Park Square, Milton Park, Abingdon, Oxon, OX14 4RN

CRC Press is an imprint of Taylor & Francis Group, LLC

Library of Congress Cataloging-in-Publication Data
Names: Rinklebe, Jörg, editor.
Title: Vanadium in soils and plants / edited by Jörg Rinklebe.
Description: First edition. | Boca Raton, FL : CRC Press, 2022. | Series: Advances in trace elements in the environment | Includes bibliographical references and index.
Identifiers: LCCN 2022009269 (print) | LCCN 2022009270 (ebook) | ISBN 9781032002293 (hbk) | ISBN 9781032002385 (pbk) | ISBN 9781003173274 (ebk)
Subjects: LCSH: Vanadium—Environmental aspects.
Classification: LCC TD196.V35 V36 2022 (print) | LCC TD196.V35 (ebook) | DDC 628.5/2—dc23/eng/20220225
LC record available at https://lccn.loc.gov/2022009269
LC ebook record available at https://lccn.loc.gov/2022009270

ISBN: 978-1-032-00229-3 (hbk)
ISBN: 978-1-032-00238-5 (pbk)
ISBN: 978-1-003-17327-4 (ebk)

DOI: 10.1201/9781003173274

Typeset in Palatino LT Pro
by Apex CoVantage, LLC

Contents

Editor vii

1 **Vanadium: Chemical Properties, Industrial Use, Global Soil Distribution, and Factors Governing the Total Content and Mobilization in Soils** 1
Sabry M. Shaheen, Ahmed Mosa and Jörg Rinklebe

2 **Critical Limits and Health Risk Assessment of Vanadium in Soils of Various Countries of the World** 33
Vasileios Antoniadis, Sabry M. Shaheen, Efi Levizou and Jörg Rinklebe

3 **Kinetics of Vanadium Sorption/Desorption in Soils** 49
Tamer A. Elbana, Wenguang Sun, Joshua Padilla and H. Magdi Selim

4 **Geochemical Fractionation and Availability of Vanadium in Soils** 73
Cho-Yin Wu, Maki Asano and Zeng-Yei Hseu

5 **Redox Chemistry of Vanadium in Soils and Sediments: Biogeochemical Factors Governing the Redox-Induced Mobilization of Vanadium in Soils** 95
Jörg Rinklebe, Vasileios Antoniadis and Sabry M. Shaheen

6 **Vanadium Speciation in Soil Aqueous and Solid Phases** 113
Worachart Wisawapipat, Yohey Hashimoto and Shan-Li Wang

7 **Remediation of Vanadium in the Soil Environment** 135
Sandun Sandanayake, S. Keerthanan, Ahamed Ashiq, Arifin Sandhi and Meththika Vithanage

8 **Potential of Biochar to Immobilize Vanadium in Contaminated Soils** 157
Ali El-Naggar, Ahmed Mosa, Avanthi D. Igalavithana, Xiao Yang, Ahmed H. El-Naggar, Sabry Shaheen, Scott X. Chang and Jörg Rinklebe

9 **Plant Uptake and Ecotoxicity of Vanadium: The Role of Soil Chemistry** 187
Jon Petter Gustafsson

10 **Vanadium in Plants: Present Scenario and Future Prospects** 205
Sheikh Mansoor, Tawseef Rehman Baba, Syed Inam ul Haq, Iqra F. Khan, Sofora Jan, Sadaf Rafiq, Jörg Rinklebe and Parvaiz Ahmad

11 **Microbial Community Responses and Bioremediation** 223
Baogang Zhang, Han Zhang and Jinxi He

12 **Vanadium in Technical Applications and Pharmaceutical Issues** 269
Dieter Rehder

Index 285

Editor

Jörg Rinklebe, PhD, is Professor for Soil- and Groundwater-Management at the University of Wuppertal, Germany. He is internationally recognized for his research in the area of biogeochemistry of trace elements in wetland soils. Dr. Rinklebe has published over 370 research papers, and he is a Globally Highly Cited Researcher. Also, he published three books, *Trace Elements in Waterlogged Soils and Sediments* (2016), *Nickel in Soils and Plants* (2018) and *Soil and Groundwater Remediation Technologies* (2020), as well as numerous book chapters. He is Editor in Chief of the international journal Environmental Pollution, *Critical Reviews in Environmental Science and Technology* (CREST) and Chief Editor for Special Issues in *Journal of Hazardous Materials*, and a guest editor of the international journals *Environment International, Chemical Engineering Journal, Science of the Total Environment, Chemosphere, Journal of Environmental Management, Applied Geochemistry* and *Environmental Geochemistry and Health*. Also, he is a member on several editorial boards (*Ecotoxicology; Geoderma; Water, Air, & Soil Pollution; Archive of Agronomy and Soil Science*) and reviewer for many leading international journals. He has organized many special symposia at various international conferences such as Biogeochemistry of Trace Elements (ICOBTE) and the International Conference on Heavy Metals in the Environment (ICHMET). He is Visiting Professor at Sejong University, Seoul, South Korea, and Guest Professor at China Jiliang University, Hangzhou, Zhejiang, China. Recently, Dr. Rinklebe was elected as President of the International Society of Trace Element Biogeochemistry (ISTEB).

1

Vanadium: Chemical Properties, Industrial Use, Global Soil Distribution, and Factors Governing the Total Content and Mobilization in Soils

Sabry M. Shaheen[1,2,3], **Ahmed Mosa**[4] **and Jörg Rinklebe**[1]

[1] *University of Wuppertal, School of Architecture and Civil Engineering, Institute of Foundation Engineering, Water- and Waste-Management, Laboratory of Soil- and Groundwater-Management, Wuppertal, Germany*

[2] *King Abdulaziz University, Faculty of Meteorology, Environment, and Arid Land Agriculture, Department of Arid Land Agriculture, Jeddah, Saudi Arabia*

[3] *University of Kafrelsheikh, Faculty of Agriculture, Department of Soil and Water Sciences, Kafr El-Sheikh, Egypt*

[4] *Mansoura University, Faculty of Agriculture, Department of Soil Sciences, Mansoura, Egypt*

CONTENTS

1.1 Introduction 2
1.2 Chemical Properties of Vanadium 3
1.3 Industrial Use of Vanadium 3
1.4 Impact of Vanadium on Human Health 5
1.5 Sources of Vanadium in the Environment 6
1.6 Global Content of Vanadium in Soils 8
    1.6.1 Total Vanadium Content 8
    1.6.2 Soil Vanadium Fate, Mobility and Availability 14
    1.6.3 Factors Governing V Content and Availability in Soils 18
        1.6.3.1 Iron/Aluminum/Manganese (Hydr)Oxides 18
        1.6.3.2 Soil Organic Matter 18
        1.6.3.3 Clay Minerals 19
        1.6.3.4 Calcium Carbonate 20
        1.6.3.5 Soil Redox Potential (E_H) 20
        1.6.3.6 Soil pH 21
        1.6.3.7 Ionic Strength 22
1.7 Conclusions 22
References 23

DOI: 10.1201/9781003173274-1

1.1 Introduction

Vanadium (V) was first discovered in 1801 and subsequently rediscovered by Nils Sefstrom in 1831 (Cintas, 2004). Vanadium occurs naturally in mineral forms; however, it was not synthesized till 1925, when J.W. Marden and M.N. Rich produced a V metal with a high purity (99.7%) through reducing V_2O_5 by calcium metal (Moskalyk & Alfantazi, 2003). Vanadium lies in Group 5B of the periodic table, and naturally occurring V comprises two isotopes: (i) the stable isotope ^{51}V (99.76%) and (ii) the weakly radioactive isotope ^{50}V (0.24%) (Gustafsson, 2019). It is resistant to acids and bases and oxidizes spontaneously at relatively high temperature (660°C). Vanadium exists in different oxidation states (from −3 to +5); however, V_2O_5 is the most common form, and the V^{4+} oxidation state persist under the presence of reducing agents (Altaf et al., 2021). It ranks among the top five transitional elements, with average concentrations of 150 mg kg^{-1} in soil, 1.8 µg L^{-1} in oceans and 1,000 ng m^3 in the atmosphere of urban areas (Aihemaiti et al., 2020).

It is well established that V is a bioactive element for human despite its uncertain essentiality. The ability to constrain protein tyrosine phosphatases might explain V's roles in a similar way to those with insulin, which promote V-based therapeutic treatments of diabetes (Driskell & Wolinsky, 2002; Panchal et al., 2017; Treviño et al., 2019). Clarkson et al.'s study (Clarkson & Rawson, 1999) also showed that V could enhance the efficacy of insulin and motivate amino acids uptake by cells. In addition, V encourages glycogen synthesis, glucose oxidation and transport into adipocytes and muscle cells (Goc, 2006). Therefore, it is suggested that V might have a key-role for athletes. According to (Toro-Román et al., 2021), physical training might enhance V concentration in serum, especially aerobic sports modalities. However, the progressive increase of global urbanization and industrialization has increased the potential exposure to high V levels.

At present, there are several V products with multiple industrial applications including V_2O_5, V_2O_3, VS_2, VN and ferrovanadium alloy (Wang, Xiang et al., 2020). Vanadium alloy is called "vitamin" in steel industries, and the international market prices of Fe-V have rallied to about $25.5/kg since producers are operating at full capacity with a progressive level of demand. Metallurgy, fossil fuel combustion, smelting, batteries, fertilizer/pesticide industries and crude oil spills are the main anthropogenic activities for V release into the ecosphere (Cao et al., 2018; Gao et al., 2021; Moskalyk & Alfantazi, 2003). The extensive mining activities caused severe contamination of V with high levels in several countries, such as Panzhihua city, China (Zhang et al., 2020), Taltal city, Chile (Reyes et al., 2020) and the North West Province of South Africa (Panichev et al., 2006). In natural seawater, V concentration lies between 34–45 nM with varied values: Baltic Sea (0.15 µg L^{-1}), East Pacific (1.35 µg L^{-1}), Peru Margin (1.8 µg L^{-1}), and the English Channel (0.75 µg L^{-1}) (Auger et al., 1999; Bauer et al., 2017; Ho et al., 2018). Accordingly,

V strongly bioaccumulates in mussels and crabs by about 10^{-5} to 10^{6}-fold greater than the concentration in seawater as a result of its high bioaccumulation factor (Seo et al., 2013). The high bioaccumulation of V causes multiple potential health risks to microorganisms, plants, animals and humans. The high uptake of V by plants causes several deteriorations in seed germination/growth, photosynthesis, nutrients uptake/translocation and oxidative stress (Chen, Liu et al., 2021). It also causes several diseases and disorders in animal health including DNA alterations, inhibition of certain enzymes, breathing problems, neurological disorders, paralyses, malformations in liver and kidneys and dysfunctions in the reproductive system of male animals and the female placenta (Desaulniers et al., 2021; Dyer & De Butte, 2022; Shrivastava et al., 2012).

The behavior of V in soil is governed by vanadate (V) sorption onto Fe/Al(hydr)oxides and vanadyl(IV) binding with natural organic matter (Gustafsson, 2019). In this chapter we present and discuss the chemical properties of vanadium and its industrial use and impacts on the environment and human health. In detail, we present and discuss the total distribution of vanadium and available content in different soils worldwide and the factors governing the total content and mobilization in soils.

1.2 Chemical Properties of Vanadium

Vanadium is a chemical element with atomic number 23. It can exist in different oxidation states, that is, −1, 0, +2, +3, +4 and +5; however, only the +3, +4 and +5 oxidation states are important in the natural environment (Shaheen et al., 2019). More details about the chemical and physical properties are included in Table 1.1.

1.3 Industrial Use of Vanadium

The "technogenic history" of V started in 1801; since then, V has become indispensable to modern industry during the past two centuries and has been widely used in several high-tech industries (Monakhov et al., 2004). As such, it is involved as a principle component in several sophisticated industries including electronics, space navigation, dyeing and nuclear industries as well as in multiple high-temperatures industrial processes (e.g., steel and iron refining) (Shaheen et al., 2019). A substantial amount of V industrial applications (about 80%) is used as ferrovanadium or as an additive in steel industries. Industrial axles, car gears, armor plates, crankshafts, springs,

TABLE 1.1
Vanadium Chemical and Physical Properties

Atomic number	23
Chemical Abstracts Service (CAS) number	7440-62-2
Atomic mass	50.94 $gmol^{-1}$
Electronegativity according to Pauling	1.6
Electrochemical equivalent	0.380 g/A/h
Density	6.1 $g.cm^{-3}$ at 20 °C
Melting point	1910 °C
Boiling point	3407 °C
Thermal neutron cross section	4.7 barns/atom
Electrode potential	−1.50 V
Van der Waals radius	0.134 nm
Ionic radius	0.074 nm (+3); 0.059 (+5)
Electro negativity	1.63
X-ray absorption edge	2.269 Å
Isotopes	5
Electronic shell	[Ar] $3d^3$ $4s^2$
Energy of first ionization	649.1 $kJ.mol^{-1}$
Energy of second ionization	1414 $kJ.mol^{-1}$
Energy of third ionization	2830 $kJ.mol^{-1}$
Energy of fourth ionization	4652 $kJ.mol^{-1}$
Discovered by	Nils Sefstrom in 1830

piston rods and cutting tools are made from V-steel alloys to produce these extremely tough tools (Monakhov et al., 2004). Vanadium is also added with titanium to create innovative alloys with outstanding strength-to-weight ratios, with several applications in aerospace industries (Yee et al., 2021). The quality of these alloys affect the mechanical properties of airplanes (jet engines, axles, crankshafts, high speed air-frames, steel alloys and other critical components) (Zhang et al., 2015). Additionally, V alloys are recognized as an attractive candidate in the structural components of nuclear reactors, given their low-neutron-absorbing properties, high thermal stress factors and radiation resistance (Nagasaka & Muroga, 2020).

In addition, vanadium oxide (V_2O_5) has been exploited as a catalyst in several industries such as sulfuric acid production (Romanovskaia et al., 2021), CO_2 methanation (Świrk et al., 2021), formaldehyde production (Laitinen et al., 2020) and the ceramic industry (Fan et al., 2021). Vanadium also has several applications in the optical domain. It is used as glass colorant to produce green or blue tint. Glass coating with vanadium dioxide (VO_2) can help in blocking infrared radiation at specific temperature (Monakhov et al., 2004). There are also several applications of V in pharmaceutical industries. Due to the similarity between vanadate and phosphate, V-based therapeutics have been exploited in several signaling pathways (Pessoa et al., 2015). Vanadium-based therapeutics are documented for their antidiabetic effects on both glucose and lipid metabolism (Jakusch & Kiss, 2017; Patel et al., 2019). Recently, anticancer properties of V-based therapeutics have been investigated, given their promising antitumor effects (Nunes et al., 2021). Interestingly, V compounds have been introduced as potential therapeutic agents for COVID-19 (Semiz, 2022).

1.4 Impact of Vanadium on Human Health

The human health risks of V compounds have received big attention during the past decade due to their widespread industrial applications (Costa Pessoa, 2015; Rehder, 2012). Vanadium toxicity varies greatly according to the nature of the V-bearing components and its oxidation state (Crans et al., 1998). The toxicity of V(+5) is greatly higher than V(+4), given the inhibitory effect of V(+5) on specific enzymes activities and the $^{+}K^{+}$-ATPase of plasma membranes (Patterson et al., 1986). Vanadium uptake by human can occur through consumption of V hyper-accumulators such *Zea mays* L., *Phaseolus vulgaris* L., and *Setaria viridis* (Chen, Liu et al., 2021). Published data also points to the potential toxicity of V via ingestion of V-containing foods and inhalation routes (Malandrino et al., 2016; Ścibior et al., 2021). Vanadium V(+5) can pass through the placental barrier to accumulate mainly in fetal membranes and fetal bones (Paternain et al., 1990). Consequently, V(+5) is categorized as a dangerous pollutant alongside with As, Hg and Pb and classified by the United Nations Environment Programme and the USEPA in the top list of toxic elements (TEs) (USEPA, 1995; ATSDR, 2012).

Pre-clinical studies also predict that V might be a main cause of anemia through its adverse effects on Fe-related proteins and Fe homeostasis (Ścibior, Hus et al., 2020). High exposure to V can cause several deteriorative impacts on human health. It has a high potential to induce toxic effects on the kidney and liver as well as neurotoxicity as secondary targets (Desaulniers et al., 2021). The International Agency for Research on Cancer categorized V_2O_5 as a possible human carcinogenic compound (International Agency for Research on the Evaluation of Carcinogenic Risks to Humans, 2006). In (Desaulniers et al., 2021), the carcinogenic effect of $NaVO_3$ might reside at low levels via the stimulation of proliferation of tumorigenic cells. In V_2O_5 factories, median values of serum V concentration of exposed and non-exposed workers were respectively 2.2 and 0.3 μg L^{-1} (Ehrlich et al., 2008). Additionally, V_2O_5-exposed workers showed oxidation of DNA bases, damage in DNA, reduction in DNA repair, and generation of micronuclei in leucocytes. Another cross-sectional study on 463 V-exposed workers and other 251 non-exposed workers showed neurobehavioral toxicity hazards associated with high doses of V exposure including neurobehavioral alteration, increase in anger-hostility and fatigue-inertia on the profile of mood states (Li et al., 2013). Vanadium emissions and vanadium-borne dusts could also cause lung diseases in humans. A comprehensive cytotoxicity and genotoxicity investigation was studied on lung cell lines A549 and BEAS-2B exposed to VO_2 nanoparticles (Xi et al., 2021). Severe damage to DNA and micronuclei was observed in A549 cells after long-term exposure to VO_2 nanoparticles, given the high production of reactive oxygen species (ROS). This high oxidative stress further causes genotoxicity, retardation in cell proliferation and cellular viability loss. Vanadium exposure via inhalation can also cause other chest diseases including bronchitis, cough, dyspnea, bronchial asthma, conjunctivitis, and pneumonia (Ścibior, Pietrzyk et al., 2020). The association of hypertension risk with V exposure was also reported. Based on a study on 1,867 participants comprising different levels in a general Chinese population, high V exposure was correlated with increased blood pressure levels and hypertension prevalence (Jiang et al., 2021). Prenatal V exposure was reported to be associated with adverse birth outcomes and fetal growth restrictions (Hu et al., 2018). Results of urinary V concentration from 1,873 Chinese mother-infant pairs showed a sex difference regarding children's anthropometric parameters, since boys were more negatively affected by high V exposure compared to girls (Li et al., 2021).

1.5 Sources of Vanadium in the Environment

Vanadium occupies the 17th place in the list of most common chemical elements, with a content of about 0.019% in the Earth's crust and 1.5 μg L^{-1} in marine resources (sea and ocean waters) (Lisichkin, 1998; Monakhov et al., 2004). In addition, V deposition's is naturally emitted into the atmosphere, with an annual release of about 8.4 tons from continental dust, volcanic debris and marine aerosol (Nriagu, 1990). Anthropogenic sources are also responsible for adding considerable amounts of V to the atmosphere (2.3×10^8 kg annually), in which most of the releases are deposited in soil (Hope, 1997). As illustrated in the global biogeochemical cycle of V (Fig. 1), there are four major V resources, specifically: (i) natural rocks weathering, (ii) mining of V-Ti magnetite ores, (iii) burning of fossil fuels and (iv) release from industrial activities (energy-intensive industries in particular) (Huang et al., 2021; Schlesinger et al., 2017; Gustafsson, 2019). Vanadium exists in more than 80 minerals in bedrock, which could be divided as: (i) sulfides (e.g., patronite), (ii) sulfates (e.g., minisragrite and cheremnykhite), (iii) silicates (e.g., roscoelite, kaolinite and sepiolite), (iv) oxides (e.g., navajoite and montroseite), (v) phosphates (e.g., vanadinite) and (vi) vanadates (e.g., chervetite, tyuyamunite and volborthite) (Boechat et al., 2000; Gehring et al., 1993, 1994). Vanadium

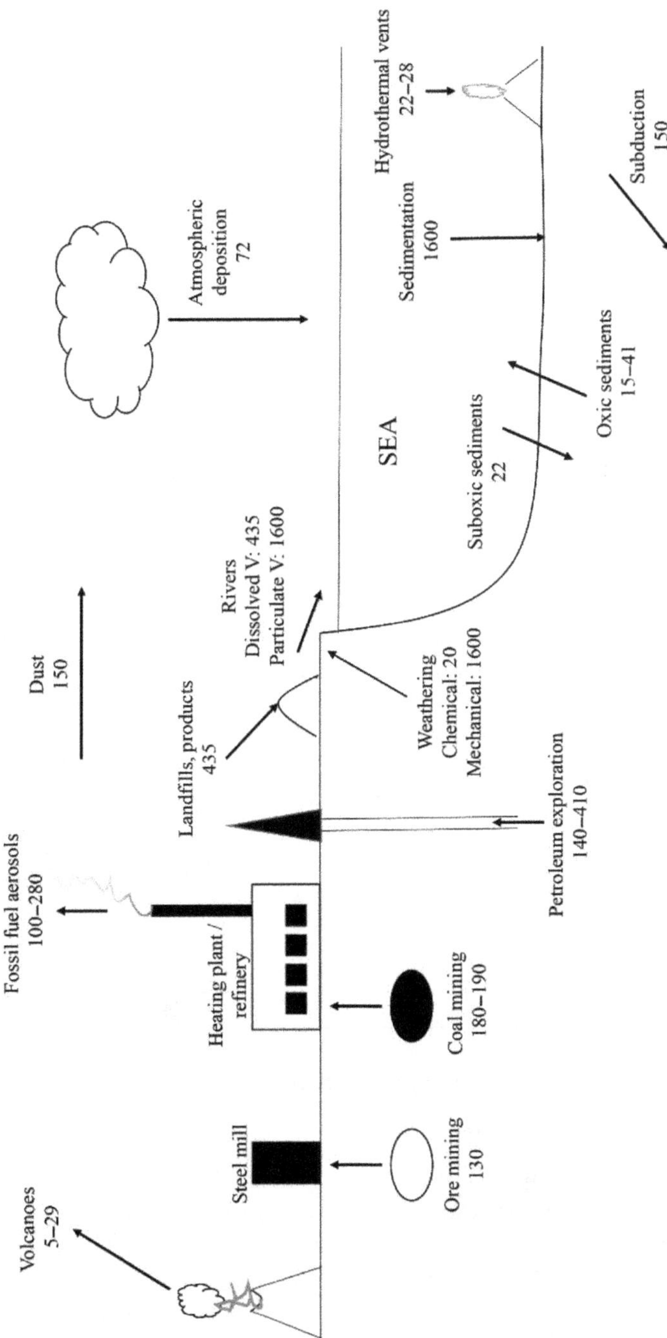

FIGURE 1.1
Global biogeochemical cycle of vanadium (after Schlesinger et al., 2017; Gustafsson, 2019). Annual fluxes are given in Gg V yr^{-1}.
Source: Reproduced from Gustafsson (2019) with permission from the publisher.

also releases into aquatic resources through natural and anthropogenic activities (mining, in particular). In this regard, elevated V levels were recorded in shallow aquifers such as 13.98 mg L^{-1} in Panzhihua, China (Zhong et al., 2015), 2.47 mg L^{-1} in the southeastern pampean region, Argentina (Fiorentino et al., 2007), 9.66 mg L^{-1} in Hubei, China (Meng et al., 2018), and 58.6 mg L^{-1} in Chisman Creek, USA (Jia et al., 2002). The oxidation states of V^{5+} and V^{4+} dominate in natural water bodies; which V^{5+} is more stable in oxidized marine bodies, but V^{4+} persists under reducing conditions. The global production of vanadium ore is about 45,000.0 tons a year, while the metal production is about 7,000 tons per year.

Different anthropogenic activities (e.g., the combustion of fossil fuels, industrial wastes, gas emissions, dust, wastewater, slag, coal and crude oil products) introduce annually about 230 Gg of V to the environment (Hope, 1997; Teng et al., 2011; Qian et al., 2014; Liu et al., 2017). For example, coal fly ash contains high content (ca. 8,000 mg/kg) of vanadium (Ketris & Yudovich, 2009; Liu et al., 2017). Therefore, these activities can be important and major sources for vanadium in soils and water (Teng et al., 2011; Qian et al., 2014).

Sedimentary rocks and the granite layers of the lithosphere, respectively, contain 171 and 623 million tons of vanadium; the hydrosphere contains 2,004 million tons; the pedosphere (soil) contains 9.3 million tons; and plants contain 3.75 million tons (Isidorov, 2001; Monakhov et al., 2004). Vanadium is constantly being redistributed in nature. Over 3 million tons of this metal have been drawn into global migratory flows, including 260,000 tons/yr in the air, 330,000 tons/yr in the ocean, 40,000 tons/yr in river waters in solutions, 2,300,000 tons/yr in river waters in suspensions and 250,000 tons/yr in eolian processes (Isidorov, 2001; Monakhov et al., 2004). As can be seen, the main component of the migration of vanadium in nature is the vanadium carried off into the ocean. More than 2 billion tons of this metal has now accumulated in the world's oceans.

1.6 Global Content of Vanadium in Soils

1.6.1 Total Vanadium Content

Vanadium exists naturally as a secondary element in V-containing minerals, but there is almost no natural mineral containing V as the principal element (Schlesinger et al., 2017). According to the statistics of the US Geological Survey, the global V reservoir is approximately 63 million tons, concentrated mainly in South Africa, China, Australia and Russia (Yuan et al., 2021). In this regard, the total worldwide output of V was about 102,365 tons in the year 2019, with about 75–85% extracted from V-Ti magnetite (Chen, 2019). It has been estimated that around 65×10^3 tons of V annually enter the ecosphere from natural resources (crustal weathering and volcanic emissions), and around 2.3 × 10^8 tons are discharged from several anthropogenic activities (Qian et al., 2014; Yuan et al., 2021).

Unlike other TEs, scant information is available regarding V behavior in soil. Additionally, several countries lack the threshold values of V in soil. In Europe, the median value of total V content in the topsoil is 60 mg kg^{-1}; however, this concentration may reach about 500 mg kg^{-1} in some soils (Larsson, 2014) (Figures 1.2a and 1.3a). Six soil profiles expressing different soil orders in Germany were studied for their total V concentration (Shaheen & Rinklebe, 2018). Results illustrated that V concentration ranged between 20.7 mg kg^{-1} in Tidalic Fluvisol and 133.1 mg kg^{-1} in Haplic Gleysol based on their wide variation in clay, carbonates and Fe/Mn oxides contents. Yet in another work, it was reported that median background values of V concentration in parent materials of soils in Germany ranged from 10–70 mg kg^{-1} (Gäbler et al., 2009). In (Guagliardi et al., 2018), total V concentration in southern Italy oscillated between 184 and 239 mg kg^{-1}, in which the highest values were recorded in peri-urban soils (av. = 211.5 mg kg^{-1}) and the lowest values were recorded in urban soils (av. = 93.5 mg kg^{-1}). In Greek soils, V concentration fluctuated between 73.2 and 155.2 mg kg^{-1}, where Entisol recorded the lowest values and Inciptisol recorded the highest values (Tsadilas & Shaheen, 2010).

The concentration of V in floodplain soils still is rarely investigated in most European countries. In view of this, alluvial soils of the Geul river showed lower V levels (45 mg kg^{-1}) relative to the Leie and Grote Beek rivers (71 and 87 mg kg^{-1}, respectively) (Cappuyns & Slabbinck, 2012). Swiss floodplains also caused elevated levels of V in Calcaric Fluvisol (58 mg kg^{-1}), closely correlated with Fe and Mn release by reductive dissolution of (oxyhydr)oxides (Abgottspon et al., 2015). In Spain, 22 representative surface soil samples were collected from the contact zone between the southern plain of the Llobregat River valley and the northeastern foothills of the Garraf massif (Tume et al., 2006). Values of V content in this study (15.2–144.9 mg kg^{-1}) were higher than the geochemical background baseline values.

Few investigations have studied V status in soils of Africa. However, the scant available information points to a noticeable variation in V contents according to the parent material and physicochemical characteristics of soil (Figures 1.2b and 1.3a). Marine deposits often contain low V concentrations compared to lacustrine deposits (El-Moselhy, 2006). According to (Tsadilas & Shaheen, 2010), the Egyptian lacustrine Entisols recorded high V concentrations (156–202 mg kg^{-1}); however, sandy marine Entisol recorded the lowest values (13–31 mg kg^{-1}). Authors also reported that soils with high clay and free iron oxide contents had the highest V values relative to other sandy ones. In another study, total V concentration in Egyptian Eutric Fluvisol, Haplic Gleysol and Sodic Fluvisols showed values higher than 66 mg kg^{-1}, suggesting their urgent remediation, according to the US Environmental Protection Agency (Shaheen & Rinklebe, 2018). In Ethiopia, wastewater irrigation in the last decades raised the total concentration of V in Vertisol and

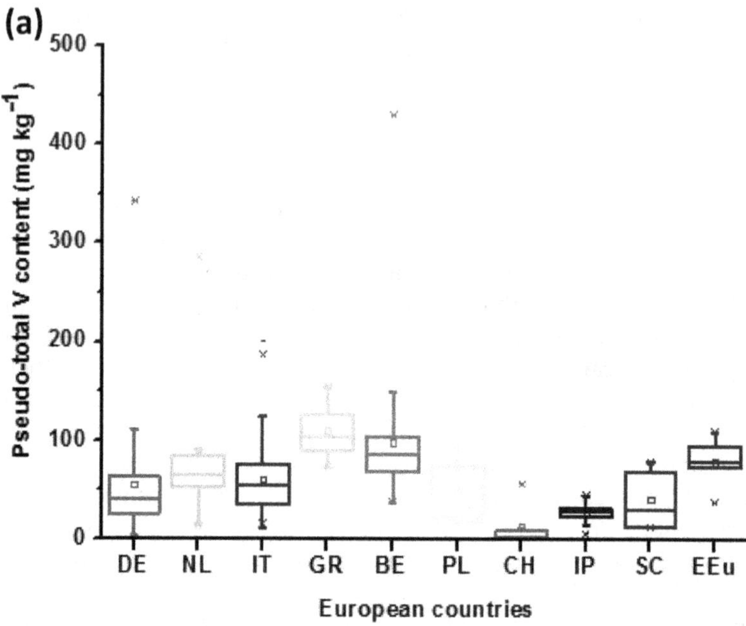

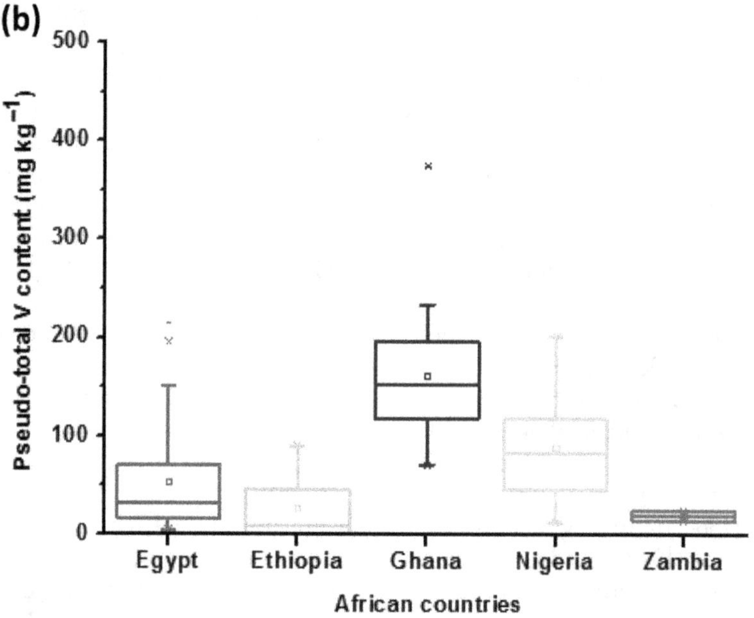

FIGURE 1.2
Total content (mg kg^{-1}) of vanadium in soils of different countries.

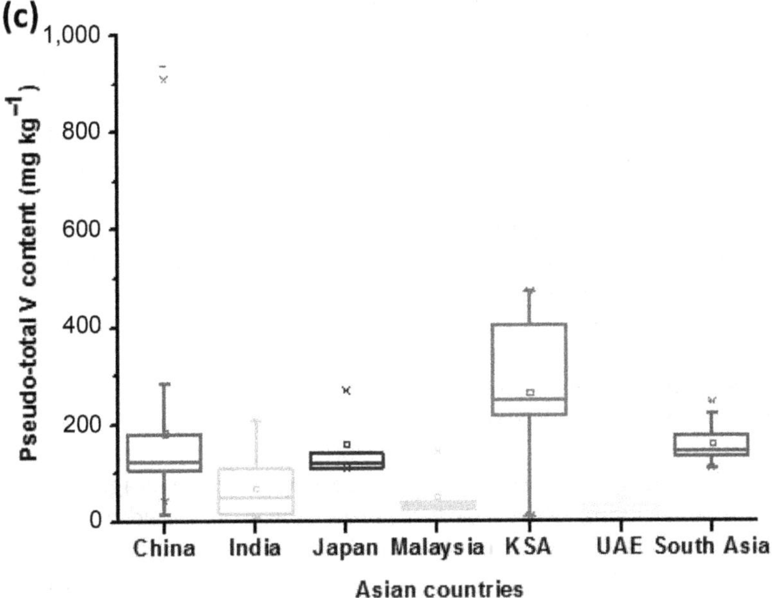
(c)
Pseudo-total V content (mg kg^{-1})
1,000
800
600
400
200
0
China
India
Japan
Malaysia
KSA
UAE
South Asia
Asian countries

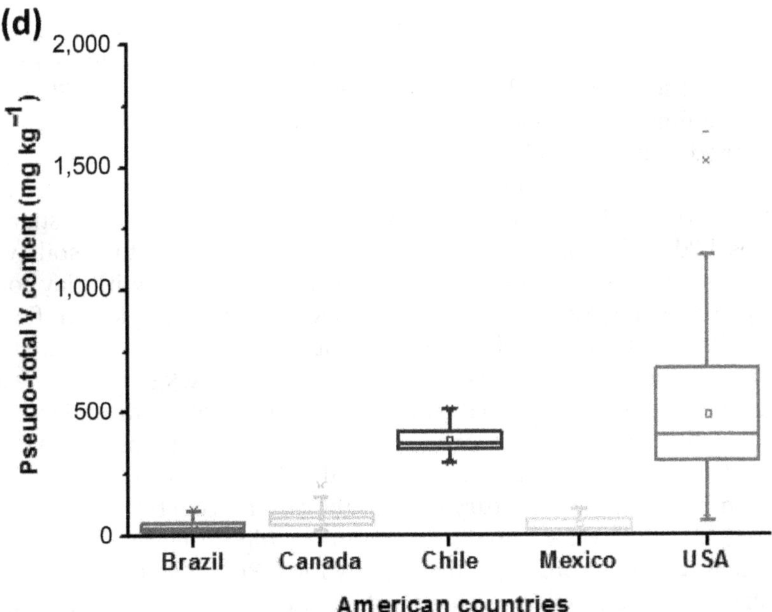
(d)
Pseudo-total V content (mg kg^{-1})
2,000
1,500
1,000
500
0
Brazil
Canada
Chile
Mexico
USA
American countries

FIGURE 1.2 (Continued)

Fluvisol (av. = 53.6 mg kg^{-1}), and a significant portion was bound with calcium carbonate (Fitamo et al., 2007). In yet another study, high values of V were recorded in some soils in Odo-Oba, southwestern Nigeria (22.0–124.0 mg kg^{-1}) (Adagunodo et al., 2018). Likewise, Ibadan, southwestern Nigeria, showed elevated V levels (median values was 84.5 mg kg^{-1}), which ranged between partially contaminated in some areas to heavily contaminated in other ones (Okonkwo et al., 2021).

In Asia, several investigations have highlighted V contamination hazards due to either geogenic and/or anthropogenic discharges (Figures 1.2c and 1.3a). In Hunan province, China, the total V content ranged from 168 to 1,538 mg kg^{-1}, in which the average concentration values in soils of wasteland, slag heap, ore pile and smelting areas were respectively 1,421, 380, 260 and 225 mg kg^{-1} (Xiao et al., 2015). Fly ash derived from coal burning was reported as the main anthropogenic source of the high V levels (100–221 mg kg^{-1}) in the topsoil of Panzhihua, China (Huang et al., 2021). Metallurgy, dyeing, electronics and textile industries caused also elevated levels of V (averaged as 54.4 mg kg^{-1}) in Ganga Plain, Uttar Pradesh, India (Srinivasa Gowd et al., 2010). In yet another study, V was determined in 78 sampling sites throughout Japan (Takeda et al., 2004). Results of this study showed that V content averaged 160 mg kg^{-1} and showed a noticeable reduction when the parent rock type shifted from mafic toward felsic. In Arabian Gulf countries, the high level of V (268 mg kg^{-1} as an average value in Al-Khobar coastline, KSA) was mainly associated with oil spills from exploration/transportation as well as the brine discharge of seawater desalination stations (Alharbi & El-Sorogy, 2017).

Like the European economy, the economy of several American countries (the United States and Canada in particular) relies on V exploration and exploitation in several sophisticated industries (Petranikova et al., 2020). Consequently, high levels of V were recorded in several investigations (Figures 1.2d and 1.3a). In Canada, total V content in Manitoba soils varies from 35 to 455 mg kg^{-1} and showed a significant correlation with clay content (averaged as 35 mg kg^{-1} in coarse-textured soils and 138 mg kg^{-1} in fine-textured soils) (Haluschak et al., 1998). Vanadium concentration in Tamaulipas State, Mexico, was also high and ranged from 19–108 mg kg^{-1} due to the long-time accumulation of industrial effluents (oil industries in particular), which settled on the bottom and incorporated into sediments (Amezcua-Allieri & Salazar-Coria, 2008). In the United States, the Brownfield site in New Jersey had been used for over a century as a freight yard for coal transportation and stocking, and this was the main reason for the high V levels (110.6 mg kg^{-1}) (Qian et al., 2014). According to the Agency for Toxic Substances and Disease Registry, the content of V in coal products ranges from 15 to 125 mg kg^{-1} (Atsdr, 2012).

New Caledonia (a pacific island 1,210 km east of Australia) has the highest cancer incidence among Pacific islands due to the significant contribution of smelting activities (Tervonen et al., 2017). To scout the toxicological hazard of TEs including V, dust wipes, roadside soil, garden soil and household

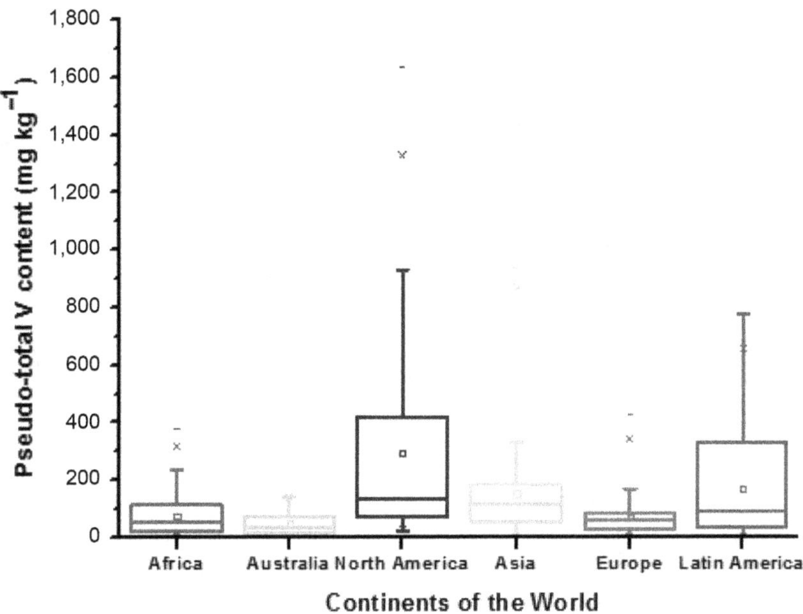

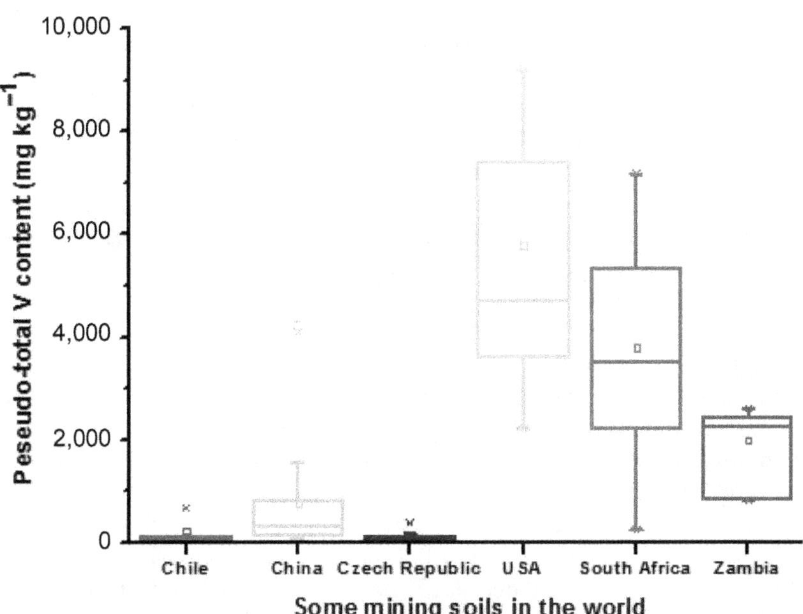

FIGURE 1.3
Pseudo-total V content in (a) continents of the world and (b) some mining soils worldwide.

vacuum dust were collected and analyzed (Fry et al., 2021). The concentration of V averaged as 77.6 mg kg^{-1}, and the risk was higher close to the smelter and in the direction of prevailing wind.

As mentioned earlier, abundant amounts of V are discharged from the world's continents into the ecosphere from mining/smelting activities (Figure 1.3b). As such, Panzhihua, China, is one of the most well-known areas for V-bearing titanomagnetite reservoirs in several mining and smelting activities (Wang, Zhang et al., 2020). It was hypothesized that soil V content varies spatiotemporally around the mines/smelters. Zhang and coworkers (Zhang et al., 2020) demonstrated that the closer to the smelter, the higher V content, with an average value of about 1,338 mg kg^{-1} in all studied locations. High levels of V (600–723 mg kg^{-1}) were also reported in abandoned agricultural land subjected to smelting wastewater from a mining area in China (stone coal-based V extraction plant, western Hunan Province) (Xiao et al., 2017).

There is a high risk of V phytotoxicity in soils close to V mines in South Africa, taking into account that 30% of the global V discharges are produced from this country (Moskalyk & Alfantazi, 2003). The content of V in the topsoil collected from a grazing soil (North West Province, South Africa) was found to be in the range of 1,570–3,600 mg kg^{-1} (Panichev et al., 2006). Also in Africa, Kabwe area, Zambia, has been ranked among the top 10 worst polluted sites in the world (Filippelli et al., 2020). In such arid areas, huge amount of dust particles are generated and cause tremendous deteriorations on the biosphere (Ettler et al., 2019). The total concentration of V in the dust fractions of <48 µm and <10 µm ranged from 84–440 mg kg^{-1} and 817–2,610 mg kg^{-1}, respectively (Ettler et al., 2020).

In the Americas, there are lots of mining areas in which V concentrations are often high. Vanadium-bearing sandstone deposits are distributed widely in western Colorado and eastern Utah as a main domestic reservoir of V discharges (Fischer, 1942). Vanadium levels in soil samples collected from Colorado, USA, ranged between 2,200 and 9,200 mg kg^{-1}, exceeding those of background sample (190 mg kg^{-1}) (Metzler & Karp, 2002). Such mining/smelting activities have been also practiced in Chile leaving behind a plethora of testimonies of dangerous disposals with elevated levels of V and other TEs. In Reyes et al.'s study (Reyes et al., 2020) V recorded high levels reaching 663 mg kg^{-1} in Taltal city, Chile.

Figure 1.2 Pseudo-total V content in (a) European countries, (b) African countries, (c) Asian countries, and (d) American countries: DE (Germany), NL (Netherlands), IT (Italy), GR (Greece), BE (Belgium), PL (Poland), CH (Switzerland), IP (Iberian peninsula), SC (Scandinavia) and EEu (Eastern Europe).

1.6.2 Soil Vanadium Fate, Mobility and Availability

Vanadium added to soils might be adsorbed by soil constituents such as clay, oxides and/or organic matter or distributed among the geochemical fractions

(Shaheen et al., 2019). The processes and factors affecting vanadium fate in soils are presented in Figure 1.4. More details about vanadium sorption and distribution among the geochemical fractions are presented and discussed in Chapters 3 and 4, respectively, in this book.

Vanadium exists in soil in different fractions including soluble and exchangeable, carbonate-occluded, Fe-Mn oxides-occluded, SOM-occluded and residual fractions. Of these, the soluble and exchangeable fraction is considered bioavailable; however, the other non-residual fractions (carbonate-, Mn oxides- and SOM- occluded fractions) could be transformed to the bioavailable form (Shaheen & Rinklebe, 2018). Several investigations have studied the availability/extractability of soil V and its relations with physicochemical properties of soil (Table 1.2). It is worth noting that extractable V content of these studies correlate not only with the total V content but also with physicochemical characters of soil. For example, extractable soil V of German soils varied between 0.06 and 8.9 mg kg^{-1}, which represents about 0.20–35% of their pseudo-total contents (median value is 32 mg kg^{-1}) (Gäbler et al., 2009). Conversely, available V concentrations in other soils (e.g., forest lands, Taiwan) account low values of that in their total contents (<4.0%), assuming the high binding capacity of their colloidal constitutions (Wu et al., 2020). According to (Luo et al., 2017), the maximum sorption capacities of V onto manural loessial and aeolian sandy soil colloids were comparable (285.7 and 238.1 mg g^{-1}, respectively). This value, however, was relatively low (41.5 mg g^{-1}) in cultivated loessial soil given the aggregated shape and the larger particle diameter of its colloids. Available V concentrations in mining and smelting areas such as the Panzhihua region, China, are normally high (17.7–95.8 mg g^{-1}) (Cao et al., 2017). Likewise, Fe-rich soils (mottled grey soils, Australia) recorded high amounts of available V (up to 45 mg g^{-1}) due to the inhibitory effect of Fe-oxides on V sequestration capacity, which maximized V extractability (Mikkonen et al., 2019).

By contrast, although marine soils contain low V contents (13–16 mg kg^{-1}), a relatively high portion of this content exists in a bioavailable form due to the high alkalinity and low colloidal content (Tsadilas & Shaheen, 2010). In according to Shaheen and Rinklebe (2018), alkaline soils exhibit a higher V mobility and availability relative to acidic soils, suggesting the high significant correlation ($r = 0.66, P < 0.001$) between soil pH and AB-DTPA-extractable V other than several TEs. This finding was further confirmed by Reijonen et al. (2016), since the bioavailability of V was found to be dictated by soil properties (i.e., pH and SOM), which favored the predominance of highly toxic V(V) species. Soil ageing reactions (alterations that occurred following prolonged periods) also showed a significant effect on V bioavailability, since V(V) dominates the soil solution of aged soils and the V(IV) form accounted for only 8.0% (Baken et al., 2012).

Several one-step extraction methods (DTPA, EDTA, $MgCl_2$, NH_4NO_3 and $MgCl_2$) have been investigated for evaluating the phytoavailability of V in soil (Feng et al., 2005). Most of these methods focused only on V species associated with certain geochemical soil components; however, these methods have ignored biological reactions in soil (soil–plant reactions in the root zone). Among them, DTPA proved a high efficiency due to its high correlation with V content in plant tissues and biological responses to these V levels (Shaheen & Rinklebe, 2018). Correspondingly, DPTA and $C_6H_8O_7$ showed a strong correlation with total and bioavailable soil V extracted via the Community Bureau of Reference (BCR) sequential extraction technique (Chen et al., 2020). Another study highlighted NH_4NO_3 as the most suitable extractant in soil–plant studies given the strong correlation between the soil extractable and plant V contents although Mehlich 3 extracted the highest V value (Pinto et al., 2015). In yet another study, Teng and coworkers (Teng et al., 2011) reported that the average values of bioavailable V in the rhizosphere of Panzhihua urban park, China, was ranked as EDTA (2.18%) > HOAc (0.67%) > $NaNO_3$ (0.43%) > HCl (0.17%).

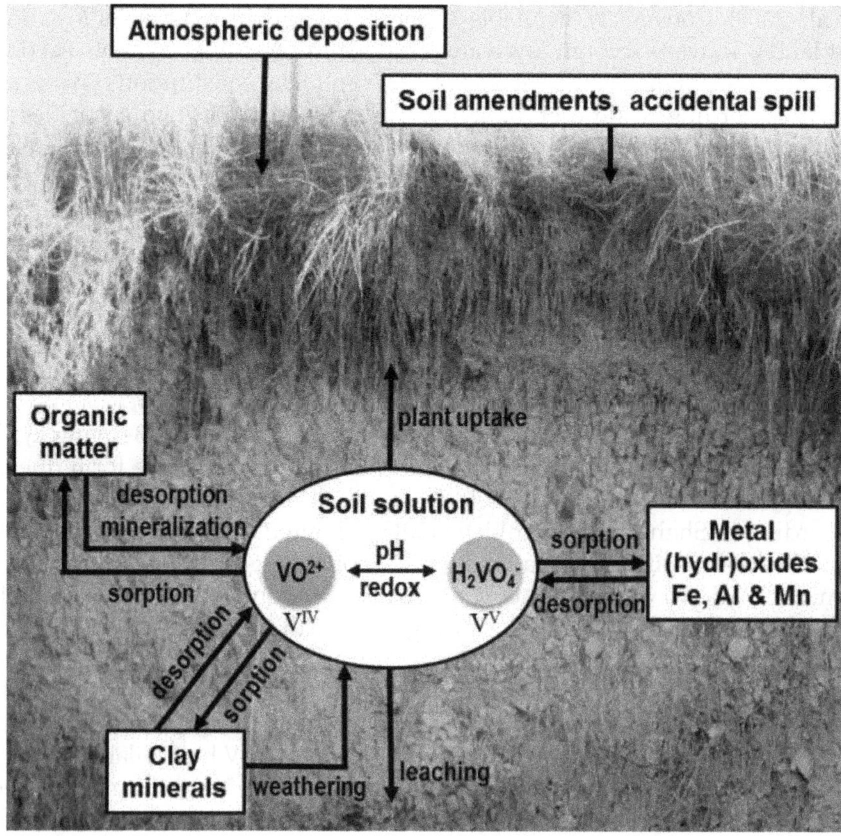

FIGURE 1.4
Processes affecting vanadium cycling in soils.
Source: Larsson, 2014.

TABLE 1.2

Soil Available V Concentration (Mg kg^{-1}), Extraction Methods and the Major Physicochemical Properties of Experimental Soils

Location	pH	Clay (%)	SOC (%)	Al/Fe/Mn oxides (g/kg)	Extraction	Available V (mg/kg)	References
Germany	3.9–7.3	0.1–54.4	0.05–9.22	0.33–268.20	Na_2-EDTA	0.06–8.9	(Gäbler et al., 2009)
Germany	5.1–7.8	0.20–56	0.23–8.26	19.6–146.14	AB-DTPA	0.04–2.14	(Shaheen & Rinklebe, 2018)
South Finland	5.4	2.0	1.7	2.93	KCl/KH_2PO_4	<0.1	(Reijonen et al., 2016)
Greece	6.3–8.1	6.14–42.6	0.33–22.84	4.12–22.15	AB-DTPA	0.5–2.3	(Tsadilas & Shaheen, 2010)
Denmark	5.2	13	/	5.9	0.2 M ammonium oxalate	7	(Baken et al., 2012)
Sweden	5.9	11		2.3		4	
Sweden	5.5	29		5.8		11	
Belgium	6.6	17		3.2		12	
Portugal	6.67–6.85	2.0–12.2	10.9–38.3	3.48–6.75	Mehlich 3	0.15–0.89	(Pinto et al., 2015)
Panzhihua region, China	5.6–8.1	1.3–34.6	0.3–20.7	/	acetic acid	0.1–1.7	(Teng et al., 2011)
Panzhihua region, China	6.8–8.2		0.58–2.13	118.3–152.4	BCR sequential extraction	17.7–95.8	(Cao et al., 2017)
Panzhihua region, China	4.03–9.42	1.2–18.3	0.08–0.20	/	BCR sequential extraction	0.19–10.5	(Chen et al., 2020)
Forest lands, Taiwan	3.60–8.51	4.97–78.3	0.08–7.90	/	0.05 M EDTA	0.11–13.1	(Wu et al., 2020)
Kuala Selangor, Malaysia	5.1–6.7	9.12	2.52–5.22	/	1.0 M MgCl2	0.74–2.11	(Ashraf et al., 2012)
Victoria, Australia	6.4–7.0	19–29	0.13–0.42	/	ammonium oxalate	0.001–45	(Mikkonen et al., 2019)
Egypt	7.87–8.94	5.38–46.43	0.09–0.70	1.02–9.19	AB-DTPA	1.51–5.77	(Tsadilas & Shaheen, 2010)
Kafr El-Sheikh, Egypt	7.87–9.33	14.1–52.2	1.22–15.58	29.6–207.42	Na_2-EDTA	0.43–4.04	(Shaheen & Rinklebe, 2018)

1.6.3 Factors Governing V Content and Availability in Soils

1.6.3.1 Iron/Aluminum/Manganese (Hydr)Oxides

Aluminum (Al), iron (Fe) and manganese (Mn) (hydr)oxides are considered as the primary reservoir of V in soil. According to (Gustafsson, 2019), V(V) is octahedrally coordinated in clay minerals in oxic soils; however, it is coordinated as surface-bound vanadate (V) onto Fe/Al (hydr)oxides. There are some other reports suggested that V^{3+} is able to enter the structure of Fe-oxides given the high similarity with Fe^{3+} in their radius (Schwertmann & Pfab, 1994). The coordination of V-Fe hydrous oxides is strong given the formation of inner-sphere bidentate complexes (Larsson, Persson et al., 2017).

The significant correlation between Al/Fe/Mn(hydr)oxides with total and/or phytoavailable V has been investigated in literature. An earlier study on the lateritic podzolic soils showed a high association between V and Fe oxides with any solution of the fine soil fraction (< 2.0 mm) (Taylor & Giles, 1970). Fe and/or Al (hydr)oxides also played a cardinal role for binding V, which derived from long-term lime application, in the mor layer of pine forest soils located in southern Sweden (Larsson et al., 2015). V K-edge XANES spectroscopy confirmed the sorption of added V onto Fe and Al hydrous oxides with minor contributions from the organically bound V(IV) (Larsson, Hadialhejazi et al., 2017). In view of this, Shaheen and coworkers (Shaheen et al., 2019) suggested that ferric (oxy)hydroxides are important component for sorption and precipitation of V as ferrous vanadate.

The role of manganese oxides on V speciation and availability has also been highlighted. According to Yang et al. (2020), upon stabilization of TEs in a mining area (Sichuan Province, China), most of V was associated with Fe and Mn oxides as main carriers. In another study, V fractionation in fluvial and lacustrine soils in the northern Nile Delta, Egypt, revealed the dominance of the residual fraction, followed by V-bound crystalline and amorphous Fe oxides and V-bound Mn oxides (Shaheen et al., 2014). This finding highlights the importance of Fe/Mn oxides in the binding of V species. Likewise, the data of V speciation in 13 different soil profiles from Egypt and Germany showed a stronger correlation between total V and free Fe/Mn oxides in Egyptian soils than in the German (Shaheen & Rinklebe, 2018).

1.6.3.2 Soil Organic Matter

Soil organic matter (SOM) acts as a carrier for storage and speciation of V in soil matrix based on its direct and indirect effects that control the bioavailability of V species. On one hand, SOM motivates the reduction of highly toxic V(V) into the less toxic V(IV). On the other hand, it serves as an adsorbent for both forms against accessibility via different biological routes (Reijonen et al., 2016). Palmer and coworkers (Palmer et al., 2013) reported that V oxides existed at reduced levels in peat soil samples collected from northern Ireland. In other reduced conditions (e.g., sediments of black shales), V exists as mixture of organically complexed vanadyl (IV) and other V(III) species (Gustafsson, 2019). This finding was further confirmed by Wu et al. (2020), since the vanadyl form might be complexed by SOM in 17 soil profiles in Taiwan.

The effect of humic substances on V sorptivity was investigated using a coexistence system of humic acid and silica to simulate organic and inorganic components of soil (Song et al., 2020). The adsorption amount of V increased by about 225% as the amount of humic acid loading increased from 0.0727 to 0.364 g L^{-1}. Organic matter creates organic inner-sphere complexes with V (particularly at pH below 6.0) via its active functional groups (Dong et al., 2021). These complexes showed weaker association with V(V) than their corresponding V(IV) complexes (Gustafsson, 2019). The quinoid moieties of humic substances act as electron shuttles in this reaction, and the reduced V(IV) form is bound onto the oxygen donor atoms at sites of active functional groups (Palmer & von Wandruszka, 2010). The oxidation of hydroquinone into quinone groups could be also considered as a potential mechanism upon V(V) reduction (Lu et al., 1998). Binding of V(IV) onto organic matter is assumed to be a mono- and bidentate complexes coordination, on which the bidentate coordination is the most dominant under most natural conditions (Gustafsson et al., 2007).

Although the active role of humic substances as reducing agents is highly acknowledged, the easily degradable organic substances showed higher efficacy in this regard (Reijonen et al., 2016). According to (Shi et al., 2016), a positive relationship was found between protein-like dissolved organic matter and labile V. Another multi-step column experiment pointed to a significant effect of low molecular weight organic acids (acetic, citric, oxalic, malic or tartaric acids) for remediation of V-contaminated soil (Zou, Gao et al., 2019).

1.6.3.3 Clay Minerals

Clay minerals are considered an important source for vanadium retention in the soil matrix. Clay-vanadium ore is one of the most important reservoirs for V recovery through destroying Al-O and Si-O bonds by fluoride addition in acid leaching (Chen, Ye et al., 2021). In clay minerals, Al^{3+} in the octahedral sheet could be substituted by V^{4+} as VO^{2+}, and V could also coordinate in sheets of clays (e.g., illite, smectites and micas) (Gehring et al., 1993).

The low contents of V are mainly associated with light-textured soils (10–60 mg kg^{-1}); however, heavy clay soils show higher V content (20–500 mg kg^{-1}) (Kabata-Pendias, 2000). In view of this, Tsadilas and Shaheen (2010) clarified that total V concentration in soils collected from Egypt and Greece was correlated significantly with clay fraction ($r = 0.73$, $P < 0.001$) suggesting that V was associated with the fine soil fraction. In addition, Shaheen and Rinklebe (2018) reported a positive significant correlation between total V concentration and clay content ($r = 0.85$, $P < 0.001$) and correlated negatively with sand fraction. This study indicated also that AB-DTPA-extractable V showed a significant correlation with clay content ($r = 0.50$, $P < 0.001$).

The high ability of clay minerals toward binding the bioavailable V confirms their high potentiality to serve as a soil stabilizer in V-contaminated soils. According to He et al. (2020), application of modified bentonite (5.0 wt%) into a calcareous purple soil resulted in stabilization rates of about 19.1, 20.5 and 37.7 for water-extractable V, pentavalent V and bioavailable V, respectively. Moreover, this stabilization retained effective for about four weeks after modified bentonite application, which reduced the risk of V translocation into aquatic resources.

1.6.3.4 Calcium Carbonate

In their studies on soil collected from Egypt and Greece, Tsadilas and Shaheen (2010) reported that total V in soil negatively correlated to $CaCO_3$ ($r = 0.35$, $P < 0.01$). The high content of Ca in soil could improve V stabilization by forming calcium orthovanadate ($Ca_3(VO_4)_2•4H_2O$) and calcium pyrovanadate ($Ca_2V_2O_7$) under strong (11.5–12.5) and mild (8.0–11.5) alkaline conditions, respectively (Aihemaiti et al., 2020). Therefore, the application of CaO to V-contaminated soil (2.0 wt%) reduced the leachable V from 18.8 up to 0.21 mg L^{-1}, with a stabilization efficiency of 98.9% (Zou, Li et al., 2019).

1.6.3.5 Soil Redox Potential (E_H)

Soil redox potential plays a crucial effect in controlling the speciation and bioavailability of redox elements including V. Soil redox potential and SOM pose the ability to reduce V(V) abiotically and interrupt its migration (Chen, Liu et al., 2021). Dissolved cationic V(IV) could be oxidized to oxyanionic V(V) under acidic oxic conditions, in which V(V) ions are sorbed to the positively charged surfaces of Fe/Al (hydr)oxides to decrease V solubility in such conditions (Frohne et al., 2015; Shaheen et al., 2016).

Flood–dry cycles can directly cause significant alterations in redox conditions of soil. In this respect, soil colloidal content tends to become reductive to enhance V solubility under flooded conditions; however, the oxic conditions increase V sorptivity onto Al/Fe/Mn hydr(oxides) and reduce V solubility (Reijonen et al., 2016; Shaheen et al., 2019). In addition, there are several reports that pointed to the potentiality of soil microorganisms in modulating soil redox conditions as an important mechanism to combat V bioavailability (Gan et al., 2020). The impact of soil redox potential on release of dissolved V was investigated under an automated biogeochemical microcosm from reducing to oxidizing conditions (–60 to +491 mV) (Shaheen et al., 2016). Results illustrated that dissolved V was significantly higher under low redox values due to the reduction of V(V) to V(IV) as well as the reduction of Fe (hydr)oxides with a subsequent release of dissolved V (Shaheen et al., 2016). The further reduction of V(IV) to V(III) requires a strong reducing agent such as sulfides, which could be the main reason for less V solubility in oxic soils having a high mount of sulfur and E_H below –180 mV (Shaheen et al., 2019). More details about the impact of soil Eh on the release of vanadium are discussed in Chapter 5 in this book.

1.6.3.6 Soil pH

Soil pH is a major factor for affecting speciation, valence state, solubility and phytotoxicity of V in soil (Aihemaiti et al., 2020). A strong active correlation between phytoavailable V and soil pH (r = 0.9916) was reported due to the negative charge of the predominant species of V (e.g., the dimer ($V_2O_7^{4-}$), the tetramer ($V_4O_{12}^{4-}$) and the pentamer ($V_5O_{15}^{5-}$)) under pH ranging between 5.0 and 8.0 (Tracey et al., 2007). Further deprotonating of V(V) species are prevalent under alkaline conditions based on values of pK_a (Heath & Howarth, 1981):

$$H_2VO_4 = HVO^{2-}{}_4 + H^+ \; pK_a = 7.1$$
$$HVO^{2-}{}_4 = HO^{3-}{}_4 + H^+ \; pK_a = 12$$

Under relatively high V concentrations (> 51 mg V L^{-1}) within pH values between 3.0 and 6.0, V(V) exists in the form of decavanadate $HVO_{10}O^{2-}{}_{28}$ (Tracey et al., 2007). On the other hand, V(IV) exists as cationic vanadyl (VO^{2+}), and above pH 4.0 it is hydrolyzed to the $VO(OH)^+$ form and $VO(OH)_2$ upon increasing the pH value (Crans et al., 1998; Reijonen et al., 2016). The solubility of organic V complexes (humic/fulvic-V complexes in particular) rather increases with an increase in the value of soil pH given the deprotonation of oxygen functional groups and increasing hydrophobicity (Mosa et al., 2021). The sorptivity of inorganic V(V) onto Fe/Al-(hydr)oxides also showed a noticeable decline as soil pH increased (above 3.0–4.0) (Naeem et al., 2007). Other reports revealed that the bioavailability of V decreased at pH 10.0 (Wang & Liu, 1999).

Several investigations have reported the high dissolution of V under alkaline soil conditions given the motivation of V(IV) oxidation to V(V) and inhibiting the active role of SOM for V(V) reduction (Blackmore et al., 1996; Shaheen & Rinklebe, 2018; Tsadilas & Shaheen, 2010). In view of this, the pH value of soil is crucial for the effectiveness of V passivators. According to Zou, Li et al. (2019), the optimum ranges of soil pH for immobilization of V in contaminated soils by $FeSO_4$ and CaO are, respectively, 4.0–5.0 and 9.5–12.5. The remarkable elevation of soil pH following passivating agents application (attapulgite, hydroxyapatite and zeolite) to a V-contaminated soil was also reported as the main mechanism for V stabilization (Yang et al., 2020). The alkaline effect of biochar application led also to increase the dissolved soil V (Yu et al., 2020). More details about the impact of soil pH on the release and solidity of vanadium are presented and discussed in Chapter 5 in this book.

1.6.3.7 Ionic Strength

Ionic strength has a crucial effect on the fate and transport of V in soil matrix. According to (Luo et al., 2017), the affinity of V(V) to natural soil colloids decreased by about 50% with the increase of ionic strength from 0.001 to 0.1M due to the reduction of zeta potential values, which minimized the thickness of the diffused double layer of soil colloids and inhibited the electrical attraction potentials. In addition, the entrance of sorbate solution into the compressed electrical diffused double layer by the high values of ionic strength could release the sorbed V into the solution (van Santen et al., 2015). This finding was also supported by Jiang et al. (2019), since the sorption of V(V) onto three agricultural soils (yellow cinnamon, manual loessial and aeolian sandy soils) decreased as ionic strength increased. This reduction was not significant with aeolian sandy soil, which possessed the lowest colloidal content. This low colloidal content diminished the noticeable alterations in the transport of V(V) in saturated quartz sand media under increasing the ionic strength values (Wang et al., 2016). In another experiment, a coexistence system was prepared from humic acid and silica to study the effect of ionic strength (0.005–1.0 mol L^{-1}) on V(V) sorptivity by a simulated soil organic and inorganic components (Song et al., 2020). Results pointed to a limited effect of ionic strength due to the reduction of effective sorption sites at high values of ionic strength.

1.7 Conclusions

We conclude that vanadium is serving as an important component of industrial activities and has bioinorganic implications to pose highly toxic hazards to humans and animals. Vanadium is an important component of certain industrial activities such as steel and iron refining, electronics, and dyeing, while its significance has also been recognized from space navigation and in the nuclear industry. Bioinorganic implications of vanadium-dependent enzymes and application of vanadium compounds in disease and cancer treatment have also recently received attention since vanadium is an essential element to human beings and animals. Vanadium can exist in different oxidation states from –1 to +5; however, only the +3, +4, and +5 oxidation states are important in the natural environment. Among vanadium compounds, vanadium (+5) in particular is toxic to pose human and environmental hazards. Vanadium (+5) is considered a dangerous pollutant. The number of people affected by vanadium pollution increased in many countries such as the United States, South Africa and China. Vanadium in soils and sediments around the world is found over a wide range of concentrations (from less than 1 mg kg^{-1} to 9,200 mg kg^{-1}). This makes the research on vanadium an important challenge from industrial, economic and environmental points of view. The total and available content of vanadium differ widely between soils worldwide, depending on the geogenic and anthropogenic sources and mining activates. The fate and mobilization of vanadium in soils are highly governed by different factors, including iron/ aluminum/ manganese (hydr)oxides, soil organic matter, clay minerals, calcium carbonate, soil redox potential, soil pH, and ionic strength. Vanadium is a redox-sensitive element, and thus its mobilization depends on the changes on soil redox potential and pH as well as the chemistry of Fe-oxides. In light of the high soil concentrations of vanadium, a release of this toxic element may increase the potential environmental risks.

References

Abgottspon, F., Bigalke, M., Wilcke, W. 2015. Fast colloidal and dissolved release of trace elements in a carbonatic soil after experimental flooding. *Geoderma*, **259**, 156–163.

Adagunodo, T.A., Sunmonu, L.A., Emetere, M.E. 2018. Heavy metals' data in soils for agricultural activities. *Data in Brief*, **18**, 1847–1855.

Aihemaiti, A., Gao, Y., Meng, Y., Chen, X., Liu, J., Xiang, H., Xu, Y., Jiang, J. 2020. Review of plant-vanadium physiological interactions, bioaccumulation, and bioremediation of vanadium-contaminated sites. *Science of the Total Environment*, **712**, 135637.

Alharbi, T., El-Sorogy, A. 2017. Assessment of metal contamination in coastal sediments of Al-Khobar area, Arabian Gulf, Saudi Arabia. *Journal of African Earth Sciences*, **129**, 458–468.

Altaf, M.M., Diao, X.-p., Shakoor, A., Imtiaz, M., Atique-ur, R., Altaf, M.A., Khan, L.U. 2021. Delineating vanadium (V) ecological distribution, its toxicant potential, and effective remediation strategies from contaminated soils. *Journal of Soil Science and Plant Nutrition*, **22**, 121–139.

Amezcua-Allieri, M.A., Salazar-Coria, L. 2008. Nickel and vanadium concentrations and its relation with sediment acute toxicity. *Bulletin of Environmental Contamination and Toxicology*, **80**(6), 555–560.

Ashraf, M.A., Maah, M.J., Yusoff, I. 2012. Chemical speciation and potential mobility of heavy metals in the soil of former tin mining catchment. *The Scientific World Journal*, **2012**.

ATSDR, U. 2012. Toxicological profile for chromium. In: *US Department of Health and Human Services, Public Health Service*. https://www.atsdr.cdc.gov/toxprofiles/tp7.pdf

Auger, Y., Bodineau, L., Leclercq, S., Wartel, M. 1999. Some aspects of vanadium and chromium chemistry in the English Channel. *Continental Shelf Research*, **19**(15–16), 2003–2018.

Baken, S., Larsson, M.A., Gustafsson, J.P., Cubadda, F., Smolders, E. 2012. Ageing of vanadium in soils and consequences for bioavailability. *European Journal of Soil Science*, **63**(6), 839–847.

Bauer, S., Blomqvist, S., Ingri, J. 2017. Distribution of dissolved and suspended particulate molybdenum, vanadium, and tungsten in the Baltic Sea. *Marine Chemistry*, **196**, 135–147.

Blackmore, D.P.T., Ellis, J., Riley, P.J. 1996. Treatment of a vanadium-containing effluent by adsorption/coprecipitation with iron oxyhydroxide. *Water Research*, **30**(10), 2512–2516.

Boechat, C.B., Eon, J.-G., Rossi, A.M., de Castro Perez, C.A., San Gil, R.A.d.S. 2000. Structure of vanadate in calcium phosphate and vanadate apatite solid solutions. *Physical Chemistry Chemical Physics*, **2**(18), 4225–4230.

Cao, L., Skyllas-Kazacos, M., Menictas, C., Noack, J. 2018. A review of electrolyte additives and impurities in vanadium redox flow batteries. *Journal of Energy Chemistry*, **27**(5), 1269–1291.

Cao, X., Diao, M., Zhang, B., Liu, H., Wang, S., Yang, M. 2017. Spatial distribution of vanadium and microbial community responses in surface soil of Panzhihua mining and smelting area, China. *Chemosphere*, **183**, 9–17.

Cappuyns, V., Slabbinck, E. 2012. Occurrence of vanadium in Belgian and European alluvial soils. *Applied and Environmental Soil Science*, **2012**.

Chen, D., 2019. Annual evaluation for vanadium industry in 2018. *Hebei Metallurgy*, **8**, 5–15.

Chen, L., Liu, J.-R., Hu, W.-F., Gao, J., Yang, J.-Y. 2021. Vanadium in soil-plant system: Source, fate, toxicity, and bioremediation. *Journal of Hazardous Materials*, **405**, 124200.

Chen, L., Wang, K.-P., Yang, J.-Y. 2020. Evaluate the potential bioavailability of vanadium in soil and vanadium titano-magnetite tailing in a mining area using BCR sequential and single extraction: A case study in Panzhihua, China. *Soil and Sediment Contamination: An International Journal*, **29**(2), 232–245.

Chen, Z., Ye, G., Xiang, P., Tao, Y., Tang, Y., Hu, Y. 2021. Effect of activator on kinetics of direct acid leaching of vanadium from clay vanadium ore. *Separation and Purification Technology*, 119937.

Cintas, P. 2004. The road to chemical names and eponyms: Discovery, priority, and credit. *Angewandte Chemie International Edition*, **43**(44), 5888–5894.

Clarkson, P.M., Rawson, E.S. 1999. Nutritional supplements to increase muscle mass. *Critical Reviews in Food Science and Nutrition*, **39**(4), 317–328.

Costa Pessoa, J. 2015. Thirty years through vanadium chemistry. *Journal of Inorganic Biochemistry*, **147**, 4–24.

Crans, D.C., Amin, S.S., Keramidas, A.D. 1998. Chemistry of relevance to vanadium in the environment. *Advances in Environmental Science and Technology-New York*, **30**, 73–96.

Desaulniers, D., Cummings-Lorbetskie, C., Leingartner, K., Xiao, G.H., Zhou, G., Parfett, C. 2021. Effects of vanadium (sodium metavanadate) and aflatoxin-B1 on cytochrome p450 activities, DNA damage and DNA methylation in human liver cell lines. *Toxicology in Vitro*, **70**, 105036.

Dong, Y., Lin, H., Zhao, Y., Gueret Yadiberet Menzembere, E.R. 2021. Remediation of vanadium-contaminated soils by the combination of natural clay mineral and humic acid. *Journal of Cleaner Production*, **279**, 123874.

Driskell, J.A., Wolinsky, I. 2002. *Nutritional assessment of athletes*. CRC Press.

Dyer, A., De Butte, M. 2022. Neurobehavioral effects of chronic low-dose vanadium administration in young male rats. *Behavioural Brain Research*, **419**, 113701.

Ehrlich, V.A., Nersesyan, A.K., Hoelzl, C., Ferk, F., Bichler, J., Valic, E., Schaffer, A., Schulte-Hermann, R., Fenech, M., Wagner, K.-H. 2008. Inhalative exposure to vanadium pentoxide causes DNA damage in workers: Results of a multiple end point study. *Environmental Health Perspectives*, **116**(12), 1689–1693.

El-Moselhy, K.M. 2006. Distribution of vanadium in bottom sediments from the marine coastal area of the Egyptian seas. *Egyptian Journal of Aquatic Research*, **32**(1), 12–21.

Ettler, V., Cihlová, M., Jarošíková, A., Mihaljevič, M., Drahota, P., Kříbek, B., Vaněk, A., Penížek, V., Sracek, O., Klementová, M. 2019. Oral bioaccessibility of metal (loid) s in dust materials from mining areas of northern Namibia. *Environment International*, **124**, 205–215.

Ettler, V., Štěpánek, D., Mihaljevič, M., Drahota, P., Jedlicka, R., Kříbek, B., Vaněk, A., Penížek, V., Sracek, O., Nyambe, I. 2020. Slag dusts from Kabwe (Zambia): Contaminant mineralogy and oral bioaccessibility. *Chemosphere*, **260**, 127642.

Fan, Z., Zhou, S., Xue, A., Li, M., Zhang, Y., Zhao, Y., Xing, W. 2021. Preparation and properties of a low-cost porous ceramic support from low-grade palygorskite clay and silicon-carbide with vanadium pentoxide additives. *Chinese Journal of Chemical Engineering*, **29**, 417–425.

Feng, M.-H., Shan, X.-Q., Zhang, S., Wen, B. 2005. A comparison of the rhizosphere-based method with DTPA, EDTA, $CaCl_2$, and $NaNO_3$ extraction methods for prediction of bioavailability of metals in soil to barley. *Environmental Pollution*, **137**(2), 231–240.

Filippelli, G., Anenberg, S., Taylor, M., van Geen, A., Khreis, H. 2020. New approaches to identifying and reducing the global burden of disease from pollution. *GeoHealth*, **4**(4), e2018GH000167.

Fiorentino, C.E., Paoloni, J.D., Sequeira, M.E., Arosteguy, P. 2007. The presence of vanadium in groundwater of southeastern extreme the pampean region Argentina: Relationship with other chemical elements. *Journal of Contaminant Hydrology*, **93**(1–4), 122–129.

Fischer, R.P. 1942. *Vanadium deposits of Colorado and Utah: A preliminary report*. US Government Printing Office.

Fitamo, D., Itana, F., Olsson, M. 2007. Total contents and sequential extraction of heavy metals in soils irrigated with wastewater, Akaki, Ethiopia. *Environmental Management*, **39**(2), 178–193.

Frohne, T., Diaz-Bone, R.A., Du Laing, G., Rinklebe, J. 2015. Impact of systematic change of redox potential on the leaching of Ba, Cr, Sr, and V from a riverine soil into water. *Journal of Soils and Sediments*, **15**(3), 623–633.

Fry, K.L., Gillings, M.M., Isley, C.F., Gunkel-Grillon, P., Taylor, M.P. 2021. Trace element contamination of soil and dust by a New Caledonian ferronickel smelter: Dispersal, enrichment, and human health risk. *Environmental Pollution*, **288**, 117593.

Gäbler, H.E., Glüh, K., Bahr, A., Utermann, J. 2009. Quantification of vanadium adsorption by German soils. *Journal of Geochemical Exploration*, **103**(1), 37–44.

Gan, C.-d., Chen, T., Yang, J.-y. 2020. Remediation of vanadium contaminated soil by alfalfa (Medicago sativa L.) combined with vanadium-resistant bacterial strain. *Environmental Technology & Innovation*, **20**, 101090.

Gao, F., Olayiwola, A.U., Liu, B., Wang, S., Du, H., Li, J., Wang, X., Chen, D., Zhang, Y. 2021. Review of vanadium production part I: Primary resources. *Mineral Processing and Extractive Metallurgy Review*, **43**(4), 1–23.

Gehring, A.U., Fry, I.V., Luster, J., Sposito, G. 1993. The chemical form of vanadium (IV) in kaolinite. *Clays and Clay Minerals,* **41**(6), 662–667.

Gehring, A.U., Fry, I.V., Luster, J., Sposito, G. 1994. Vanadium in sepiolite: A redox-indicator for an ancient closed brine System in the Madrid Basin, central Spain. *Geochimica et cosmochimica acta,* **58**(16), 3345–3351.

Goc, A. 2006. Biological activity of vanadium compounds. *Central European Journal of Biology,* **1**(3), 314–332.

Guagliardi, I., Cicchella, D., De Rosa, R., Ricca, N., Buttafuoco, G. 2018. Geochemical sources of vanadium in soils: Evidences in a southern Italy area. *Journal of Geochemical Exploration,* **184**, 358–364.

Gustafsson, J.P. 2019. Vanadium geochemistry in the biogeosphere—speciation, solid-solution interactions, and ecotoxicity. *Applied Geochemistry,* **102**, 1–25.

Gustafsson, J.P., Persson, I., Kleja, D.B., Van Schaik, J.W.J. 2007. Binding of iron (III) to organic soils: EXAFS spectroscopy and chemical equilibrium modeling. *Environmental Science & Technology,* **41**(4), 1232–1237.

Haluschak, P., Eilers, R.G., Mills, G.F., Grift, S. 1998. Status of Selected Trace Elements in Agricultural Soils of Southern Manitoba. Technical Report 1998-6E Land Resource Unit. Brandon Research Centre. Research Branch. Agriculture and Agri-Food Canada.

He, W.-y., Yang, J.-y., Li, J., Ai, Y.-w., Li, J.-x. 2020. Stabilization of vanadium in calcareous purple soil using modified Na-bentonites. *Journal of Cleaner Production,* **268**, 121978.

Heath, E., Howarth, O.W. 1981. Vanadium-51 and oxygen-17 nuclear magnetic resonance study of vanadate (V) equilibria and kinetics. *Journal of the Chemical Society, Dalton Transactions,* 5, 1105–1110.

Ho, P., Lee, J.-M., Heller, M.I., Lam, P.J., Shiller, A.M. 2018. The distribution of dissolved and particulate Mo and V along the US GEOTRACES East Pacific Zonal Transect (GP16): The roles of oxides and biogenic particles in their distributions in the oxygen deficient zone and the hydrothermal plume. *Marine Chemistry,* **201**, 242–255.

Hope, B.K. 1997. An assessment of the global impact of anthropogenic vanadium. *Biogeochemistry,* **37**(1), 1–13.

Hu, J., Peng, Y., Zheng, T., Zhang, B., Liu, W., Wu, C., Jiang, M., Braun, J.M., Liu, S., Buka, S.L. 2018. Effects of trimester-specific exposure to vanadium on ultrasound measures of fetal growth and birth size: A longitudinal prospective prenatal cohort study. *The Lancet Planetary Health,* **2**(10), e427–e437.

Huang, Y., Long, Z., Zhou, D., Wang, L., He, P., Zhang, G., Hughes, S.S., Yu, H., Huang, F. 2021. Fingerprinting vanadium in soils based on speciation characteristics and isotope compositions. *Science of the Total Environment,* **791**, 148240.

IARC Working Group on the Evaluation of Carcinogenic Risks to Humans. 2006. *Cobalt in hard metals and cobalt sulfate, gallium arsenide, indium phosphide and vanadium pentoxide.* International Agency for Research on Cancer (IARC monographs on the evaluation of carcinogenic risks to humans, No. 86.) Available from: https://www.ncbi.nlm.nih.gov/books/NBK321688/

Isidorov, V.A., 2001. *Ecological chemistry: Textbook for higher educational institutions [in Russian].* Khimizdat, St. Petersburg.

Jakusch, T., Kiss, T. 2017. In vitro study of the antidiabetic behavior of vanadium compounds. *Coordination Chemistry Reviews,* **351**, 118–126.

Jia, L., Anthony, E.J., Charland, J.P. 2002. Investigation of vanadium compounds in ashes from a CFBC firing 100 petroleum coke. *Energy & Fuels,* **16**(2), 397–403.

Jiang, S., Zhou, S., Liu, H., Peng, C., Zhang, X., Zhou, H., Wang, Z., Lu, Q. 2021. Concentrations of vanadium in urine with hypertension prevalence and blood pressure levels. *Ecotoxicology and Environmental Safety*, **213**, 112028.

Jiang, Y., Yin, X., Luo, X., Yu, L., Sun, H., Wang, N., Mathews, S. 2019. Sorption of vanadium (V) on three typical agricultural soil of the Loess Plateau, China. *Environmental Pollutants and Bioavailability*, **31**(1), 120–130.

Kabata-Pendias, A. 2000. *Trace elements in soils and plants*. CRC Press.

Ketris, M.P., Yudovich, Y.E. 2009. Estimations of clarkes for carbonaceous biolithes: World averages for trace element contents in black shales and coals. *International Journal of Coal Geology*, **78**, 135–148.

Laitinen, T., Ojala, S., Cousin, R., Koivikko, N., Poupin, C., El Assal, Z., Aho, A., Keiski, R.L. 2020. Activity, selectivity, and stability of vanadium catalysts in formaldehyde production from emissions of volatile organic compounds. *Journal of Industrial and Engineering Chemistry*, **83**, 375–386.

Larsson, MA. 2014. Vanadium in soils. PhD Thesis. Swedish University of Agricultural Sciences, Uppsala, Sweden.

Larsson, M.A., D'Amato, M., Cubadda, F., Raggi, A., Öborn, I., Kleja, D.B., Gustafsson, J.P. 2015. Long-term fate and transformations of vanadium in a pine forest soil with added converter lime. *Geoderma*, **259–260**, 271–278.

Larsson, M.A., Hadialhejazi, G., Gustafsson, J.P. 2017. Vanadium sorption by mineral soils: Development of a predictive model. *Chemosphere*, **168**, 925–932.

Larsson, M.A., Persson, I., Sjöstedt, C., Gustafsson, J.P. 2017. Vanadate complexation to ferrihydrite: X-ray absorption spectroscopy and CD-MUSIC modelling. *Environmental Chemistry*, **14**(3), 141–150.

Li, C., Wu, C., Zhang, J., Li, Y., Zhang, B., Zhou, A., Liu, W., Chen, Z., Li, R., Cao, Z., Xia, W., Xu, S. 2021. Associations of prenatal exposure to vanadium with early-childhood growth: A prospective prenatal cohort study. *Journal of Hazardous Materials*, **411**, 125102.

Li, H., Zhou, D., Zhang, Q., Feng, C., Zheng, W., He, K., Lan, Y. 2013. Vanadium exposure-induced neurobehavioral alterations among Chinese workers. *NeuroToxicology*, **36**, 49–54.

Lisichkin, G.V. 1998. The raw-materials crisis and the problem of recovering metals from seawater. *Soros Education Journal*, **6**, 65–70.

Liu, Y., Liu, G., Qu, Q., Qi, C., Sun, R., Liu, H. 2017. Geochemistry of vanadium (V) in Chinese coals. *Environmental Geochemistry and Health*, **39**, 967–986.

Lu, X., Johnson, W.D., Hook, J. 1998. Reaction of vanadate with aquatic humic substances: An ESR and 51V NMR study. *Environmental Science & Technology*, **32**(15), 2257–2263.

Luo, X., Yu, L., Wang, C., Yin, X., Mosa, A., Lv, J., Sun, H. 2017. Sorption of vanadium (V) onto natural soil colloids under various solution pH and ionic strength conditions. *Chemosphere*, **169**, 609–617.

Malandrino, P., Russo, M., Ronchi, A., Minoia, C., Cataldo, D., Regalbuto, C., Giordano, C., Attard, M., Squatrito, S., Trimarchi, F. 2016. Increased thyroid cancer incidence in a basaltic volcanic area is associated with non-anthropogenic pollution and biocontamination. *Endocrine*, **53**(2), 471–479.

Meng, R., Chen, T., Zhang, Y., Lu, W., Liu, Y., Lu, T., Liu, Y., Wang, H. 2018. Development, modification, and application of low-cost and available biochar derived from corn straw for the removal of vanadium (v) from aqueous solution and real contaminated groundwater. *RSC Advances*, **8**(38), 21480–21494.

Metzler, D.R., Karp, K.E. 2002. Performance report on the operations of a pilot plant for the treatment of vanadium in ground water—New Rifle UMTRA Site, Rifle, Colorado, USA. In: *Uranium in the aquatic environment*. Springer, pp. 541–552.

Mikkonen, H.G., van de Graaff, R., Collins, R.N., Dasika, R., Wallis, C.J., Howard, D.L., Reichman, S.M. 2019. Immobilisation of geogenic arsenic and vanadium in iron-rich sediments and iron stone deposits. *Science of The Total Environment*, **654**, 1072–1081.

Monakhov, I.N., Khromov, S.V., Chernousov, P.I., Yusfin, Y.S. 2004. The flow of vanadium- bearing materials in industry. *Metallurgist*, **48**, 381–385.

Mosa, A., Taha, A.A., Elsaeid, M. 2021. In-situ and ex-situ remediation of potentially toxic elements by humic acid extracted from different feedstocks: Experimental observations on a contaminated soil subjected to long-term irrigation with sewage effluents. *Environmental Technology & Innovation*, **23**, 101599.

Moskalyk, R.R., Alfantazi, A.M. 2003. Processing of vanadium: A review. *Minerals Engineering*, **16**(9), 793–805.

Naeem, A., Westerhoff, P., Mustafa, S. 2007. Vanadium removal by metal (hydr) oxide adsorbents. *Water research*, **41**(7), 1596–1602.

Nagasaka, T., Muroga, T. 2020, June 1. Vanadium for nuclear systems☆. In: *Comprehensive nuclear materials*, (Eds.) R.J.M. Konings, R.E. Stoller (2nd ed.). Elsevier, pp. 1–18.

Nriagu, J.O. 1990. Global metal pollution: Poisoning the biosphere? *Environment: Science and Policy for Sustainable Development*, **32**(7), 7–33.

Nunes, P., Correia, I., Cavaco, I., Marques, F., Pinheiro, T., Avecilla, F., Pessoa, J.C. 2021. Therapeutic potential of vanadium complexes with 1,10-phenanthroline ligands, quo vadis? Fate of complexes in cell media and cancer cells. *Journal of Inorganic Biochemistry*, **217**, 111350.

Okonkwo, S.I., Idakwo, S.O., Ameh, E.G. 2021. Heavy metal contamination and ecological risk assessment of soils around the pegmatite mining sites at Olode area, Ibadan southwestern Nigeria. *Environmental Nanotechnology, Monitoring & Management*, **15**, 100424.

Palmer, N.E., von Wandruszka, R. 2010. Humic acids as reducing agents: The involvement of quinoid moieties in arsenate reduction. *Environmental Science and Pollution Research*, **17**(7), 1362–1370.

Palmer, S., Ofterdinger, U., McKinley, J.M., Cox, S., Barsby, A. 2013. Correlation analysis as a tool to investigate the bioaccessibility of nickel, vanadium and zinc in Northern Ireland soils. *Environmental Geochemistry and Health*, **35**(5), 569–584.

Panchal, S.K., Wanyonyi, S., Brown, L. 2017. Selenium, vanadium, and chromium as micronutrients to improve metabolic syndrome. *Current Hypertension Reports*, **19**(3), 10.

Panichev, N., Mandiwana, K., Moema, D., Molatlhegi, R., Ngobeni, P. 2006. Distribution of vanadium (V) species between soil and plants in the vicinity of vanadium mine. *Journal of Hazardous Materials*, **137**(2), 649–653.

Patel, N., Prajapati, A.K., Jadeja, R.N., Patel, R.N., Patel, S.K., Gupta, V.K., Tripathi, I.P., Dwivedi, N. 2019. Model investigations for vanadium-protein interactions: Synthesis, characterization and antidiabetic properties. *Inorganica Chimica Acta*, **493**, 20–28.

Paternain, J.L., Domingo, J.L., Gomez, M., Ortega, A., Corbella, J. 1990. Developmental toxicity of vanadium in mice after oral administration. *Journal of Applied Toxicology*, **10**(3), 181–186.

Patterson, B.W., Hansard 2nd, S.L., Ammerman, C.B., Henry, P.R., Zech, L.A., Fisher, W.R. 1986. Kinetic model of whole-body vanadium metabolism: studies in sheep. *American Journal of Physiology-Regulatory, Integrative and Comparative Physiology*, **251**(2), R325–R332.

Pessoa, J.C., Etcheverry, S., Gambino, D. 2015. Vanadium compounds in medicine. *Coordination Chemistry Reviews*, **301–302**, 24–48.

Petranikova, M., Tkaczyk, A.H., Bartl, A., Amato, A., Lapkovskis, V., Tunsu, C. 2020. Vanadium sustainability in the context of innovative recycling and sourcing development. *Waste Management*, **113**, 521–544.

Pinto, E., Almeida, A.A., Ferreira, I.M.P.L.V.O. 2015. Assessment of metal(loid)s phytoavailability in intensive agricultural soils by the application of single extractions to rhizosphere soil. *Ecotoxicology and Environmental Safety*, **113**, 418–424.

Qian, Y., Gallagher, F.J., Feng, H., Wu, M., Zhu, Q. 2014. Vanadium uptake and translocation in dominant plant species on an urban coastal brownfield site. *Science of the Total Environment*, **476**, 696–704.

Rehder, D. 2012. The potentiality of vanadium in medicinal applications – Review. *Future Medicinal Chemistry*, **4**, 1823–1837.

Reijonen, I., Metzler, M., Hartikainen, H. 2016. Impact of soil pH and organic matter on the chemical bioavailability of vanadium species: The underlying basis for risk assessment. *Environmental Pollution*, **210**, 371–379.

Reyes, A., Thiombane, M., Panico, A., Daniele, L., Lima, A., Di Bonito, M., De Vivo, B. 2020. Source patterns of potentially toxic elements (PTEs) and mining activity contamination level in soils of Taltal city (northern Chile). *Environmental Geochemistry and Health*, **42**(8), 2573–2594.

Romanovskaia, E., Romanovski, V., Kwapinski, W., Kurilo, I. 2021. Selective recovery of vanadium pentoxide from spent catalysts of sulfuric acid production: Sustainable approach. *Hydrometallurgy*, **200**, 105568.

Schlesinger, W.H., Klein, E.M., Vengosh, A. 2017. Global biogeochemical cycle of vanadium. *Proceedings of the National Academy of Sciences*, **114**(52), E11092–E11100.

Schwertmann, U.T., Pfab, G. 1994. Structural vanadium in synthetic goethite. *Geochimica et Cosmochimica Acta*, **58**(20), 4349–4352.

Ścibior, A., Hus, I., Mańko, J., Jawniak, D. 2020. Evaluation of the level of selected iron-related proteins/receptors in the liver of rats during separate/combined vanadium and magnesium administration. *Journal of Trace Elements in Medicine and Biology*, **61**, 126550.

Ścibior, A., Pietrzyk, Ł., Plewa, Z., Skiba, A. 2020. Vanadium: Risks and possible benefits in the light of a comprehensive overview of its pharmacotoxicological mechanisms and multi-applications with a summary of further research trends. *Journal of Trace Elements in Medicine and Biology*, **61**, 126508.

Ścibior, A., Wnuk, E., Gołębiowska, D. 2021. Wild animals in studies on vanadium bioaccumulation—Potential animal models of environmental vanadium contamination: A comprehensive overview with a Polish accent. *Science of the Total Environment*, **785**, 147205.

Semiz, S. 2022. Vanadium as potential therapeutic agent for COVID-19: A focus on its antiviral, antiinflamatory, and antihyperglycemic effects. *Journal of Trace Elements in Medicine and Biology*, **69**, 126887.

Seo, D., Vasconcellos, M.B.A., Catharino, M.G.M., Moreira, E.G., de Sousa, E.C.P.M., Saiki, M. 2013. Vanadium determination in *Perna* mussels (Linnaeus, 1758: Mollusca, Bivalvia) by instrumental neutron activation analysis using the passive biomonitoring in the Santos coast, Brazil. *Journal of Radioanalytical and Nuclear Chemistry*, **296**(1), 459–463.

Shaheen, S.M., Alessi, D.S., Tack, F.M.G., Ok, Y.S., Kim, K.-H., Gustafsson, J.P., Sparks, D.L., Rinklebe, J. 2019. Redox chemistry of vanadium in soils and sediments: Interactions with colloidal materials, mobilization, speciation, and relevant environmental implications- A review. *Advances in Colloid and Interface Science*, **265**, 1–13.

Shaheen, S.M., Rinklebe, J. 2018. Vanadium in thirteen different soil profiles originating from Germany and Egypt: Geochemical fractionation and potential mobilization. *Applied Geochemistry*, **88**, 288–301.

Shaheen, S.M., Rinklebe, J., Frohne, T., White, J.R., DeLaune, R.D. 2014. Biogeochemical factors governing cobalt, nickel, selenium, and vanadium dynamics in periodically flooded Egyptian North Nile Delta rice soils. *Soil Science Society of America Journal*, **78**(3), 1065–1078.

Shaheen, S.M., Rinklebe, J., Frohne, T., White, J.R., DeLaune, R.D. 2016. Redox effects on release kinetics of arsenic, cadmium, cobalt, and vanadium in Wax Lake Deltaic freshwater marsh soils. *Chemosphere*, **150**, 740–748.

Shi, Y.X., Mangal, V., Guéguen, C. 2016. Influence of dissolved organic matter on dissolved vanadium speciation in the Churchill River estuary (Manitoba, Canada). *Chemosphere*, **154**, 367–374.

Shrivastava, S., Jadon, A., Shukla, S., Mathur, R. 2012. Reversal of vanadium-induced toxicity by combination therapy of tiferron and α-tocopherol in rat during pregnancy and their fetuses. *Therapies*, **67**(2), 173–182.

Song, Q.-Y., Liu, M., Lu, J., Liao, Y.-L., Chen, L., Yang, J.-Y. 2020. Adsorption and desorption characteristics of vanadium (V) on coexisting humic acid and silica. *Water, Air, & Soil Pollution*, **231**(9), 460.

Srinivasa Gowd, S., Ramakrishna Reddy, M., Govil, P.K. 2010. Assessment of heavy metal contamination in soils at Jajmau (Kanpur) and Unnao industrial areas of the Ganga Plain, Uttar Pradesh, India. *Journal of Hazardous Materials*, **174**(1), 113–121.

Świrk, K., Summa, P., Wierzbicki, D., Motak, M., Da Costa, P. 2021. Vanadium promoted Ni(Mg,Al)O hydrotalcite-derived catalysts for CO_2 methanation. *International Journal of Hydrogen Energy*, **46**(34), 17776–17783.

Takeda, A., Kimura, K., Yamasaki, S.-I. 2004. Analysis of 57 elements in Japanese soils, with special reference to soil group and agricultural use. *Geoderma*, **119**(3–4), 291–307.

Taylor, R.M., Giles, J.B. 1970. The association of vanadium and molybdenum with iron oxides in soils. *Journal of Soil Science*, **21**(2), 203–215.

Teng, Y., Yang, J., Sun, Z., Wang, J., Zuo, R., Zheng, J. 2011. Environmental vanadium distribution, mobility and bioaccumulation in different land-use Districts in Panzhihua Region, SW China. *Environmental Monitoring and Assessment*, **176**(1), 605–620.

Tervonen, H., Foliaki, S., Bray, F., Roder, D. 2017. Cancer epidemiology in the small nations of Pacific Islands. *Cancer Epidemiology*, **50**, 184–192.

Toro-Román, V., Bartolomé, I., Siquier-Coll, J., Alves, J., Grijota, F.J., Muñoz, D., Maynar-Mariño, M. 2021. Serum vanadium concentrations in different sports modalities. *Journal of Trace Elements in Medicine and Biology*, **68**, 126808.

Tracey, A.S., Willsky, G.R., Takeuchi, E.S. 2007. *Vanadium: Chemistry, biochemistry, pharmacology and practical applications*. CRC Press.

Treviño, S., Díaz, A., Sánchez-Lara, E., Sanchez-Gaytan, B.L., Perez-Aguilar, J.M., González-Vergara, E. 2019. Vanadium in biological action: Chemical, pharmacological aspects, and metabolic implications in diabetes mellitus. *Biological Trace Element Research*, **188**(1), 68–98.

Tsadilas, C.D., Shaheen, S.M. 2010. Distribution of total and ammonium bicarbonate-DTPA-extractable soil vanadium from Greece and Egypt and their correlation to soil properties. *Soil Science*, **175**(11), 535–543.

Tume, P., Bech, J., Longan, L., Tume, L., Reverter, F., Sepulveda, B. 2006. Trace elements in natural surface soils in Sant Climent (Catalonia, Spain). *Ecological Engineering*, **27**(2), 145–152.

United States Environmental Protection Agency (USEPA). 1995. *Ecological soil screening levels for vanadium*. Office of Solid Waste and Emergency Responses.

van Santen, R.A., Tranca, I., Hensen, E.J.M. 2015. Theory of surface chemistry and reactivity of reducible oxides. *Catalysis Today*, **244**, 63–84.

Wang, J.F., Liu, Z. 1999. Effect of vanadium on the growth of soybean seedlings. *Plant and Soil*, **216**(1), 47–51.

Wang, S., Zhang, B., Li, T., Li, Z., Fu, J. 2020. Soil vanadium (V)-reducing related bacteria drive community response to vanadium pollution from a smelting plant over multiple gradients. *Environment International*, **138**, 105630.

Wang, X., Xiang, J., Ling, J., Huang, Q., Lv, X. 2020. Comprehensive utilization of vanadium extraction tailings: A brief review. *Energy Technology 2020: Recycling, Carbon Dioxide Management, and Other Technologies*, 327–334.

Wang, Y., Yin, X., Sun, H., Wang, C. 2016. Transport of vanadium (V) in saturated porous media: Effects of pH, ionic-strength and clay mineral. *Chemical Speciation & Bioavailability*, **28**(1–4), 7–12.

Wu, C.-Y., Asano, M., Hashimoto, Y., Rinklebe, J., Shaheen, S.M., Wang, S.-L., Hseu, Z.-Y. 2020. Evaluating vanadium bioavailability to cabbage in rural soils using geochemical and micro-spectroscopic techniques. *Environmental Pollution*, **258**, 113699.

Xi, W.-S., Li, J.-B., Liu, Y.-Y., Wu, H., Cao, A., Wang, H. 2021. Cytotoxicity and genotoxicity of low-dose vanadium dioxide nanoparticles to lung cells following long-term exposure. *Toxicology*, **459**, 152859.

Xiao, X.-Y., Jiang, Z., Guo, Z., Wang, M., Zhu, H., Han, X. 2017. Effect of simulated acid rain on leaching and transformation of vanadium in paddy soils from stone coal smelting area. *Process Safety and Environmental Protection*, **109**, 697–703.

Xiao, X.-Y., Miao, Y., Guo, Z.-H., Jiang, Z.-C., Liu, Y.-N., Xia, C.A.O. 2015. Soil vanadium pollution and microbial response characteristics from stone coal smelting district. *Transactions of Nonferrous Metals Society of China*, **25**(4), 1271–1278.

Yang, J., Gao, X., Li, J., Zuo, R., Wang, J., Song, L., Wang, G. 2020. The stabilization process in the remediation of vanadium-contaminated soil by attapulgite, zeolite and hydroxyapatite. *Ecological Engineering*, **156**, 105975.

Yee, S.L., Wan, H., Chen, M., Li, L., Li, J., Ma, X. 2021. Development of a cleaner route for aluminum-vanadium alloy production. *Journal of Materials Research and Technology*, **16**, 187–193.

Yu, Y.-Q., Li, J.-X., Liao, Y.-L., Yang, J.-Y. 2020. Effectiveness, stabilization, and potential feasible analysis of a biochar material on simultaneous remediation and quality improvement of vanadium contaminated soil. *Journal of Cleaner Production*, **277**, 123506.

Yuan, R., Li, S., Che, Y., He, J., Song, J., Yang, B. 2021. A critical review on extraction and refining of vanadium metal. *International Journal of Refractory Metals and Hard Materials*, **101**, 105696.

Zhang, G., Feng, K., Li, Y., Yue, H. 2015. Effects of sintering process on preparing iron-based friction material directly from vanadium-bearing titanomagnetite concentrates. *Materials & Design*, **86**, 616–620.

Zhang, H., Zhang, B., Wang, S., Chen, J., Jiang, B., Xing, Y. 2020. Spatiotemporal vanadium distribution in soils with microbial community dynamics at vanadium smelting site. *Environmental Pollution*, **265**, 114782.

Zhong, C., Huang, Y., Ni, S.-J., Lu, H.-L., Liu, Z.-D. 2015. The distribution and influence factors of species vanadium in shallow groundwater near the slag field of Baguan river in Panzhihua area. *Computing Techniques for Geophysical and Geochemical Exploration*, **37**(2), 263–266.

Zou, Q., Gao, Y., Yi, S., Jiang, J., Aihemaiti, A., Li, D.A., Yang, M. 2019. Multi-step column leaching using low-molecular-weight organic acids for remediating vanadium-and chromium-contaminated soil. *Environmental Science and Pollution Research*, **26**(15), 15406–15413.

Zou, Q., Li, D.A., Jiang, J., Aihemaiti, A., Gao, Y., Liu, N., Liu, J. 2019. Geochemical simulation of the stabilization process of vanadium-contaminated soil remediated with calcium oxide and ferrous sulfate. *Ecotoxicology and Environmental Safety*, **174**, 498–505.

2

Critical Limits and Health Risk Assessment of Vanadium in Soils of Various Countries of the World

Vasileios Antoniadis[1], Sabry M. Shaheen[2], Efi Levizou[1] and Jörg Rinklebe[2]

[1] *University of Thessaly, Department of Agriculture Crop Production and Rural Environment, Volos, Greece*

[2] *University of Wuppertal, School of Architecture and Civil Engineering, Institute of Foundation Engineering, Water- and Waste-Management, Laboratory of Soil- and Groundwater-Management, Wuppertal, Germany*

CONTENTS

2.1 Introduction ... 33
2.2 Regulation Limits ... 34
  2.2.1 The Necessity of Setting Up Limits ... 34
  2.2.2 Discrepancies in Trace Element Regulation Limits ... 35
    2.2.2.1 Land Uses ... 35
    2.2.2.2 Names of Regulation Categories ... 35
    2.2.2.3 Countries with No Soil Regulation ... 36
    2.2.2.4 Vanadium Limits in Water ... 37
2.3 Vanadium-Based Human Health Risk Assessment ... 37
2.4 Conclusions ... 41
References ... 45

2.1 Introduction

Vanadium is a trace element of metallic nature which is considered to be a nonessential trace element to living organisms. Its natural "background" total soil concentration is reported as 129 mg kg^{-1} (Kabata-Pendias, 2011, p. 24), but it is often found at elevated concentrations, even if naturally occurring. Due to its very low mobility from soil to plants (Ochoa et al., 2020) and due to the fact that there are analytical difficulties in measuring it, V is an often-overlooked

DOI: 10.1201/9781003173274-2

element in studies concerning soil contamination (Antoniadis, Golia et al., 2019). However, V does have some vital industrial usages, for example, in the production of certain steel and aluminum alloys; it also is a byproduct of uranium mining (Chen et al., 2021). It is thus an element reported at elevated concentrations in various soil contamination studies, especially when soils around industrial or mining sites are studied (Rinklebe et al., 2019).

Given these considerations, unlike other frequently studied potentially toxic elements, V is often overlooked by countries and legal entities concerning maximum allowable limits. As a result, only a few countries have set up limits for soils concerning V. These limits are not only few in number but also highly diverse. Thus there is a need for reviewing all such V limits, by collecting and presenting them under one roof in a comprehensive manner. This is very important given the fact that some countries have recognized that V contamination is an important issue: this element has recently been included in the new German Soil Conservation Law (currently a draft). Also there is a need to critique the set values as to whether they are suitable or not entirely protective. One way to approach this is to apply a human health risk assessment (HRA) based on the three major human exposure pathways – direct soil particles ingestion, dermal contact and inhalation. Having as a base the limit values set by various countries, one can assess if health risk is low (thus limits are indeed protective) or high (limits need to be lower).

Thus, the aim of this chapter is to (a) present the V maximum allowable limit values in a comprehensive way and (b) to critique them based on the HRA approach as to whether they are sufficiently protective.

2.2 Regulation Limits

2.2.1 The Necessity of Setting Up Limits

Various countries impose V limits, assuming a certain level of critical exposure to such a nonessential element. Such limits should be land use-specific, as we will see later; also, different age groups do not have the same sensitivity, with children being the most vulnerable (Shaheen, Antoniadis et al., 2020), and among adults, typically women are more sensitive than men (Shaheen, El-Naggar et al., 2020). Limits must be sufficiently low so that they may be protective. However, lower limits tend to be more restrictive; thus, social and industrial restrictions may unnecessarily be imposed and law enforcement may then be more costly. On the other hand, if limits are overly relaxed, this would lead to health risks being realized (Antoniadis, Shaheen et al., 2019). Hence there is a need to propose correct limit values that would be both sufficiently protective and not unnecessarily relaxed.

2.2.2 Discrepancies in Trace Element Regulation Limits

2.2.2.1 Land Uses

V limits are usually set up depending on land use; residential areas usually have the lowest limits. This is understandable since in such areas humans spend most of their time, and also there is activity of the more vulnerable members of the society, that is, children. There are some countries that also set limit values for industrial land use (usually identified in the same category as "commercial"), where values are typically one order of magnitude higher than those set for residential areas. In some countries, there is an indication of an even more sensitive land use than residential, that is, amenity areas such as parks and playgrounds; in such land uses, limits are even lower due to the risk of children playing and thus directly coming into contact with soil. In Europe (Table 2.1), V is regulated in Italy (90 mg kg^{-1}) and the Czech Republic (450 mg kg^{-1}) concerning residential areas, and only in Canada and South Africa (300 mg kg^{-1}) outside of Europe. In Canada, there are different limits for the different provinces (Alberta = 1; Ontario = 86 mg kg^{-1}) and for the central government (130 mg kg^{-1}; Table 2.1). As for industrial land use, apart from Italy (250) and the Czech Republic (550), V is also regulated in Poland (300); in countries outside of Europe, apart from Canada and South Africa, V is also regulated in Australia (7,200 mg kg^{-1}). There is also a land use concerning agriculture, set by France (560) and Poland (300), as well as Canada (130) and Australia (550), while there are a number of countries that regulate V without land use specifications (Denmark = 42; Sweden = 200; Netherlands and Finland = 250; Lithuania = 450; Slovakia = 500; all above-mentioned units in mg kg^{-1}). As for Germany, there is a draft document recording precautionary values for three soil categories, not based on land uses but rather based on particle size distribution, with sand having a limit of V = 30, and two other listed categories ("Silt/Loam," and "Clay") having V = 100 mg kg^{-1}.

2.2.2.2 Names of Regulation Categories

One source of confusion in legal limits may well be the fact that different names that imply different meaning are being used; for V, the residential land use limit of Italy is the "Screening Value," while that of the Czech Republic is "Action Value" and of Canada is "Soil Quality Guidelines." In Australia, the industrial land use limit value bears the name "Health Investigations Levels," while for agricultural use in France the name is "Impact Statement Value," and in Poland "Soil Quality Standards." In other countries that have set up limits without any land use specifications, names also vary greatly, with Denmark having "Trigger Value," the Netherlands and Lithuania "Guideline Value" and Slovakia "Value for Decontamination Measures." This differentiation in limit names might be the source of great confusion, as it may

TABLE 2.1

Limits for V for Various Land Uses (Residential, Industrial, Agricultural, and Non-Specified Use) in Various Countries

		Country	Value	References
Residential	Europe	Italy (SV)	90	Carlon (2007)
		Czech R. (AV)	450	Carlon (2007)
	Non-Europe	Ontario-Ca (SS)	86	Ontario (2011)
		Alberta-Ca (Res/Park)	1	Alberta (2016)
		Canada (SQG)	130	CCME (2018)
		S. Africa (StandRes)	300	MoE (2018)
Industrial	Europe	Italy (SV)	250	Carlon (2007)
		Poland (SQS, Deep/LHC)	300	Carlon (2007)
		Czech R. (AV)	550	Carlon (2007)
	Non-Europe	Canada (SQG)	130	CCME (2018)
		Australia (HIL)	7,200	DoEC (2010)
		S. Africa (Comm/Ind)	2,600	GGS (2012)
Agricultural	Europe	France (ISV, Gard/playgr)	560	Carlon (2007)
		Poland (SQS, Deep/LHC)	300	Carlon (2007)
	Non-Europe	Canada (SQG)	130	CCME (2018)
		Australia (HIL, Res/Gard)	550	DoEC (2010)
Use non-specified	Europe	Denmark (TV)	42	DEPE (2002)
		Sweden (GV, Non-sens)	200	INSURE (2017)
		Slovakia (VDM)	500	Carlon (2007)
		Netherlands (IV)	250	MvV (2000)
		Lithuania (IV)	450	CF (2009)
		Finland (GV, High)	250	MotE (2007)
		Germany[a] (PV)	100	Germany (2012)

a. Draft document.

be falsely expected that common names should bear common meaning or should indicate the same rate of alarm, while this is not necessarily the case.

2.2.2.3 Countries with No Soil Regulation

Undoubtedly the greatest problem concerning V management is the fact that most countries even in the developed world fail to take the element into account concerning regulation limits. Thus, it is not just an issue of V limit values varying greatly among countries with similar climate (e.g., Italy and Czechia in residential soils) but rather the lack of any legal concern for V in many countries. The possible reason for that presumably lies in the behavior of V in soil and its rather low mobility to plants and thus into the human food chain. However, such certainties could be overturned if cases of V contamination arise, since there are currently no legal barriers to protect human health. In other words, V should indeed be among the regulated elements even if no acute toxicities are currently reported, so that countries may be protected if such crises arise.

2.2.2.4 Vanadium Limits in Water

Similar regulatory discrepancies to those in soil exist in the case of V in water. However, in water there are only three countries that regulate V, to the best of our knowledge: Lithuania, the Netherlands and South Africa. Similar to the soil limits, category names, even among those three countries, are totally dissimilar, and this is also the case with the values themselves, ranging from as low as 1.2 to 200 μg L^{-1} (Table B1, Appendix). Due to the lack of sufficient data, we will not discuss V further concerning health risks derived from water consumption. However, we do recognize the necessity of performing such assessments concerning elements like V, which are not studied very often concerning contamination of water bodies as related to human health.

2.3 Vanadium-Based Human Health Risk Assessment

Table 2.2 shows the soil intake values via ingestion, dermal absorption and inhalation (in μg V per day per kg body weight); the values, for clarity, are magnified by 10^6 times. The values of the factors used in the HRA equations are given in Table B2 (Appendix). The values have been calculated based on the equations given in the Appendix and conform with the scenario that the maximum allowable concentrations of soil V have been attained. It is evident that the primary exposure pathway is that of ingestion, with the other two being a rather insignificant portion compared to it. Then, the sum of soil intake (indicated as $\Sigma(D)$, also magnified by a factor of 10^6) is also calculated. In Figure 2.1 we present the resulting hazard quotient (HQ) values (i.e., as divided by the RfD value for V, equal to 5.04×10^{-3} mg kg^{-1} day^{-1}). The analysis we made distinguished between adults in residential areas and children in these areas. Also between industrial workers and workers in agriculture (in which sections, by definition, no children are included). As for the countries that have not specified land uses, we took the worst-case scenario, that

TABLE 2.2

Average Daily Dose concerning the Pathways of Human Exposure (in μg V per day per kg body weight) via Ingestion (D-Ing), Dermal Contact (D-Derm), and Inhalation (D-Inh), and the Resultant HQ as an Index of Human Health Risk

			D-Ing (×10⁶)	D-Derm (×10⁶)	D-Inh (×10⁶)	Σ(D) (×10⁶)
Residential-Adult						
Italy	SV	Res/publ	123.29	0.49	0.02	123.80
Czech	AV	Res	616.44	2.46	0.09	618.99
Ontario-Ca	SS		117.81	0.47	0.02	118.30
Alberta-Ca	Res/Park		1.37	0.01	0.00	1.38
Canada	SQG	Res	178.08	0.71	0.03	178.82
S. Africa		Stand Res	410.96	1.64	0.06	412.66
Denmark[a]	TV		57.53	0.23	0.01	57.77
Sweden[a]	GV	Non-sens	273.97	1.09	0.04	275.11
Slovakia[a]	VDM		684.93	2.73	0.10	687.77
Netherland[a]	IV		342.47	1.37	0.05	343.88
Lithuania[a]	IV		616.44	2.46	0.09	618.99
Germany[a, b]	PV	Silt/Loam/ Clay	136.99	0.55	0.02	137.55
Finland[a]	GV		342.47	1.37	0.05	343.88
Residential-Child						
Italy	SV	Res/publ	1150.68	3.22	NA	1153.91
Czech	AV	Res	5753.42	16.11	NA	5769.53
Ontario-Ca	SS		1099.54	3.08	NA	1102.62
Alberta-Ca	Res/Park		12.79	0.04	NA	12.82
Canada	SQG	Res	1662.10	4.65	NA	1666.75
S. Africa	1	Stand Res	3835.62	10.74	NA	3846.36
Denmark[a]	TV		536.99	1.50	NA	538.49
Sweden[a]	GV	Non-sens	2557.08	7.16	NA	2564.24
Slovakia[a]	VDM		6392.69	17.90	NA	6410.59
Netherland[a]	IV		3196.35	8.95	NA	3205.30
Lithuania[a]	IV		5753.42	16.11	NA	5769.53
Germany[a, b]	PV	Silt/Loam/ Clay	1278.54	3.58	NA	1282.12
Finland[a]	GV		3196.35	8.95	NA	3205.30
Industrial-Worker						
Italy	SV	Comm/Ind	78.63	0.04	0.03	78.70
Poland		Ind/deep/ LHC	94.35	0.05	0.03	94.44
Czech	AV	Ind	172.98	0.09	0.06	173.13
Canada	SQG	Ind	40.89	0.02	0.01	40.92
Australia	HIL	Ind/Comm	2264.47	1.22	0.78	2266.47

(Continued)

TABLE 2.2 (Continued)

			D-Ing ($\times 10^6$)	D-Derm ($\times 10^6$)	D-Inh ($\times 10^6$)	Σ(D) ($\times 10^6$)
S. Africa		Comm/Ind	817.72	0.44	0.28	818.45
Agriculture-Worker						
France	ISV	Gard/ playgr	547.95	2.11	0.08	550.14
Poland		Agr/deep/ LHC	293.54	1.13	0.04	294.72
Canada	SQG	Agr	127.20	0.49	0.02	127.71
Australia	HIL	Res/gard	538.16	2.07	0.08	540.31

Note: The average daily dose values are given in $\times 10^6$ for clarity.
a. Soil use not specified, thus values are classified as for residential use.
b. Draft document
NA: Not applicable

is, we used the values in the case of them being in a residential area (thus we distinguished in the previously mentioned countries two age groups). Our analysis reveals that the HQ values for adults in residential areas are always well below the threshold value of 1.0, with the maximum being 0.136 for the Netherlands, followed by 0.122 (for the Czech Republic and Lithuania; Figure 2.1). However, concerning children, HQ for the Netherlands was 1.268 and those of Czechia and Lithuania 1.142, indicating considerable health risk for children. In the other two land uses, HQ was still well below the 1.0 threshold.

This shows that the V limit values are overly protective concerning industrial areas and agricultural lands, meaning that it is feasible for authorities to examine the possibility of increasing the limit values for such zones. However, concerning residential areas, it is evident that not all age groups are protected, with children being exposed to health risk if maximum allowable levels are approached in the Netherlands, the Czech Republic and Lithuania. It may thus be advisable for a reconsideration of the regulations concerning V in the direction of them being decreased to include the protection of the vulnerable age groups. In Table 2.3 we back-calculated the maximum values of soil concentration of V that would result in HQ = 1.0. It is exhibited that V in industrial zones could be set as high as more than 16,000 mg kg^{-1} without any significant risk to the employees, and likewise the limit could be relaxed to as high as 5,000 mg kg^{-1} in agricultural lands. However, in residential areas, if children are to be sufficiently protected, V legal limits should be made sure not to exceed 400 mg kg^{-1}.

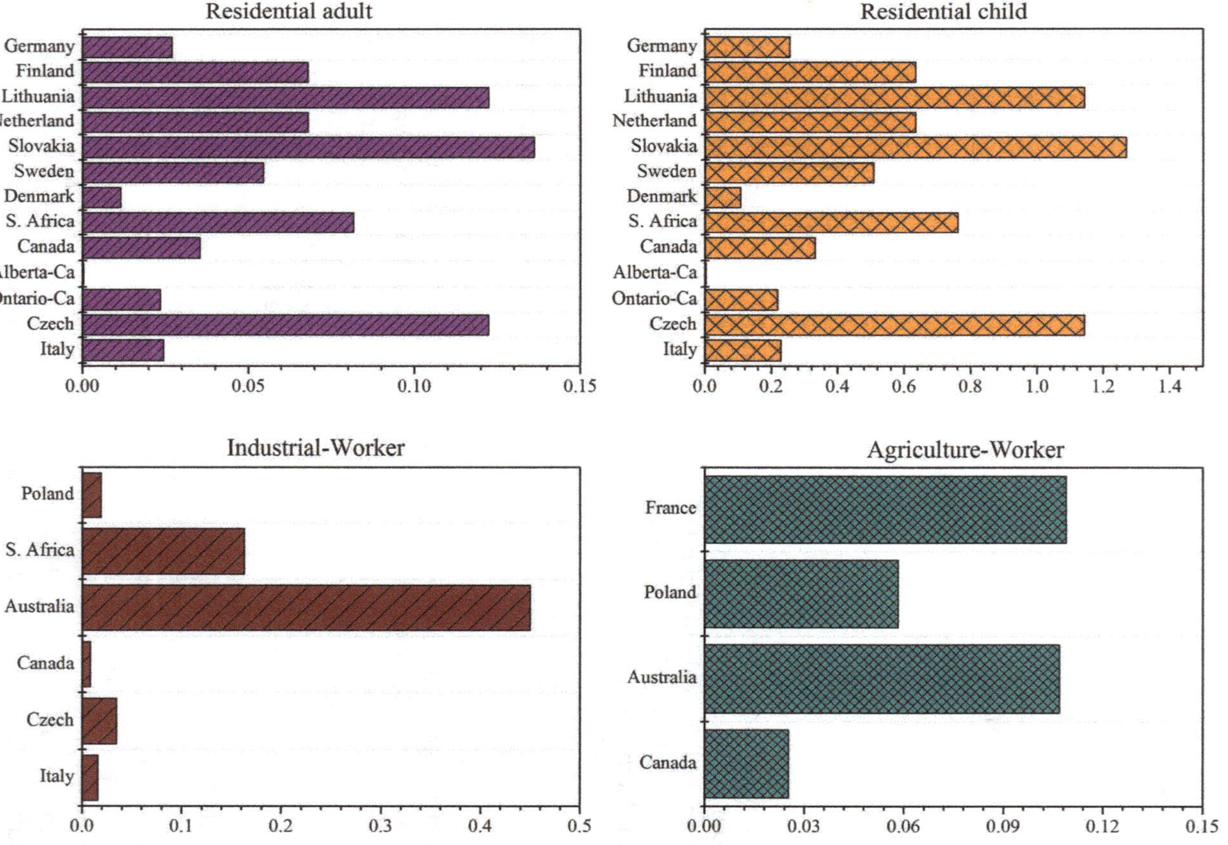

FIGURE 2.1
Hazard quotient (HQ) of V in various countries, land uses and person ages.

TABLE 2.3

Maximum Allowable Levels of V in Soil So That HQ Derived from All Three Human Exposure Pathways Would Be Equal to 1.0 (the Upper Limit of No Induced Health Risk) according to Age Groups and Land Uses

Age Group	Occupation/ Land Use	Total Soil V That Would Result in HQ = 1.0
Adult	Residential	3,333
Child	Residential	400
Adult	Industrial	16,667
Adult	Agriculture [a]	5,000

a. Concerns only exposure concerning a worker in agriculture, not the exposure via the consumption of V-containing plants.

2.4 Conclusions

In this chapter, we showed, to the best of our knowledge, all regulation limits concerning V in soil and water around the globe under one roof. There is a great difficulty when dealing with these values that has to do with (a) the immense diversity in values set up by different countries and (b) the differences in names given to the values even when they serve the same purpose (i.e., even when they represent limits for the same land uses). We recognize the fact that any attempt at normalizing theses discrepancies is a rather herculean task; however, this is a necessity, since soil and water management in some cases may be trans-boundary (e.g., when a uniform pedogenetic area may lie across country boundaries or common pollution sources may affect soils and water bodies across countries uniformly). The HRA analysis we performed followed a scenario in which soils under certain land uses (the same ones specified in the V limits) have the maximum allowable concentrations; this analysis revealed that the set limits are rather overprotective (i.e., too low) concerning industrial sites and agricultural lands, but in residential areas, vulnerable age groups (here, children) are unprotected by the limits set in three European countries. We propose, based on our HRA analysis, that a sufficiently protective limit for soil V would be 400 mg kg^{-1} in residential areas in order to cover sensitive groups of people. We recognize that such a probabilistic approach is by no means conclusive, as a more in-depth approach should also take into account properties and characteristics that are now unaccounted for, such as soil pH, clay, oxides and organic matter content and the geochemical fractionation of V under specific conditions, as well as plant species involved.

Abbreviations

SV	Screening Value
IV	Intervention Value
TV	Trigger Value
AV	Action Value
SS	Stratified Site Condition Standards in Non-potable Ground Water Conditions
Res/Park	Residential/Parks
SQG	Soil Quality Guidelines
StandRes	Value for "Standard Residential Area"
SQS	Soil Quality Standards
Deep/LHC	Agricultural Use/Deep Soil (depth >15 m with low hydraulic conductivity (<10^{-7} m sec^{-1}))
HIL	Health Investigation Levels
Comm/Ind	Commercial/Industrial
ISV	Impact Statement Value
Gard/playgr	Garden/Playground
Res/Gard	Residence with Garden
Non-sens	Non-sensitive Soils
VDM	Value for Decontamination Measures
High	High Value in a Range of Values
MAV	Maximum Allowable Value
TVSW	Target Value, Shallow Watertable
ROL	Recommended Operational Limit
TestV	Test Value
PV	Precautionary Value

Appendix A. Theory and Equations of Health Risk Assessment

In the health risk assessment (HRA) approach, there are usually three possible pathways examined: (a) ingestion (insertion of soil particles into the human body via the gastrointestinal tract), (b) dermal absorption (insertion of soil particles via absorption caused by dermal contact) and (c) inhalation (insertion of soil particles via the inhalation system – i.e., inhaled towards the lungs).

First the intake quantity, *D*, must be calculated in mg V per kg bodyweight per day, with the quantity of ingestion being D_{Ing}, that of dermal contact D_{Derm} and that of inhalation D_{Inh} (USEPA, 2002):

$$D_{S-Ing} = C_S \cdot \frac{IngR \cdot EF \cdot ED}{BW \cdot AT} \cdot 10^{-6} \tag{1}$$

$$D_{Derm} = C_S \cdot \frac{SA \cdot SAF \cdot ABS \cdot EF \cdot ED}{BW \cdot AT} \cdot 10^{-6} \tag{2}$$

$$D_{Inh} = C_S \cdot \frac{InhR \cdot EF \cdot ED}{PEF \cdot BW \cdot AT} \tag{3}$$

where C_S is soil V (mg kg^{-1}); IngR = Ingestion rate (mg soil d^{-1}); EF = Exposure frequency (d yr^{-1}); ED = Exposure duration (yr); BW = Body weight (kg); AT = Averaging time (d); SA = Surface area (cm^2); SAF = Skin adherence factor (mg soil cm^{-2}); ABS = Dermal absorption factor (unitless); InhR = Inhalation rate (m^3 d^{-1}); PEF = Particle emission factor (m^3 kg^{-1}). Values used here are shown in Table B2 (Appendix) and were all obtained by USEPA (2002).

Risk was then assessed by the hazard quotient (HQ), the ratio of D over reference dose (RfD), i.e., the maximum allowable quantity of V, equal to 5.04×10^{-3} mg V per kg bodyweight per day. If *D* is higher than RfD, i.e., HQ > 1, this indicates considerable health risk; if HQ < 1, then there is no significant risk.

Full details concerning the HRA equations may be found elsewhere (Antoniadis, Shaheen et al., 2019). In this work, we obtained as C_S the values of the regulation limits of V to test the scenario in which a person ingests, absorbs dermally and inhales air-borne soil particles containing V concentrations equal to the maximum allowable concentrations, and thus assess if the set limits are sufficiently protective.

Appendix B. Supplementary Tables

TABLE B1

Vanadium Regulations Maximum Allowable Limits in Water (Values in μg L^{-1})

Country	Category	Value
Lithuania	MAC	100
Holland-GW	TVSW	1.2
South Africa	ROL	200
Germany	TestV	4

TABLE B2

Values of the Various Factors Used in the Equations concerning Human Health Risk in This Work Obtained from USEPA (2002) Unless Otherwise Indicated

Route to Human	Factor (units)	Exposure Scenarios		
		Residential-Adult	Residential-Child	Non-residence (outdoor worker)
Ingestion				
	IngR soil (mg d^{-1}) [water (L d^{-1})]	100 [2]	200 [0.8]	100
Dermal				
	SA (cm^2)	5700	2800	3300
	SAF (mg cm^{-2})	0.07	0.2	0.07
	ABS (unitless)	As: 0.03 All else: 10^{-3}	As: 0.03 All else: 10^{-3}	As: 0.03 All else: 10^{-3}
Inhalation				
	InhR (m^3 d^{-1})	20	NA	20
	PEF[b] (m^3 kg^{-1})	1.36×10^9	NA	1.36×10^9
	EF (d yr^{-1})	350	350	225 (agr'ture: 250)
Common factors in all routes				
	ED (yr)	30	6	25 (agr'ture: 50)
	BW (kg)	70	15	
	AT (d)	ED × 365	ED × 365	ED × 365
	LT (yr)	70	–	70

IngR = Ingestion rate; SA = Surface area; SAF = Skin adherence factor; ABS = Dermal absorption factor; InhR = Inhalation rate; PEF = Particle emission factor; EF = Exposure frequency; ED = Exposure duration; BW = Body weight; AT = Averaging time.

References

Alberta Government, 2016. Alberta tier 1: Soil and groundwater remediation guidelines. Available at www.alberta.ca/part-one-soil-and-groundwater-remediation.aspx?utm_source=redirector (Last access on 24 April 2021).

Antoniadis, V., Golia, E.E., Liu, Y., Wang, S., Shaheen, S.M., Rinklebe, J., 2019. Soil and maize contamination by trace elements and associated health risk assessment in the industrial area of Volos, Greece. *Environment International* 124, 79–88.

Antoniadis, V., Shaheen, S.M., Levizou, E., Shahid, M., Niazi, N.B., Vithanage, M., Ok, Y.S., Bolan, N., Rinklebe, J., 2019. A critical prospective analysis of the potential toxicity of trace element regulation limits in soils worldwide: Are they protective concerning health risk assessment?—A review. *Environment International* 127, 819–847.

Carlon, C., 2007. Derivation methods of soil screening values in Europe: A review and evaluation of national procedures towards harmonization. Available at https://esdac.jrc.ec.europa.eu/content/derivation-methods-soil-screening-values-europe-review-and-evaluation-national-procedures (Last access on 24 April 2021).

CCME (Canadian Council of Ministers of the Environment), 2018. Soil quality guidelines for the protection of environmental and human health. Available at http://st-ts.ccme.ca/en/ (Last access on 24 April 2021).

Chen, L., Liu, J.-R., Hu, W.-F., Gao, J., Yang, J.-Y., 2021. Vanadium in soil-plant system: Source, fate, toxicity, and bioremediation. *Journal of Hazardous Materials* 405, 124200.

CF (Common Forum—Working Document), 2009. Compilation of standards for contamination of surface water, groundwater, sediments and soil. Available at www.commonforum.eu/Documents/WorkingDocument/CFWD_Standards.pdf (Last access on 24 April 2021).

DEPE (Danish Environmental Protection Agency), 2002. Environmental guidelines no. 7, 2002. Vejledning fra Miljøstyrelsen. Guidelines on remediation of contaminated sites. Available at https://www2.mst.dk/udgiv/publications/2002/87-7972-280-6/pdf/87-7972-281–4.pdf (Last access on 16 September 2018).

DoEC (Department of Environment and Conservation, Australia), 2010. Contaminated sites management series assessment levels for soil, sediment and water, version 4, revision 1, February 2010. Available at www.der.wa.gov.au/images/documents/your-environment/contaminated-sites/guidelines/2009641_-_assessment_levels_for_soil_sediment_and_water_-_web.pdf (Last access on 24 April 2021).

Germany, 2012. Draft document of the 31st of October 2012: Verordnung zur Festlegung von Anforderungen für das Einbringen oder das Einleiten von Stoffen in das Grundwasser, an den Einbau von Ersatzstoffen und für die Verwendung von Boden und bodenähnlichem Material [Ordinance laying down requirements for introducing or the discharge of substances into the groundwater, the incorporation of substitutes and the use of soil and soil-like material]. Referat WA III 3 WA III 3–73103–1/0 RefL. Pastor, Bonn, den (Last access on 31 October 2012).

GGS (Government Gazette Staatskoerant, South Africa), 2012. National environmental management: Waste act (59/2008): Draft national norms and standards for the remediation of contaminated land and soil quality: For public comments no. 35160. Available at https://cer.org.za/wp-content/uploads/2010/03/national-environmental-management-waste-act-59–2008-national-norms-and-standards-for-the-remediation-of-contaminated-land-and-soil-quality_20140502-GGN-37603–00331.pdf (Last access on 24 April 2021).

INSURE, 2017. EQS limit and guideline values for contaminated sites—report 2017. Available at www.meteo.lv/fs/CKFinderJava/userfiles/files/EQS_limit_and_guideline_values.pdf (Last access on 24 April 2021).

Kabata-Pendias, A., 2011. *Trace Elements in Soils and Plants*, 4th Edition. CRC Press and Taylor and Francis Group, Boca Raton.

MotE (Ministry of the Environment, Finland), 2007. Government decree on the assessment of soil contamination and remediation needs 214/2007 of March 1, 2007. Available at www.finlex.fi/en/laki/kaannokset/2007/en20070214.pdf. (Last access on 24 April 2021).

MoE (Ministry of Environment, Korea), 2018. Soil contaminants and control limits. Available at http://eng.me.go.kr/eng/web/index.do?menuld=311 (Last access on 3 June 2019).

MvV (Ministerie von Volkshuissvesting), 2000. Dutch target and intervention values, 2000 (the New Dutch List). Annexes: Circular on target values and intervention values for soil remediation. Available at www.esdat.net/Environmental%20 Standards/Dutch/annexS_I2000Dutch%20Environmental%20Standards.pdf (Last access on 24 April 2021).

Ochoa, M., Tierra, W., Tupuna-Yerovi, D.S., Guanoluisa, D., Otero, X.L., Ruales, J. 2020. Assessment of cadmium and lead contamination in rice farming soils and rice (*Oryza sativa* L.) from Guayas province in Ecuador. *Environmental Pollution* 260, 11450.

Ontario, 2011. Soil, groundwater, and sediment standards for use under part XV.1 of the environmental protection act. Available at www.ontario.ca/page/soil-ground-water-and-sediment-standards-use-under-part-xv1-environmental-protection-act (Last access on 24 April 2021).

Rinklebe, J., Antoniadis, V., Shaheen, S.M., Rosche, O., Altermann, M. 2019. Health risk assessment of potentially toxic element with the aid of indices on the example of soils along the Central Elbe River as a model region. *Environment International* 126, 76–88.

Shaheen, S.M., Antoniadis, V., Kwon, E., Song, H., Wang, S.-L., Hseu, Z.-Y., Rinklebe, J. 2020. Soil contamination by potentially toxic elements and the associated human health risk in geo- and anthropogenic contaminated soils: A case study from the temperate region (Germany) and the arid region (Egypt). *Environmental Pollution* 262, 114312.

Shaheen, S.M., El-Naggar, A., Antoniadis, V., Moghanm, F.S., Zhang, Z., Tsang, D.C.W., Ok, Y.S., Rinklebe, J. 2020. Release of toxic elements in fishpond sediments under dynamic redox conditions: Assessing the potential environmental risk for a safe management of fisheries systems and degraded waterlogged sediments. *Journal of Environmental Management* 225, 109778.

USEPA (United States Environment Protection Agency), 2002. OSWER 9355.4–24 December 2002. Supplemental guidance for developing soil screening levels for superfund sites. office of emergency and remedial response, U.S. Environmental Protection Agency, Washington, DC, 20460. Available at https://nepis.epa.gov/Exe/ZyNET.exe/91003IJK.TXT?ZyActionD=ZyDocument&Client=EPA&Index=2000+Thru+2005&Docs=&Query=&Time=&EndTime=&SearchMethod=1&Toc

Restrict=n&Toc=&TocEntry=&QField=&QFieldYear=&QFieldMonth=&QField Day=&IntQFieldOp=0&ExtQFieldOp=0&XmlQuery=&File=D%3A%5Czyfiles %5CIndex%20Data%5C00thru05%5CTxt%5C00000023%5C91003IJK.txt&User= ANONYMOUS&Password=anonymous&SortMethod=h%7C-&MaximumDoc uments=1&FuzzyDegree=0&ImageQuality=r75g8/r75g8/x150y150g16/i425&D isplay=hpfr&DefSeekPage=x&SearchBack=ZyActionL&Back=ZyActionS&Back Desc=Results%20page&MaximumPages=1&ZyEntry=1&SeekPage=x&ZyPURL (Last access on 24 April 2021).

3

Kinetics of Vanadium Sorption/Desorption in Soils

Tamer A. Elbana[1], Wenguang Sun[2], Joshua Padilla[3] and H. Magdi Selim

[1] *Soils and Water Use Department, National Research Centre, Cairo, Egypt*

[2] *Nebraska Water Center, Robert B. Daugherty Water for Food Global Institute, University of Nebraska, Lincoln, Nebraska, USA*

[3] *School of Plant, Environmental, and Soil Sciences, Louisiana State University, Baton Rouge, Louisiana, USA*

CONTENTS

3.1 Introduction 49
3.2 Soil/Solution Interaction 52
  3.2.1 Influence of Soil pH 52
  3.2.2 Influence of Phosphate 53
  3.2.3 Influence of Iron Oxides 55
  3.2.4 Influence of Clay Minerals and Organic Substances 56
3.3 Sorption/Desorption in Soils 56
3.4 Distribution Coefficients and Sorption Capacity of Soils 57
3.5 Retention/Release Hysteresis 58
3.6 Equilibrium and Kinetics Modeling 59
3.7 Prospects of Vanadium Kinetics Studies 65
3.8 Acknowledgment 66
References 67

3.1 Introduction

The formation of attractive colors of vanadium (V) species in solutions is related to its various oxidation states. The valence state of V varies widely from –2 to +5, where V forms numerous cationic, anionic and organometallic compounds. In soil, V (+3, +4, +5) occurs in mineral lattices, and it exists as free vanadate anions (+5) in soil solution (Madejón, 2013). Oxocation such as VO^{2+} is dominant in acidic soils, whereas anions such as VO_3^-, HVO_4^{2-}, and $H_2VO_4^-$ occur

DOI: 10.1201/9781003173274-3

in nonacidic soils (Kabata-Pendias and Sadurski, 2004). However, plants absorb soluble tetravalent (oxocation vanadyl: VO^{2+}) and pentavalent (oxoanion vanadate: HVO_4^{2-} and $H_2VO_4^-$) species (Welch, 1973; Hopkins et al., 1977; Morrell et al., 1986). Vanadate toxicity is attributed to its similarity with o-phosphate and its ability to substitute for P (Hurlbut and Klein, 1977; Gustafsson, 2019). A worldwide concentration of V in agricultural soils is reported in a range of 15 mg kg^{-1} to 250 mg kg^{-1} (Teng et al., 2011) and 10 mg kg^{-1} to 500 mg kg^{-1}, with an average of 60 mg kg^{-1} (Kabata-Pendias and Mukherjee, 2007). Vanadium ecological soil screening levels of 7.8 mg kg^{-1} and 280 mg kg^{-1} are derived for avian and mammalian species, respectively (United States Environmental Protection Agency, 2005). However, accidental V release endangers the soil environment and living organisms. In 2010, the spill of red mud slurry containing a high V concentration of 870 mg kg^{-1} polluted soils and natural water resources of Hungary (Ruyters et al., 2011). Mayes et al., 2016 recommended the necessity of long-term monitoring of vanadate in such polluted soils due to its potential toxic effect. Improper use of a slag containing 3% V on a farm in northern Sweden caused toxicity and death of 23 heifers and depression of milk production for the surviving cows (Frank et al., 1996). While V is involved in nitrogen fixation bacteria and enzymes, high V concentrations cause plant damage due to its oxidative effects (Imtiaz et al., 2015).

Understanding the chemical behavior of V in soils is essential to quantify its fate in the environment. Natural minerals (e.g., vanadinite and roscoelite), fossil fuel combustion, refining process of petroleum products and steel production, among other industrial activities, are various sources of V contamination into agricultural soils (Madejón, 2013). The behavior of V in soils varies based on its speciation under different soil conditions (soil aeration, water saturation and alkalization, and among others). In contaminated soils adjacent to a vanadium mine, Panichev et al. (2006) reported a range from 51% to 68% of the total V (3505 mg kg^{-1} and 2200 mg kg^{-1}, respectively) occurred as V (+4), whereas the respective percentage for V (+5) represented 29% to 51%. However, they stated that in the growing grass, 52% and 41% of the total V (7.5 mg kg^{-1} and 10.9 mg kg^{-1}, respectively) were found in the form V(+4), whereas the respective values for V(+5) were 35% and 54%. In three neutral or weakly alkaline soils, Yang et al. (2021) found that 87.5% to 95.4% of total V occurred as V(+4). Furthermore, they reported ranges of 6% to 14 and 5% to 9% of V were associated with reducible and oxidizable fractions, respectively. At the same time, less than 0.1% of V occurred in the acid-soluble form. Based on the high-performance liquid chromatography (HPLC) separation procedure, Kuo et al. (2007) explored V speciation in soil samples with total V in the range of 72.8 to 101.0 mg kg^{-1}. Their results revealed that V(+4) fraction varied from 49% to 64%, whereas V(+5) varied from 36% to 51%. However, Baken et al. (2012) treated three acidic soils (pH: 5.5–6.6) with V(+5), which then aged for periods of 5 to 11 months. They found that V(+4) concentration in soil extracts using 10 mM $CaCl_2$ ranged from 0.06 and 0.14 mg L^{-1}, which accounted for less than 8% of the total V, whereas V(+5) varied between 0.59 and 3.02 mg L^{-1}. The V(+4) exhibits a high ability to form complexes at low pH with organic and inorganic ligands such as $VOH(\text{malonate})^{+}$, $VO(\text{oxalate})_2^{2-}$, $VOH_2PO_4^{+}$, $VOOHCO_3^{-}$, $H_2V_2O_4^{2+}$, and $VOCl^{+}$ whereas at high pH, V exists predominantly as reactive tetrahedral vanadate(+5) (Gustafsson, 2019).

Moreover, V forms associations with different soil constituents such as OM, Fe/Al (hydr)oxides, carbonate and various clay minerals. For example, Shaheen and Rinklebe (2018) assessed V fractionation in Egyptian and German soils, where V was below detection limits in the acid-soluble fraction. However, they found that 27%, 8% and 65% of V were associated with reducible, oxidizable and residual fractions, respectively, in the Egyptian soils, while the respective values in the German soils were 19%, 6% and 76%. Their results indicated a high potential risk of V mobility in alkaline soils. Cappuyns and Slabbinck (2012) reported a significant positive correlation between V concentration and organic carbon, clay, Al and Fe contents in soils. In addition, they found that Al, Fe and Mg concentrations in soil can be applied as independent variables to predict V concentration. A strong positive linear V-Fe relationship ($r^2 > 0.82$) was found for moderately well-drained to well-drained soil profiles representing Alfisols, Aridisols, Entisols and Ultisols soils that can be applied to distinguish between anthropogenic and natural V (Aide, 2005). In addition, Poledniok and Buhl (2003) performed V sequential extraction for two different soils collected from industrial and agricultural regions. Their results indicated that similar V contents were found in the exchangeable form, representing 42% and 12% in agricultural and industrial soils, respectively. However, the dominant fractions of V were associated with OM (39%) and the residual fraction (32%) in the industrial soil. Such results indicated the role of OM in controlling V mobility in the contaminated soils. Furthermore, Cappuyns and Swennen (2014) quantified a neglected concentration of the acid-soluble V form and <10% for the oxidizable fraction, whereas reducible and residual fractions accounted for 80% to 90% of the total V. They concluded a limited V release (< 5%) based on $CaCl_2$, ammonium-EDTA and acetic acid extractions from the oxidized sediments at a pH range of 4 to 8. However, V exhibited high mobility in saturated quartz sand columns and a pH range of 4 to 8. Specifically, a pulse of 50 mg L^{-1} V resulted in effluent concentrations exceeding those of the influent (Wang et al., 2016). Moreover, Reijonen et al. (2016) emphasized the effect of adding low levels (50 mg kg^{-1} and 100 mg kg^{-1}) and high concentration (500 mg kg^{-1}) of V to soils on the fractionation of V on soil components. Specifically, they found that <11% was an easily soluble fraction: 30% to 68% of V was sorbed on OM, and 8% to 35% was sorbed on Al/Fe oxides for the low level. When V was added at the high concentrations, they reported that 31% was an easily soluble fraction, 9% to 77% was bound to OM, and 12% to 28% was sorbed by Al/Fe oxides. The reactivity of V in soils is complex because it is an element that is sensitive to oxidation and pH conditions that allow V to be involved in a wide range of reactions. The sorption/desorption reactions control the solubility of V under oxic, and alkaline conditions depend on amorphous oxide contents and pH, whereas V precipitation reactions likely occur under anoxic circumstances (Wright et al., 2014).

In this chapter, we discuss numerous factors affecting the reactivity of V in soils, such as pH and the influence of phosphate, clay minerals, (hydr)oxides and organic substances. Furthermore, we review how these factors control equilibrium and time-dependent V sorption/desorption in bulk soils as well as with individual soil constituents. Moreover, modeling V reactivity and kinetics were reviewed, and an example of sensitivity analysis of sorption kinetics was carried out using a multi-reaction simulation model.

3.2 Soil/Solution Interaction

A literature review reveals that soil pH, iron oxides, clay minerals and OM contents control the distribution of V between solid and liquid soil phases (Chen et al., 2021). In addition, anions in the soil solution, such as phosphate ions, induce specific behavior of V reactivity on soil surfaces. In the following sections, we will discuss in detail these influences on the V reactivity in soils.

3.2.1 Influence of Soil pH

Soil pH is a crucial parameter that controls V fate and behavior in soils. Hydrogen ions influence the distribution of V among various species in the soil solution, as shown in the following examples:

$$VO^{2+} + 2H^{+} \leftrightarrow V^{3+} + H_2O$$

$$VO_2^{+} + 2H^{+} \leftrightarrow VO^{2+} + H_2O$$

$$VO(OH)^{+} + H^{+} \leftrightarrow VO^{2+} + H_2O$$

$$VO(OH)_{2(S)} + H^{+} \leftrightarrow VO(OH)^{+} + H_2OHVO_4^{2-} + H^{+} \leftrightarrow H_2VO_4^{-}$$

$$H_2VO_4^{-} + H^{+} \leftrightarrow H_3VO_4$$

Reijonen et al. (2016) studied the impact of soil pH (4 to 6.9) on the V fractionation in soils. They found that increasing the pH raised the easily soluble fraction in the surface soil that characterized by high OM contents

(OM = 3.2%; Fe oxide= 1683 mg kg^{-1}) and reduced the OM affinity for V for both the surface and subsurface (low OM of 0.5% and Fe oxide = 784 mg kg^{-1}) soils. Such behavior is likely attributable to more electrostatic repulsion between V anions and an increasingly negatively charged OM surface. Moreover, Mikkonen and Tummavuori (1994) quantified the influence of altering the pH on V retention for three non-alkaline soils over a pH range of 2.3 to 7.1. They found that the maximum V sorption occurred at pH 4, where more than 70% of the added V was retained on soils. However, Luo et al. (2017) studied the effect of changing the pH on V retention by three alkaline soils. They reported that increasing pH from 4 to 8 increased the affinity of different colloidal systems to retain V, whereas following a pH of 8 to 9 range, the V-retention decreased due to the competition between OH^- ions and V(+5) species. In fact, V(+5) species are dominant under alkaline environments where the dissociation constants (pK_a) of $H_2VO_4^-$ and HVO_4^{2-} are 7.1 and 12, respectively (Heath and Howarth, 1981).

Several studies revealed that V exhibits pH-dependent mobility in soils. For instance, Cappuyns and Swennen (2014) illustrated that the V leaching curve exhibited a V-shape where minimum concentration occurred at pH range 4 to 6, whereas high V leaching was observed beyond this pH range. Moreover, in quartz sand-packed columns, Wang et al. (2016) reported that mobility of V increased and the relative concentration on the effluent increased from 1.26 to 1.42 by increasing the pH from 4 to 8. Such increase in V mobility is associated with the reduction of V sorption on soil surfaces. Wright et al. (2014) simulated the data of adsorption isotherm experiments using geochemical modeling (PHREEQC) with the MINTEQ.v4 database. Their study revealed that $H_2VO_4^-$ represented > 80% of V (+5), whereas HVO_4^{2-} was < 20% in the experimental solution (pH of 7.4 to 8.2). Furthermore, they found that V concentration in the groundwater exhibited a positive correlation with pH under oxic conditions and a negative correlation with pH under suboxic conditions due to the availability of negatively charged surfaces.

3.2.2 Influence of Phosphate

Numerous researchers investigated the influence of phosphate on V mobility in the soil environment (Molina et al., 2009; Sun et al., 2019; Liao and Yang, 2020). Specifically, Molina et al. (2009) attributed V accumulation in soils to the application of phosphate fertilizers. Similarly, Liao and Yang (2020) showed that nano-hydroxyapatite, $Ca_5(PO_4)_3(OH)$, was able to decrease the mobility and availability of V in soils by increasing V sorption on reactive soil sites. However, Mikkonen and Tummavuori (1994) observed a linear correlation between the sorbed V on soils and the dissolved phosphate in the filtrates at a pH range of 5 to 7. They explained this correlation as a result of the competitive sorption of $H_2VO_4^-$ on the same sites for phosphate on soil surfaces. Nevertheless, a formation of phosphovanadates was not detected after 72-hour soil equilibration at natural pH with 0.01 M of V solution (Mikkonen and Tummavuori, 1994). The influence of phosphate on vanadate sorption can be expected due to the chemical similarity of these two anions. Blackmore et al. (1996) showed that vanadate exhibited stronger sorption than phosphate, whereas high concentrations of phosphate reduced vanadate retention. In agreement with that, Larsson (2014) explored the influence of adding 200 μM of phosphate with 50 μM of vanadate solution on the ferrihydrite sorption of V. Their results indicated that phosphate reduced V retention on sorption sites. Moreover, Larsson (2014) obtained an excellent simulation of V adsorption in such a system using the CD-MUSIC model that revealed the significant role of OM and Al (hydr)oxides in V retention on soil surfaces.

The influence of phosphate ions on V sorption was assessed comprehensively on acidic clay soil (Sun et al., 2019). Specifically, the results of the sorption batch experiments indicated that V sorption was nonlinear with S_{max} of 54.2 mmol kg^{-1} and 49.5 mmol kg^{-1} for single V and phosphate, respectively. The addition of 1.61 and 3.23 mmol L^{-1} phosphate had little effect on V sorption, indicating a minimal competition between P and V for sorption sites (See Figure 3.1) However, competition between the two species was more apparent during stirred-flow experiments in which the presence of P increased the V recovery in the effluent (Figure 3.1). Moreover, Selim and Amacher (2001) presented phosphate's influence on V retention based on miscible displace-

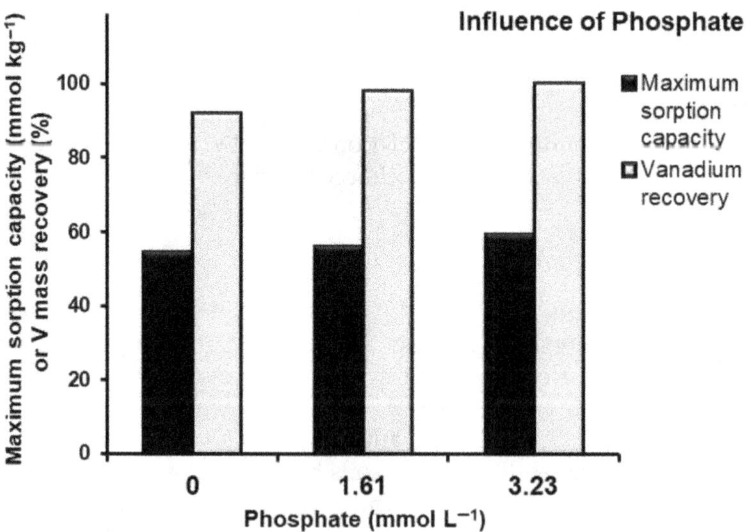

FIGURE 3.1
The influence of phosphate on V mass recovery and V maximum sorption capacity, S_{max} from stir flow and batch experiments, respectively.
Source: Created based on published data from Sun et al., 2019.

ment experiments. Their results revealed that both phosphate and V exhibited nonlinear/kinetic sorption behavior in montmorillonitic (with 1.4% OM) and kaolinitic (with 0.74% OM) soils. However, both were acidic soils (pH of 5.9 and 5.6, respectively); the montmorillonitic soil exhibited higher phosphate retention than V, whereas kaolinitic soil sorbed higher V than phosphate. Introducing a pulse of phosphate displaced a considerable amount of the already-sorbed V from the kaolinitic soil column, indicating competition between the two species for sorption sites; no such evidence of competition was observed in the montmorillonitic soil column. Such results indicate the influence of phosphate on V sorption, and release from soils is dependent on the dominant clay mineralogy.

3.2.3 Influence of Iron Oxides

Vanadium is a redox-sensitive element that exhibits a high potential to react with (hydr)oxides in soils, dependent on soil pH and the chemical reactivity of the (hydr)oxide. Wällstedt et al. (2010) found that the concentration of V in natural waters is positively correlated to iron concentration, indicating a general association between the two. Moreover, soils under oxic environments exhibit a linear positive relation between V and Fe concentrations (Aide, 2005). Adsorption to metal (hydr)oxides is considered one of the mechanisms that control the fate of V in the environment. Under batch experimental conditions, Gäbler et al. (2009) found that (hydr)oxides content controls V sorption behavior in soils where a high correlation between the Fe, Mn and Al (hydr)oxides contents and sorption parameters was reported. Previous studies have indicated soil properties, including metal (hydr)oxides content, organic matter and pH, as governing factors of V retention and transport in soils (Mikkonen and Tummavuori, 1994; Imtiaz et al., 2015).

Several studies have confirmed that maximum adsorption capacities of V(+5) on iron oxides were achieved at pH 3–4 (Peacock and Sherman, 2004; Naeem et al., 2007). Larsson et al. (2017) found that the affinity of ferrihydrite surfaces to sorb vanadate decreased with increasing pH. Such a decline in sorption is probably because of increasing negative surface charge at high pH. Their analyses of extended X-ray absorption fine structure (EXAFS) spectra revealed that V(+5) sorbed as an edge-sharing bidentate-ferrihydrite complex (V-Fe distance of 2.8 Å at pH 3.6–9.4) and V kept the tetrahedral V(+5) nature. This behavior has been attributed to the strong adsorption of vanadate (+5) on colloidal iron (hydr)oxides. Furthermore, EXAFS analyses revealed that sorption of V(+5) on Fe(+3) (hydr)oxides is strong and involves the formation of inner-sphere bidentate complexes (Peacock and Sherman, 2004). The high concentration of the dissolved V under anoxic conditions can be attributed to the rise of V solubility as well as to the reductive dissolution of Fe (hydr)oxides under such circumstances (Shaheen et al., 2016). In general, redox and mineral adsorption reactions control the sorption of V onto iron (hydr)oxides such as goethite surfaces (Wu et al., 2015).

3.2.4 Influence of Clay Minerals and Organic Substances

Inorganic sandy soils tend to exhibit a lower affinity for V than finer textured soils (Imtiaz et al., 2015). Treating sandy soils with modified clay minerals could increase V sorption. For instance, He et al. (2020) showed that Na-bentonite with ferrous modification exhibited a high ability to remediate V-contaminated calcareous soils by reducing the mobility of V. Furthermore, treating the soil with kaolinite increased sorption and fixation of V in saturated columns of quartz sand more effectively than adding montmorillonite (Wang et al., 2016). Agnieszka and Barbara (2012) showed that the mobile V-fractions (the exchangeable and the associated with carbonates fractions) in soils correlated significantly positively with soil pH and sand content but were negatively associated with organic carbon. However, in an organic soil containing 0.2 g L^{-1} and 0.6 g L^{-1} of fulvic and humic acids, respectively, Gustafsson (2019) showed the influence of pH and the presence of Fe(+3) on the adsorption of vanadyl (+4). Gustafsson (2019) simulated the adsorption using the Stockholm Humic Model, which revealed the increase of V(+4) adsorption by increasing the pH from 2 to 6, and a negative influence of Fe(+3) competition was observed on this pH range.

Considering the soil texture or OM content alone cannot determine V adsorption on soils from diluted solutions (Mikkonen and Tummavuori, 1994). For example, Teng et al. (2011) found no clear-cut association between V fractionation and soil pH, texture or organic carbon content. In contrast, Tyler (2004) investigated the vertical distribution of trace elements in a Haplic Podzol soil profile and found that the surface layer, higher in OM content, retained relatively large concentrations of V. Moreover, Cappuyns and Slabbinck (2012) reported that V contents in floodplain soils correlated significantly and positively with clay content and organic carbon. Moreover, they predicted the released V from these Belgian soils using a regression model formulated based on clay, organic carbon and phosphorus contents. In opposition to the notion that OM has a high affinity for V, Reijonen et al. (2016) found that soils with high OM contents exhibited higher easily soluble V than soils containing low OM. Their results revealed the role of OM in reducing V(+5) to V(+4) as well as the competition between V(+5) and the dissolved organic carbon (DOC) for the ligand exchange sites on soil surfaces. Such a positive association between the dissolved V and DOC was confirmed at estuary environments in which V-rich sediments interact with oxic water (Shi et al., 2016).

3.3 Sorption/Desorption in Soils

Sorption isotherms represent the distribution of a chemical between solid (sorbed) and solution phases in soil. Such a relationship between the sorbed amount on solid (S) and solute concentration in the soil solution (C) can be expressed mathematically as $S = f(C)$. Linear (Equation 3.1) and nonlinear (Freundlich, Equation 3.2, and Langmuir, Equation 3.3) models are commonly applied to describe sorption isotherms of trace element in soils (Sparks, 2003; Shaheen et al., 2009; Elbana et al., 2018; Zhu et al., 2020; Gonzalez-Rodriguez and Fernandez-Marcos, 2021).

$$S = k_d C \tag{3.1}$$

$$S = k_f C^n \tag{3.2}$$

$$S = \frac{S_{max} k_l C}{1 + k_l C} \tag{3.3}$$

where $k_d, k_f,$ and k_l, are linear distribution, Freundlich and Langmuir coefficients, respectively. The parameter S_{max} is the maximum sorption capacity, and n is the nonlinear Freundlich parameter. The linear and Freundlich sorption isotherms are empirical models, whereas the Langmuir equation can be driven based on the first-order kinetics principle (Tóth, 2002). Mathematically, the Langmuir equation limits the sorption on solid phase by maximum sorption capacity, whereas linear and Freundlich models do not put a ceiling for the sorption capacity mathematically.

3.4 Distribution Coefficients and Sorption Capacity of Soils

The distribution of a trace element between solid and liquid soil phases depends on the properties of soil colloids and the chemical features of the sorbed element itself (Shaheen et al., 2013). Numerous studies revealed the nonlinearity of V sorption isotherms. For example, Gäbler et al. (2009) found that different soil types exhibited various affinities to retain V. Specifically, they studied the sorption behavior of V on 30 German soils with the Freundlich isotherm. Their results revealed that sorption isotherms were mainly nonlinear, with n values that varied between 0.33 and 1, whereas the value of k_f ranged over two orders of magnitude between $10^{2.27}$ and $10^{4.63}$. Moreover, Luo et al. (2017) concluded that sorption isotherms of V on alkaline soils were nonlinear with n values varied between 0.51 and 0.89, whereas S_{max} ranged between 41.5 mg g^{-1} and 285.7 mg g^{-1} and was found to be pH-dependent.

The nonlinear sorption of V can be attributed to different affinities of the various soil constituents toward V or the availability of sorption sites. For instance, Gan et al. (2020) reported that V showed a nonlinear sorption behavior with an n value of 0.74 and S_{max} of 81.0 mg kg^{-1} on silica. However, humic acid colloids exhibited a high V sorption capacity of 861.2 mg kg^{-1} with a nonlinear parameter, n of 0.19 (Chen et al., 2020). However, in a humic acid and silica system, Song et al. (2020) reported nonlinear V sorption with n of 0.23 and S_{max} of 166.7 mg kg^{-1}. Moreover, the size of particles can influence the sorption behavior and reactivity of the soil solid phase. For example, Wehrli et al. (1990) reported complete vanadyl sorption on nanosize aluminum oxide particles after <90 seconds at pH> 5. Considering multiple reaction sites would be recommended to model the sorption behavior of trace elements in soil surfaces due to the heterogeneity of the soil system. Under low contamination and diluted solutions, a nonlinear sorption isotherm can be attributed as a composite curve of different linear sections with decreasing slopes (Selim, 2014).

3.5 Retention/Release Hysteresis

Irreversibility of the sorption reaction is referred to as sorption/desorption hysteresis. Each soil constituent exhibits a different affinity and retention strength to sorb and release V. For instance, Yang et al. (2020) showed that less than 9% of the sorbed V on colloidal kaolinite (median particle size of 0.627 μm) was able to be released during desorption, where 71% of the desorbed V was recovered during the first eight hours. Such behavior is indicative of the V kinetics and time-dependent release. Gan et al. (2020) found that silica was effective adsorbent to retain V under acidic conditions with low desorption of 5.4% of the total sorbed V. Moreover, they reported that the adsorption/desorption of V on silica was concentration-dependent. Specifically, sorbed V is associated with high energy sites at low input concentrations, whereas it becomes more associated with low energy (more reversible) sites at high input concentrations. Such behavior resulted in the presence of retention release hysteresis. In agreement with Gan et al. (2020), Chen et al. (2020) reported a concentration-dependence of V sorption/desorption on colloidal humic substances where increasing initial V concentrations in solution increased the sorbed as well as the desorbed V. However, humic colloids exhibited lower V desorption than sorption, with a desorption fraction in a range of 0.8%–1.7% of the total sorbed V. Song et al. (2020) concluded that V exhibited high sorption and less desorption under the acidic conditions in a coexisting system of organic (humic acid) and inorganic soil components such as silica. The dependency of sorption/desorption processes of V on pH and initial concentration was confirmed by numerous studies (Chen et al., 2020; Gan et al., 2020; Song et al., 2020).

Hysteresis of V sorption/desorption is anticipated in soil systems where multiple ions exist simultaneously. Gonzalez-Rodriguez and Fernandez-Marcos (2021) reported that volcanic soils could sorb >90% of the added vanadate with S_{max} of 37.5 mmol kg^{-1} due to their high contents of amorphous Al and Fe. However, treating those soils with 1 mM phosphate resulted in very limited V release (<1%), indicating hysteresis. Sun et al. (2019) demonstrated that V exhibited higher retention strength than phosphate. Specifically, based on sorption/desorption, the released V amounts from clayey acidic soil were found to be 2.3% and 9.9% of that sorbed, with initial V concentrations of 0.49 mM and 0.98 mM, respectively. However, the released P was 17.7% and 20.0% of the total sorbed phosphate associated, with initial phosphate concentrations of 0.49 mM and 0.98 mM, respectively. Moreover, their results indicated that V desorption increased to 28% of the total sorbed in the presence of high phosphate concentration (3.2 mM). Such results confirm V retention/release hysteresis in agricultural soils.

3.6 Equilibrium and Kinetics Modeling

Environmental scientists are interested in quantifying the change of contaminant concentration, such as the change of V concentration in soil solution as a function of time. Researchers applied numerous mathematical expressions to estimate the rate of sorption/desorption reactions that control the fate of V in soils. The equilibrium of a reaction can be realized at the equalization of the forward and reverse reaction rates. For a reaction:

$$aA + bB \leftrightarrow cC + dD$$

Here, equilibrium constant (K_{eq}) is the ratio between the forward rate constant (K_{for}) and reverse rate constant (K_{rev}) at equilibrium, as shown in Equation 3.4.

$$K_{eq} = \frac{[C]^c[D]^d}{[A]^a[B]^b} = \frac{K_{for}}{K_{rev}} \tag{3.4}$$

where [] indicates the activity of each reactant or product of the chemical reaction. On the other hand, modeling sorption/desorption kinetics aims to quantify the change of reaction rate versus time. A zero-order reaction is applied when the reaction rate does not depend on solute concentration (Equation 3.5). However, the first-order reaction (Equation 3.6) and the second-order reaction (Equation 3.7) are applied when the reaction rate is proportional to V concentration in solution [V] to the first power and second power, respectively.

$$\frac{d[V]}{dt} = \pm k_0 \tag{3.5}$$

$$\frac{d[V]}{dt} = \pm k_1 [V] \tag{3.6}$$

$$\frac{d[V]}{dt} = \pm k_2 [V]^2 \tag{3.7}$$

where k_0, k_1, and k_2 are the rate constants for zero, first, and second-order reactions, respectively. For concentration and time units in mol L^{-1} and s, the units of k_0, k_1, and k_2 are mol L^{-1} s^{-1}, s^{-1}, and L mol^{-1} s^{-1}, respectively. Furthermore, the n-order reaction can be simulated according to Equation 3.8 where the rate constant, k_n, units is L^{n-1} mol^{1-n} s^{-1}:

$$\frac{d[V]}{dt} = \pm k_n [V]^n \tag{3.8}$$

Vanadium sorption/desorption is a heterogeneous reaction that occurs at solid and liquid soil interfaces. This reaction can be reversible, including the forward and the reversible reactions that can occur as a consecutive reaction (A↔B↔C) or in parallel reactions (A↔B and B↔C). Such possibility can be explained through a multi-reaction model that involves more than one reaction type (exchangeable, outer- and inner-sphere complexes and precipitation) or including more than one soil compartment such as OM, carbonates, oxides and clay minerals, and among others (see Figure 3.2).

The kinetic behavior of V reactions controls its fate in soils. Wehrli et al. (1990) found that sorption of V(+4) on aluminum oxide was a relatively fast reaction followed by a pH-dependent slow reaction. The reaction of V with soil constituents can be reversible or irreversible and/or time-dependent. For example, Gan et al. (2020) investigated the kinetics of V adsorption on silica. Their results revealed that V sorption was rate limited, where V quickly sorbed during the first hour, followed by a gradual decrease to reach equilibrium in ten hours. They attributed the kinetic behavior to the diffusion process and the reduced available reaction sites with time. The external and intra-particle diffusion reactions control V sorption, where higher OM content was associated with the lower diffusion rate constant (Luo et al., 2017). In agreement with that, Chen et al. (2020) obtained good simulations of the kinetic behavior of V sorption on humic acid colloids using pseudo first- and second-order models. Their results indicated that both chemical process and diffusion control V sorption on humic substances. Moreover, Song et al. 2020 showed that chemisorption was a controlling mechanism for V sorption in a mixed system of silica and humic acid; the pseudo-second order model

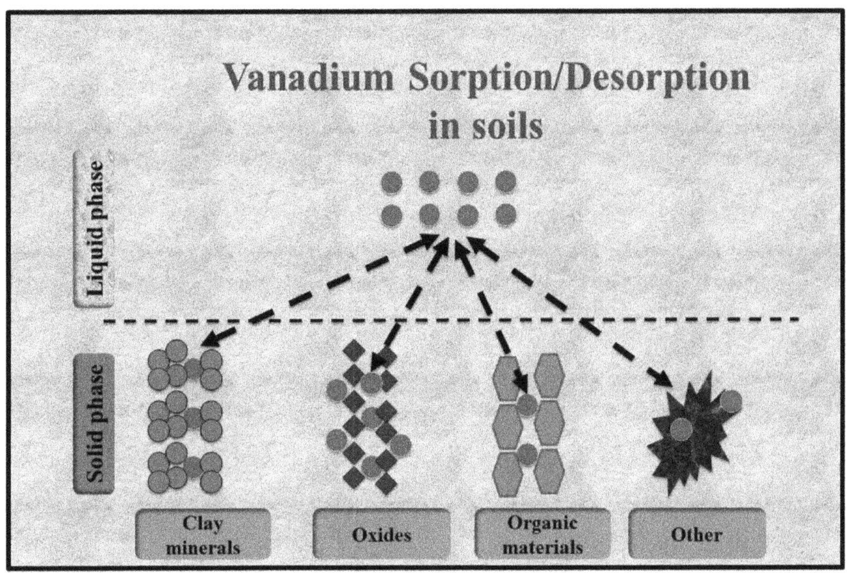

FIGURE 3.2
Schematic diagram of vanadium sorption/desorption reactions in soils.

successfully presented a more reliable simulation of sorption/desorption data than a pseudo first-order model.

The slow sorption process/step can explain the kinetics of chemical release from soils. A decrease of V concentration in soil solution versus time was observed after spiking soil with V(+4) (Martin and Kaplan, 1998). Likewise, Baken et al. (2012) investigated the change of the extractable V using 0.01 M $CaCl_2$ after spiking four acidic soils (pH: 5.2 to 6.6) with vanadate solutions during an ageing period of 100 days. They found that V became more sorbed over time of ageing where the ratio between soluble V concentrations from freshly spiked (14 days) and from the aged soil (100 days) varied between 1.6 and 2.5, whereas the rate constant of the reversible first-order kinetic model varied from 0.030 day^{-1} to 0.078 day^{-1}. Such kinetic sorption behavior was also reported for alkaline soils. For instance, Luo et al. (2017) studied the sorption of V(+5) onto three alkaline soils and found that soils exhibited a rapid initial rate of V sorption in the early stage, followed by slow retention. Their results indicated that the rate of V sorption was limited by particle diffusion, and the pseudo second-order model simulated V kinetic data better than the pseudo first-order model. Good fitting of sorption kinetic data using the second-order model indicates the role of the chemisorption in V retention on soil constituents (Song et al., 2020).

Batch, stir-flow and miscible displacement experiments are effective techniques for exploring chemical kinetics in soils. Sun et al. (2019) studied V sorption/desorption kinetics in acidic clay soil (Sharkey soil, La). The results of batch experiments revealed that V sorption was time-dependent, where the sorbed V increased as a function of time. Furthermore, they found that V desorption was not fully reversible. Therefore, they successfully simulated the kinetic data of V sorption/desorption using the multi-reaction model (MRM) version that assumed three types of available sites for sorption/desorption i) instantaneous equilibrium, S_{eq} (Equation 3.9), ii) reversible kinetic, S_k (Equation 3.10), and iii) consecutive irreversible, S_{irr} (Equation 3.11):

$$S_{eq} = K_{eq}\left(\theta/\rho\right)C^{n} \qquad \text{(Equation 3.9)}$$

$$\frac{\partial S_k}{\partial t} = k_1\left(\theta/\rho\right)C^{n} - \left(k_2 + k_3\right)S_k \qquad \text{(Equation 3.10)}$$

$$\frac{\partial S_{irr}}{\partial t} = k_3 S_k \qquad \text{(Equation 3.11)}$$

where θ and ρ are soil moisture content and bulk density of the soil, respectively. The parameters k_1 and k_2 are the reaction rate constants for forward and reverse kinetic reactions, respectively. k_3 is the rate constant associated with the consecutive-irreversible reaction. The MRM is an available-free online software as a part of the Chem-Transport software package (Selim, 2016). Such application of the multi-reaction model provides a comprehensive understanding of the fate of contaminants in the soil environment. Sun et al. (2019) successfully modeled V stir-flow kinetic data using the same values of K_{eq}, k_1, k_2 and k_3 obtained based on the batch experiment without further fitting. The stir-flow results confirmed that non-equilibrium reaction was the main V sorption/desorption mechanism on the acidic clayey soil. Based upon miscible displacement column experiments, Selim and Amacher (2001) found that V breakthrough curves (BTCs) for two acidic clayey soils – (i) Sharkey (montmorillonitic soil) and (ii) Cecil (kaolinitic soil) – exhibited extensive tailing during the leaching of V. Such results indicated non-equilibrium and kinetic behavior of V in those soils.

In the following section, we presented a hypothetical example for modeling V sorption/desorption using MRM based on the following situation:

- In a batch experiment, 3 g soil (ρ = 1.30 Mg m^{-3}) was reacted with a 30 ml of V solution (5, 10, 20, 50, and 100 mg L^{-1}). Use an average value of 0.65 for the exponential parameter, n, and assume 10 days for the sorption period to simulate sorption kinetics using MRM.

- Consider the first-order reaction model for scenario A, including a reversible kinetic site k_1 of 0.4 h^{-1} and k_2 of 0.02 h^{-1}, in addition to a consecutive irreversible site with k_3 of 0.005 h^{-1}. Then, Run MRM sensitivity analysis for
  - Examining the change of solute concentration as a function of time
  - Examining the case of changing K_2 to 0.2 h^{-1} (scenario B).
  - Examining the case of considering equilibrium reaction site with K_{eq} of 10 (scenario C)
  - Examining scenario D for the case of including an additional reaction site (S_{d_irr}) that for a direct irreversible reaction between solution and solid phases where $S_{d\_irr} = k_{irr}\ \Theta\ C$

Results for the sensitivity analyses are shown in Figures 3.3–3.6. The MRM manipulates different data sets associated with various initial concentrations (from 5 to 100 mg L^{-1}) as one set using only one set of experimental conditions and reaction rate constants (n, ρ, Θ, k_1, k_2, k_3, among others). Figure 3.3 shows MRM output for the change of solute concentration as a

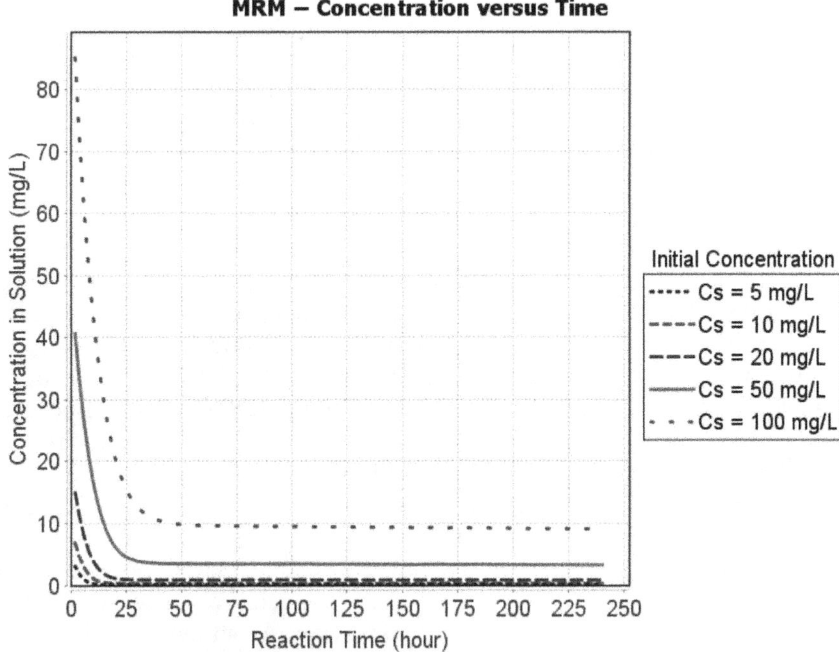

FIGURE 3.3
The MRM simulation of solute concentration as a function of time for scenario A (a reversible kinetic site k_1 of 0.4 h^{-1} and k_2 of 0.02 h^{-1}, in addition to a consecutive irreversible site with k_3 of 0.005 h^{-1}).

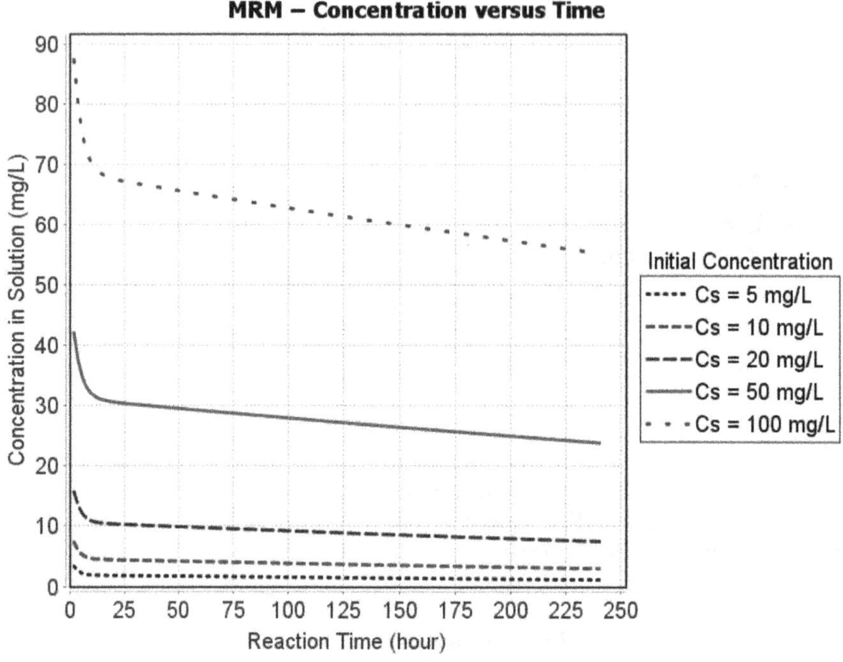

FIGURE 3.4
The MRM simulation of solute concentration as a function of time for scenario B (a reversible kinetic site k_1 of 0.4 h^{-1} and k_2 of 0.2 h^{-1}, in addition to a consecutive irreversible site with k_3 of 0.005 h^{-1}).

function of time in scenario A. Results revealed the continuous decrease of the soluble V for each initial concentration associated with increasing the sorbed amount by time. Figure 3.4 shows the influence of increasing the backward rate constant K_2 ten folds (from 0.02 to 0.2 h^{-1}, scenario B). Such a scenario can exist due to a change of soil pH or the influence of competitive ions. Sorption of V is expected to be reduced as a result of increasing the reversibility. Both scenarios C and D represent considering an additional reaction site. Specifically, scenario C considered reducing kinetic and time-dependent reactions by considering instantaneous equilibrium reaction, S_{eq} (see Equation 3.4). However, the sensitivity analysis of considering an additional direct irreversible reaction site is expected to increase sorption and reduce solute concentration in the solution. As shown in Figures 3.3 to 3.6, such multi-reaction models that produced reasonable simulations provide a practical approach to assess different options and strategies to

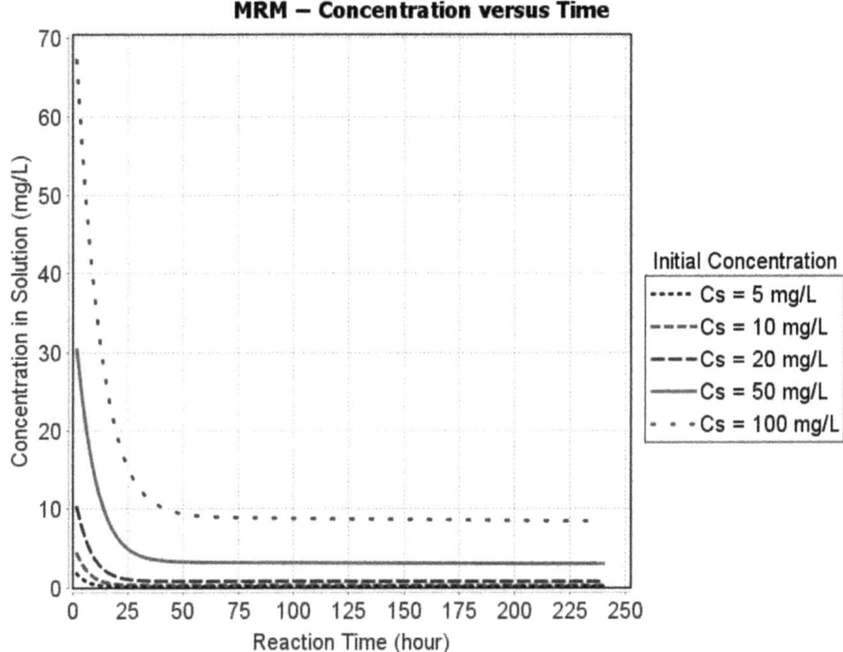

FIGURE 3.5
The MRM simulation of solute concentration as a function of time for scenario C (a reversible kinetic site k_1 of 0.4 h^{-1} and k_2 of 0.02 h^{-1}, in addition to a consecutive irreversible site with k_3 of 0.005 h^{-1} and an equilibrium reaction with K_{eq} of 10).

remediate soils and quantify risk assessment of environmental pollution problems.

3.7 Prospects of Vanadium Kinetics Studies

Vanadium fate and behavior in soils associate with numerous environmental conditions that control V reactivity and the interactions with natural resources. Toxicity and environmental risk of V pollution will continue as driving factors for further research in the future. Our literature review of V kinetic studies revealed the existence of extensive and valuable studies on V sorption/desorption kinetics for V as a single element in soil systems. However, few studies considered the competitive systems, including multiple-ion. Furthermore,

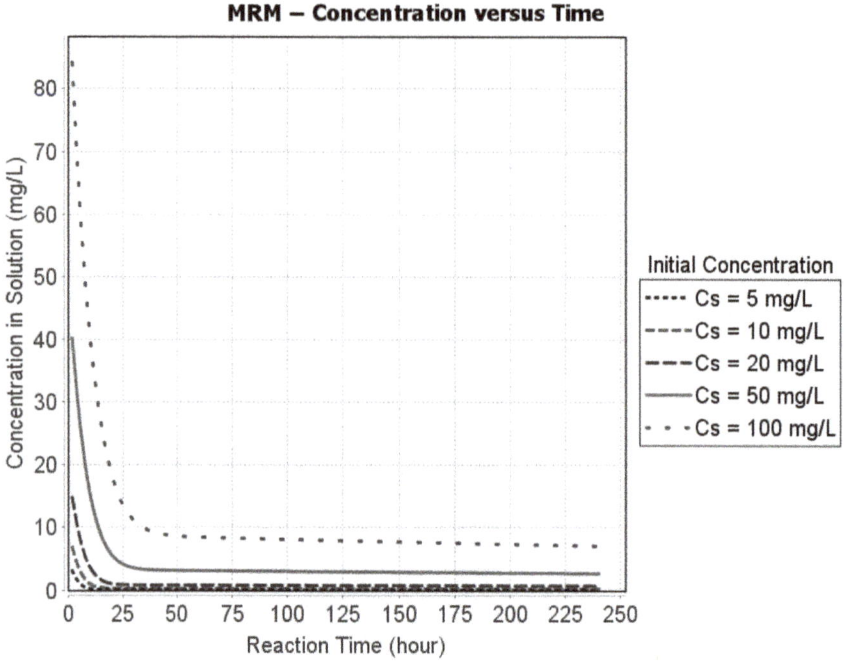

FIGURE 3.6
The MRM simulation of solute concentration change as a function of time for scenario C (a reversible kinetic site k_1 of 0.4 h^{-1} and k_2 of 0.02 h^{-1}, in addition to a consecutive irreversible site with k_3 of 0.005 h^{-1} and direct irreversible reaction site with k_{irr} of 0.005 h^{-1}).

there is a need to explore V kinetic behavior in heterogeneous soil systems and investigate the interconnected influence of various organic and inorganic soil constituents on V sorption behavior. Additional advanced research is needed to quantify V kinetics under redox reactions under field and large-scale conditions to fully understand V reactivity. Moreover, comprehensive research is needed for modeling the kinetic V release from contaminated soils under dynamic environment circumstances. Furthermore, research is needed on coupling the precise reaction characterization techniques and modeling V time-dependent behavior in different soils.

3.8 Acknowledgment

The authors would like to dedicate this book chapter to the soul of Prof. Dr. H.M. Selim, who passed away November 2020 in Baton Rouge, La.

References

Agnieszka, J., G. Barbara. 2012. Chromium, nickel and vanadium mobility in soils derived from fluvioglacial sands. *Journal of Hazardous Materials*. 237–238: 315–322. http://dx.doi.org/10.1016/j.jhazmat.2012.08.048.

Aide, M. 2005. Geochemical assessment of iron and vanadium relationships in oxic soil environments. *Soil & Sediment Contamination*. 14: 403–416. https://doi.org/10.1080/15320380500180382.

Baken, S., M.A. Larsson, J.P. Gustafsson, F. Cubadda, E. Smolders. 2012. Ageing of vanadium in soils and consequences for bioavailability. *European Journal of Soil Science*. 63(6): 839–847. https://doi.org/10.1111/j.1365-2389.2012.01491.x.

Blackmore, D.P.T., J. Ellis, P.J. Riley. 1996. Treatment of a vanadium-containing effluent by adsorption/coprecipitation with iron oxyhydroxide. *Water Research*. 30(10): 2512–2516.

Cappuyns, V., E. Slabbinck. 2012. Occurrence of vanadium in Belgian and European alluvial soils. *Applied and Environmental Soil Science*. Article ID 979501. https://doi.org/10.1155/2012/979501.

Cappuyns, V., R. Swennen. 2014. Release of vanadium from oxidized sediments: Insights from different extraction and leaching procedures. *Environmental Science and Pollution Research*. 21:2272–2282. https://doi.org/10.1007/s11356-013-2149-0.

Chen, L., J. Liu, W. Hu, J. Gao, J. Yang. 2021. Vanadium in soil-plant system: Source, fate, toxicity, and bioremediation. *Journal of Hazardous Materials*. 405: 124200. https://doi.org/10.1016/j.jhazmat.2020.124200.

Chen, L., Y. Zhu, H. Luo, J. Yang. 2020. Characteristic of adsorption, desorption, and co-transport of vanadium on humic acid colloid. *Ecotoxicology and Environmental Safety*. 190: 110087. https://doi.org/10.1016/j.ecoenv.2019.110087.

Elbana, T.A., H.M. Selim, N. Akrami, A. Newman, S. Shaheen, J. Rinklebe. 2018. Freundlich sorption parameters for cadmium, copper, nickel, lead, and zinc for different soils: Influence of kinetics. *Geoderma*. 324: 80–88. https://doi.org/10.1016/j.geoderma.2018.03.019.

Frank, A., A. Madej, V. Galgan, L.R. Petersson. 1996. Vanadium poisoning of cattle with basic slag. Concentrations in tissues from poisoned animals and from a reference, slaughter-house material. *Science of the Total Environment*. 181(1): 73–92. https://doi.org/10.1016/0048-9697(95)04962-2.

Gäbler, H.E., K. Gluh, A. Bahr, J. Utermann. 2009. Quantification of vanadium adsorption by German soils. *Journal of Geochemical Exploration*. 103: 37–44. https://doi.org/10.1016/j.gexplo.2009.05.002.

Gan, C., M. Liu, J. Lu, J. Yang. 2020. Adsorption and desorption characteristics of vanadium (V) on silica. *Water Air Soil Pollution*. 231: 10. https://doi.org/10.1007/s11270-019-4377-5.

Gonzalez-Rodriguez, S., M.L., Fernandez-Marcos. 2021. Sorption and desorption of vanadate, arsenate and chromate by two volcanic soils of equatorial Africa. *Soil System*. 5(2): 22. https://doi.org/10.3390/soilsystems5020022.

Gustafsson, J.P. 2019. Vanadium geochemistry in the biogeosphere—speciation, solid-solution interactions, and ecotoxicity. *Applied Geochemistry*. 102: 1–25. https://doi.org/10.1016/j.apgeochem.2018.12.027.

He, W., J. Yang, J. Li, Y. Ai, J. Li. 2020. Stabilization of vanadium in calcareous purple soil using modified Na-bentonites. *Journal of Cleaner Production*. 268: 121978. https://doi.org/10.1016/j.jclepro.2020.121978.

Heath, E., O.W. Howarth. 1981. Vanadium-51 and oxygen-17 nuclear magnetic resonance study of vanadate (V) equilibria and kinetics. *Journal of the Chemical Society, Dalton Transactions*. 5: 1105–1110. https://doi.org/10.1039/DT9810001105.

Hopkins, L.L., H.L. Cannon, A.T. Musch, R.M. Welch, F.H. Neilsen. 1977. Vanadium Geochem. *Environmental*. 2: 93–107.

Hurlbut, C.S., C. Klein. 1977. *Manual of Mineralogy*. John Wiley and Sons, New York.

Imtiaz, M., M.S. Rizwan, S. Xiong, H. Li, M. Ashraf, S.M. Shahzad, M. Shahzad, M. Rizwan, S. Tu. 2015. Vanadium, recent advancements and research prospect: A review. *Environment International*. 80(2015): 79–88. https://doi.org/10.1016/j.envint.2015.03.018.

Kabata-Pendias, A., A.B. Mukherjee. 2007. *Trace Elements from Soil to Human*. Springer, Berlin and Heidelberg. https://doi.org/10.1007/978-3-540-32714-1_12.

Kabata-Pendias, A., W. Sadurski. 2004. Trace elements and compounds in soil. In: E. Merian, M. Anke, M. Ihnat, M. Stoepppler (Eds.), *Elements and Their Compounds in the Environment*, 2nd ed. Wiley-VCH, Weinheim, pp 79–99.

Kuo, C.Y., S.J. Jiang, A.C. Sahayam. 2007. Speciation of chromium and vanadium in environmental samples using HPLC-DRC-ICP-MS. *Journal of Analytical Atomic Spectrometry*. 22(6): 636–641. https://doi.org/10.1039/B701112A.

Larsson, M.A. 2014. Vanadium in soils chemistry and ecotoxicity. Doctoral Thesis. Swedish University of Agricultural Sciences. ISBN 978-91-576-8153-9.

Larsson, M.A., I. Persson, C. Sjöstedt, J.P. Gustafsson. 2017. Vanadate complexation to ferrihydrite: X-ray absorption spectroscopy and CD-MUSIC modelling. *Environmental Chemistry*. 14(3): 141–150. https://doi.org/10.1071/EN16174.

Liao, Y., J. Yang. 2020. Remediation of vanadium contaminated soil by nano-hydroxyapatite. *Journal of Soils and Sediments*. 20: 1534–1544. https://doi.org/10.1007/s11368-019-02522-0.

Luo, X., L. Yu, C. Wang, X. Yin, A. Mosa, J. Lv, H. Sun. 2017. Sorption of vanadium (V) onto natural soil colloids under various solution pH and ionic strength conditions. *Chemosphere*. 169: 609–617. http://dx.doi.org/10.1016/j.chemosphere.2016.11.105.

Madejón, P. 2013. Vanadium. In: B. Alloway (Eds.), *Heavy Metals in Soils. Environmental Pollution*, Vol. 22. Springer, Dordrecht. https://doi.org/10.1007/978-94-007-4470-7_27.

Martin, H.W., D.I. Kaplan. 1998. Temporal changes in cadmium, thallium, and vanadium mobility in soil and phytoavailability under field conditions. *Water Air & Soil Pollution*. 101: 399–410. https://doi.org/10.1023/A:1004906313547.

Mayes, W.M., I.T. Burke, H.I. Gomes, et al. 2016. Advances in understanding environmental risks of red mud after the Ajka Spill, Hungary. *Journal of Sustainable Metallurgy*. 2: 332–343. https://doi.org/10.1007/s40831-016-0050-z.

Mikkonen, A., J. Tummavuori. 1994. Retention of vanadium (V) by three Finnish mineral soils. *European Journal of Soil Science*. 45: 361–368.

Molina, M., F. AbuRto, R. Calderón, M. Cazanga, M. Escudey. 2009. Trace element composition of selected fertilizers used in Chile: Phosphorus fertilizers as a source of long-term soil contamination. *Soil and Sediment Contamination: An International Journal*. 18: 497–511. https://doi.org/10.1080/15320380902962320.

Morrell, B.G., N.W. Lepp, D.A. Phipps. 1986. Vanadium uptake by higher plants: Some recent developments. *Environ Geochem Health*. 8: 14–18. https://doi.org/10.1007/BF02280116.

Naeem, A., P. Westerhoff, S. Mustafa. 2007. Vanadium removal by metal (hydr) oxide adsorbents. *Water Research*. 41: 1596–1602. https://doi.org/10.1016/j.watres.2007.01.002.

Panichev, N., K. Mandiwana, D. Moema, R. Molatlhegi, P. Ngobeni. 2006. Distribution of vanadium(V) species between soil and plants in the vicinity of vanadium mine. *Journal of Hazardous Materials*. A137: 649–683. https://doi.org/10.1016/j.jhazmat.2006.03.006.

Peacock, C.L., D.M. Sherman. 2004. Vanadium (V) adsorption onto goethite (a-FeOOH) at pH 1.5 to 12: A surface complexation model based on ab initio molecular geometries and EXAFS spectroscopy. *Geochimica et Cosmochimica Acta*. 68: 1723–1733.

Poledniok, J., F. Buhl. 2003. Speciation of vanadium in soil. *Talanta*. 59: 1–8. https://doi.org/10.1016/S0039-9140(02)00322-3.

Reijonen, I., M. Metzler, H. Hartikainen. 2016. Impact of soil pH and organic matter on the chemical bioavailability of vanadium species: The underlying basis for risk assessment. *Environmental Pollution*. 210: 371–379. http://dx.doi.org/10.1016/j.envpol.2015.12.046.

Ruyters, S., J. MErtens, E. Vassilieva, B. Dehandschutter, A. Poffijn, E. Smolders. 2011. The red mud accident in Ajka (Hungary): Plant toxicity and trace metal bioavailability in red mud contaminated soil. *Environmental Science & Technology*. 45(4): 1616–1622. https://doi.org/10.1021/es104000m.

Selim, H.M. 2014. On the nonlinearity of sorption isotherms of solutes in soils. *Soil Science*. 179: 237–241. https://doi.org/10.1097/SS.0000000000000063.

Selim, H.M. 2016. *Chem_Transport Software Models for Chemical Kinetic Retention and Transport in Soils and Geological Media User's Manual. School of Plant.* Environmental and Soil Science, LSU-Agcenter. www.spess.lsu.edu/chem_transport/. Accessed date: June 2021.

Selim, H.M., M.C. Amacher. 2001. Sorption and release of heavy metals in soils: Nonlinear kinetics. In: H.M. Selim, D.L. Sparks (Eds.), *Heavy Metals Release in Soils*. CRC Press, Boca Raton, FL, pp. 1–29.

Shaheen, S.M., J. Rinklebe. 2018. Vanadium in thirteen different soil profiles originating from Germany and Egypt: Geochemical fractionation and potential mobilization. *Applied Geochemistry*. 88: 288–301. http://dx.doi.org/10.1016/j.apgeochem.2017.02.010.

Shaheen, S.M., J. Rinklebe, T., Frohne, J.R. White, R.D. DeLaune. 2016. Redox effects on release kinetics of arsenic, cadmium, cobalt, and vanadium in Wax Lake Deltaic freshwater marsh soils. *Chemosphere*. 150: 740–748. http://dx.doi.org/10.1016/j.chemosphere.2015.12.043.

Shaheen, S.M., C.D. Tsadilas, T. Mitsibonas, M. Tzouvalekas. 2009. Distribution coefficient of copper in different soils from Egypt and Greece. *Communications in Soil Science and Plant Analysis*. 40: 1–6, 214–226. https://doi.org/10.1080/00103620802625625.

Shaheen, S.M., C.D. Tsadilas, J. Rinklebe. 2013. A review of the distribution coefficients of trace elements in soils: Influence of sorption system, element characteristics, and soil colloidal properties. *Advances in Colloid and Interface Science*. 201–202: 43–56. http://dx.doi.org/10.1016/j.cis.2013.10.005.

Shi, Y.X., V. Mangal, C. Guéguen. 2016. Influence of dissolved organic matter on dissolved vanadium speciation in the Churchill River estuary (Manitoba, Canada). *Chemosphere*. 154: 367–374. http://dx.doi.org/10.1016/j.chemosphere.2016.03.124.

Song, Q., M. Liu, J. Lu, Y. Liao, L. Chen, J. Yang. 2020. Adsorption and desorption characteristics of vanadium (V) on coexisting humic acid and silica. *Water Air Soil Pollution*. 231: 460. https://doi.org/10.1007/s11270-020-04839-w.

Sparks, D.L. 2003. *Environmental Soil Chemistry*, 2nd ed. Academic Press, San Diego, CA, p. 352.

Sun, W., X. Li, J. Padilla, T.A. Elbana, H.M. Selim. 2019. The influence of phosphate on the adsorption—desorption kinetics of vanadium in an acidic Soil. *Journal of Environmental Quality*. 48(3): 686–693. https://doi.org/10.2134/jeq2018.08.0316.

Teng, Y., Yang, J., Sun, Z., J. Wang, R. Zuo, J. Zheng. 2011. Environmental vanadium distribution, mobility and bioaccumulation in different land-use Districts in Panzhihua Region, SW China. *Environmental Monitoring and Assessment*. 176: 605–620. https://doi.org/10.1007/s10661-010-1607-0.

Tóth, J. 2002. Uniform and thermodynamically consistent interpretation of adsorption isotherms. In: J. Tóth (Ed.), *Adsorption Theory: Modeling and Analysis*, Vol. 107, Surfactant Science Series. Marcel Dekker, New York, pp 1–104. https://doi.org/10.1201/b12439.

Tyler, G. 2004. Vertical distribution of major, minor, and rare elements in a Haplic Podzol. *Geoderma*. 119: 277–290. https://doi.org/10.1016/j.geoderma.2003.08.005.

United States Environmental Protection Agency (USEPA). 2005. *Ecological Soil Screening Levels for Vanadium, Interim Final*. Office of Solid Waste and Emergency Response. OWSER Directive 9285.7–75. Washington, DC.

Wällstedt, T., L. Bjorkvald, J.P. Gustafsson. 2010. Increasing concentrations of arsenic and vanadium in (southern) Swedish streams. *Applied Geochemistry*. 25: 1162–1175. https://doi.org/10.1016/j.apgeochem.2010.05.002.

Wang, Y., X. Yin, H. Sun, C. Wang. 2016. Transport of vanadium (V) in saturated porous media: Effects of pH, ionic-strength and clay mineral. *Chemical Speciation & Bioavailability*. 28:1–4, 7–12. https://doi.org/10.1080/09542299.2015.1133238.

Wehrli, B., S. Ibric, W. Stumm. 1990. Adsorption kinetics of vanadyl (IV) and chromium (III) to aluminum oxide: Evidence for a two-step mechanism. *Colloids and Surfaces*. 51: 77–88. https://doi.org/10.1016/0166-6622(90)80133-O.

Welch, R.M. 1973. Vanadium uptake by plants absorption kinetics and the effects of pH, metabolic inhibitors, and other anions and cations. *Plant Physiology*. 51(5): 828–832. https://doi.org/10.1104/pp.51.5.828.

Wright, M.T., K.G. Stollenwerk, K. Belitz. 2014. Assessing the solubility controls on vanadium in groundwater, northeastern San Joaquin Valley, CA. *Applied Geochemistry*. 48: 41–52. http://dx.doi.org/10.1016/j.apgeochem.2014.06.025.

Wu, F., T. Qin, X. Li, Y. Liu, J. Huang, Z. Wu. 2015. First-principles investigation of vanadium isotope fractionation in solution and during adsorption. *Earth and Planetary Science Letters*. 426: 216–224. http://dx.doi.org/10.1016/j.epsl.2015.06.048.

Yang, J., X. Gao, J. Li, R. Zuo, J. Wang, L. Song. 2021. The distribution and speciation characteristics of vanadium in typical cultivated soils. *International Journal of Environmental Analytical Chemistry*. https://doi.org/10.1080/03067319.2021.1890057.

Yang, J., H. Luo, Y. Zhu, Y. Yu, W. He, Z. Wu, B. Wang. 2020. Adsorption-desorption and co-migration of vanadium on colloidal kaolinite. *Environmental Science and Pollution Research*. 27: 17910–17922. https://doi.org/10.1007/s11356-020-07845-x.

Zhu, H., X. Xiao, Z. Guo, C. Peng, X. Wang, A. Yang. 2020. Characteristics and behaviour of vanadium (V) adsorption on goethite and birnessite. *Environmental Earth Sciences*. 79: 240. https://doi.org/10.1007/s12665-020-08992-7.

4

Geochemical Fractionation and Availability of Vanadium in Soils

Cho-Yin Wu[1], Maki Asano[2] and Zeng-Yei Hseu[3]

[1] *Department of Agricultural Chemistry, National Taiwan University, Taipei, Taiwan*

[2] *School of Life and Environmental Sciences, University of Tsukuba, Tsukuba, Japan*

[3] *Department of Agricultural Chemistry, National Taiwan University, Taipei, Taiwan*

CONTENTS

4.1 Introduction 73
 4.1.1 Background 73
 4.1.2 Natural and Anthropogenic Sources of Vanadium 74
 4.1.3 Soil Component Sinks of Vanadium 75
4.2 Extractability of Vanadium 77
 4.2.1 Sequential Extraction 77
 4.2.2 Single Extraction 81
4.3 Bioavailability of Vanadium to Plants 83
 4.3.1 Vanadium Content in Plants 83
 4.3.2 Soil Factors Controlling Availability of Vanadium to Plants 83
References 88

4.1 Introduction

4.1.1 Background

According to Largo Resources (2019), global vanadium consumption in industry increased over 221.4% from 2012 to 2019. Such an increase leads to a large amount of vanadium released into the environment through anthropogenic activities mining, smelting, refining, fertilizer and soil amendment application (Poggioli et al., 2001; Imtiaz et al., 2016; Park et al., 2016; Chen et al., 2020). In consideration of applying efficient environmental assessment and food safety evaluation, the exploration of vanadium geochemical fractionation and availability is important to human health. Therefore, the aim of this chapter is to illustrate the sources, extractability and bioavailability of vanadium in soils.

DOI: 10.1201/9781003173274-4

4.1.2 Natural and Anthropogenic Sources of Vanadium

Vanadium is the fifth most abundant element among all transitional metals and ranks 22nd among all discovered elements in the Earth's crust, with an average content of 97 mg/kg, which is two times higher than copper and 10 times higher than lead (Nriagu, 1998; McDonough and Sun, 1995). Vanadium exists in over 80 minerals, which can be divided into following groups: sulfides (e.g., patronite and sulvanite), sulfates (e.g., minisragrite and cheremnykhite), silicates (e.g., roscoelite), oxides (e.g., navajoite and montroseite), phosphates (e.g., vanadinite) and vanadates (e.g., chervetite, tyuyamunite, carnotite and volborthite) (Nriagu, 1998). In addition, vanadium associates in nearly 70 minerals such as pyroxenes, hornblende, biotite and magnetite as an admixture due to the sorption properties of vanadium oxides (Kabata-Pendias, 2011).

Among various oxidation states of vanadium, such as +2, +3, +4, and +5, the ionic radius of V^{3+} (64.0 pm) is close to that of Fe^{3+} (64.5 pm), and thus vanadium is able to incorporate into a great amount of Fe-bearing mineral phases because V^{3+} is the common oxidation state in nature (Schwertmann and Pfab, 1996; Gustafsson, 2019). Therefore, around 92% of vanadium is found in basic rocks such as basalt, gabbro, amphibolite and eclogite, with vanadium concentration ranged from 200 to 300 mg/kg, whereas 7% is found in acidic and neutral rocks such as granite, with vanadium concentration ranged from 5 to 80 mg/kg (Hurlbut and Klein, 1977; Nriagu, 1998; Huang et al., 2015). Table 4.1 lists vanadium concentration in soils from different countries in which the great diversity of vanadium levels is regarded as the inheritance from their parent materials.

Approximate 230 Gg of vanadium is released into the environment each year through anthropogenic activities to cause undeniable soil contamination (Hope, 1997). Mining and high-temperature industrial activities are regarded as the major sources of vanadium pollution; for example, soils from the adjacent area to a vanadium mine in South Africa contained vanadium ranging from 1,570 to 3,600 mg/kg (Panichev et al., 2006). Another example is the urban park, agricultural area, smelting area and mine in Panzhihua City, China, where the soils contained vanadium ranging from 94.0–183.6 mg/kg, 71.7–227.2 mg/kg, 208.1–938.4 mg/kg and 111.6–591.2 mg/kg, respectively (Teng, Yang, Sunet al., 2011). The recovery of vanadium in the refining industry is lower than 70%, and around 30–40% of vanadium is discharged into the environment by gas emission, dust, wastewater and slag (Hope, 1997; Jiao and Teng, 2008); for example, the soil vanadium level was significantly higher in Salamanca (64–666 mg/kg), a city in Mexico surrounded by refinery, thermoelectric and chemical industries, than that in the rural area nearby (11–126 mg/kg) (Hernandez and Rodriguez, 2012). Furthermore, considerable vanadium levels were reported in soils from fossil oil combustion and leaching areas (Su et al., 2008; Qian et al., 2014). Additionally, application of phosphate fertilizers and soil amendments that contain rock phosphate (10–1,000 mg V/kg),

TABLE 4.1
Vanadium Concentration in Soils from Different Countries

Countries	Concentration (mg/kg)	Reference
Baltic States	18	Reimann et al. (2003)
Belgian	69	Cappuyns and Slabbinck (2012)
Canada	31.7–180.1	Mermut et al. (1996)
China	87.36	Yang, Teng et al. (2017)
Denmark	31	Baken et al. (2012)
Egypt	37.4–122.1	Shaheen and Rinklebe (2018)
Finland	12.0–13.0	Reijonen et al. (2016)
Germany	20.7–133.1	Shaheen and Rinklebe (2018)
Greece	23–158	Tsadilas and Shaheen (2010)
Hungary	15.2–42.0	Ovari et al. (2001)
Ireland	52.2	Zhang et al. (2008)
Italy	58	Manta et al. (2001)
Italy	34–276	Cicchella et al. (2015); Guagliardi et al. (2018)
Japan	94–250	Takeda et al. (2004)
Lithuania	38	Salminen and Gregorauskiene (2000)
Northern England	103	Rawlins et al. (2002)
Poland	18.39	Dudka and Market (1992)
Portugal	32	Ferreira et al. (2001)
Russia	79–91	Protasova and Kopayeva (1985)
Spain	6.01–144.9	Granero and Domingo (2002), Tume et al. (2006)
Sri Lanka	29.3–810	Chandrajith et al. (2005), Jayawardana et al. (2015)
Sweden	27–58	Baken et al. (2012)
Taiwan	35.4–475	Wu et al. (2020)
Thailand	210–591	Wisawapipat and Kretzschmar (2017)
Turkey	74	Yay et al. (2008)
USA	36–150	Govindaraju (1994)

superphosphate (50–2,000 mg V/kg), and basic slag (1000–5,000 mg V/kg) can also contribute considerable vanadium contamination in the ecosystem (van Zinderen Bakker and Jaworski, 1980; Yang, Teng et al., 2017).

4.1.3 Soil Component Sinks of Vanadium

Vanadium in soils is either dissolved into soil solution or retained by Fe/Al (hydr)oxides, clay minerals and organic matters (OM) as reported by Wisawapipat and Kretzschmar (2017) and Wu et al. (2020). Larsson, Hadialhejazi et al. (2017) mentioned a strong adsorption of pentavalent vanadium to ferrihydrite was observed between pH 4 to 10, in which the adsorption clearly decreased below pH 4 because of the predominance of cations (i.e., VO_2^+) by lowering pH. Aside from the adsorption mechanism, V(III) is able to substitute Fe(III) in goethite due to the similarity between ionic radii of Fe and V (Taylor and Giles, 1970; Aide, 2005). The effects of Fe oxides on vanadium sequestration during pedogenetic processes were demonstrated by the authors in this chapter, showing that the dithionite-citrate-bicarbonate (DCB) extractable vanadium positively and significantly correlated with the DCB-extractable iron ($r = 0.62$, $p < 0.001$) (Figure 4.1). From the raw data by the authors in this chapter, a Fe nodule was identified in the subsurface horizon of a basaltic soil based on the brighter appearance in the backscattered electron (BSE) mode of electron probe micro-analysis (EPMA); moreover, a strong and visible correlation between vanadium and Fe comparing to that with Si, Al and Mn in the EPMA mapping of a Fe nodule was observed (Figure 4.2), supporting the sequestered vanadium by pedogenic Fe oxides.

Gehring et al. (1994) reported the substitution of Al(III) by V(IV) in octahedral sites of clay minerals as VO^{2+} form, which is in good agreement with the close relationship between vanadium sorption and amorphous aluminosilicates elucidated by Burke et al. (2012) and Larsson, Persson et al. (2017). Furthermore, Wisawapipat and Kretzschmar (2017) reported on the mechanism of vanadium fixation in highly weathered soils, where it is adsorbed onto kaolinite, goethite, ferrihydrite and gibbsite as V(V) and structurally

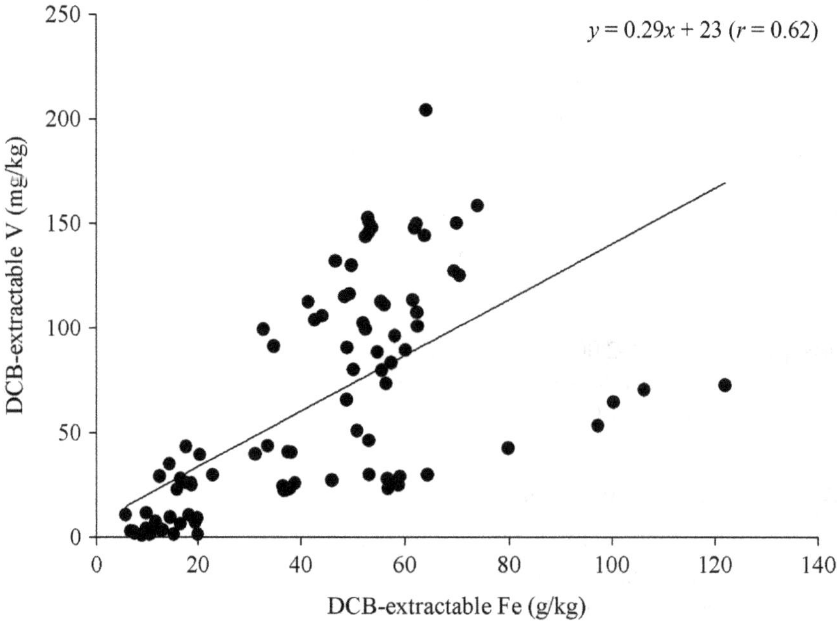

FIGURE 4.1
Relationship between DCB-extractable vanadium and Fe in soils.

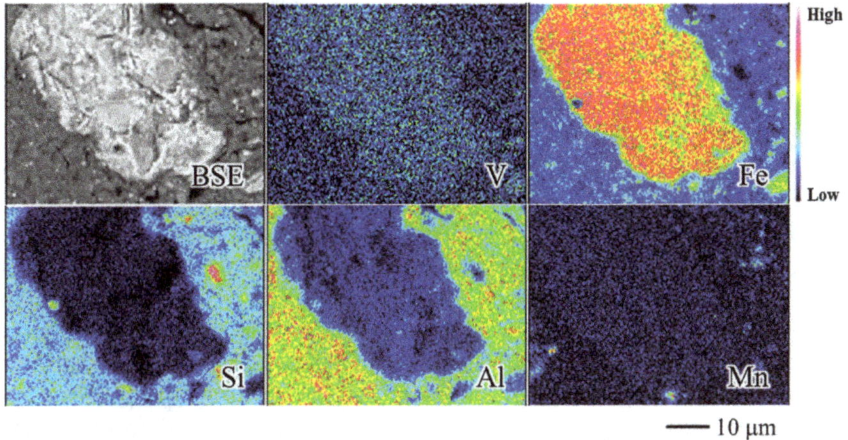

FIGURE 4.2
The BSE image corresponding to the electron mappings of V, Fe, Si, Al and Mn obtained by EPMA in a Fe nodule from the subsurface horizon of a basaltic soil.

incorporated in kaolinite as V(IV). V(IV) existed as a cation species (Crans et al., 1998), which led to a preferential retention of V(IV) by the functional groups of soil OM in comparison with the anion V(V) species (Larsson et al., 2015; Reijonen et al., 2016). V(V) is also able to form complexes with various organic molecules, yet the association is usually weaker than that with V(IV) and generally induces the reduction of V(V) to V(IV) (Gustafsson, 2019).

4.2 Extractability of Vanadium

4.2.1 Sequential Extraction

As a potentially toxic element (PTE), the available pool of vanadium in the biogeosphere has received a great amount of attention for environment quality and food safety (Huang et al., 2015; Imtiaz et al., 2015; Gustafsson, 2019; Shaheen et al., 2019; Chen et al., 2020). Wu et al. (2020) determined the geochemical fractions of vanadium in the rural soils from Taiwan using the Commission of the European Communities Bureau of Reference (BCR) sequential extraction method. They found that the residual fraction (81.5–98.9%) played a predominant role in their studied soils, followed by the reducible fraction (0.91–9.71%) and the oxidizable fraction (ND-16.7%), and the acid-soluble fraction of vanadium was negligible (ND-0.12%) (Table 4.2). Moreover, the potentially mobile fractions (PMF) – which meant all the non-residual fractions including acid soluble, Fe/Mn oxides-associated and organically bound fractions of vanadium – mainly consisted of reducible fractions, implying the predominance that Fe-oxides play in the immobilization of vanadium during the pedogenetic process. The oxidizable fraction of vanadium was especially enriched in the surface soils of most pedons, illustrating the accumulation of OM in the surface soils and the retention of vanadium by OM (Table 4.2). Shaheen and Rinklebe (2018) reported the predominance of residual fraction in slightly developed soils such as Fluvisols, Luvisols and Gleysols, which accounted for 35.3–81.8% of total vanadium by the BCR sequential extraction method. Nevertheless, in soils with extremely high degrees of weathering such as Oxisols, the dominant fraction was the Fe-oxides binding fraction (23–54%) followed by the acid soluble fraction (2–36%) and the residual fraction (10–38%) (Wisawapipat and Kretzschmar, 2017). The PMF of vanadium increased along with the progressive soil weathering, indicating that vanadium is gradually released from primary minerals along the pedogenetic processes and is further sequestered by soil components, especially by Fe-oxides in tropical regions.

TABLE 4.2

Chemical Fractions (mg/kg) of Vanadium and Their Proportions of Total Vanadium in Soils (reorganized from Wu et al., 2020)

Soil	Depth cm	Horizon	F1	F2	F3	F4
Ustorthent	0–15	A	ND	3.87 (2.08%)	28.7 (15.5%)	153 (82.5%)
	15–26	AC1	ND	12.9 (5.75%)	2.36 (1.05%)	210 (93.2%)
	26–36	AC2	ND	22.8 (8.57%)	0.07 (0.03%)	244 (91.4%)
	>36	AC3	ND	20.0 (6.84%)	ND	272 (93.2%)
Humudept	0–17	A1	0.06 (0.08%)	1.18 (1.50%)	ND	76.4 (98.4%)
	17–30	A2	0.05 (0.06%)	1.57 (1.70%)	ND	88.6 (98.2%)
	30–40	AB	0.03 (0.03%)	1.60 (1.50%)	ND	103 (98.4%)
	40–60	Bw1	0.06 (0.05%)	2.71 (2.10%)	ND	128 (97.9%)
	60–75	Bw2	0.06 (0.05%)	3.30 (2.90%)	ND	110 (97.0%)
	>75	C	0.07 (0.06%)	3.67 (3.20%)	ND	109 (96.7%)
Haplustalf	0–15	Ap	ND	2.49 (1.81%)	23.0 (16.7%)	112 (81.5%)
	15–25	AB	ND	13.8 (8.79%)	0.79 (0.50%)	143 (90.7%)
	25–50	BA	ND	15.9 (9.71%)	0.17 (0.11%)	148 (90.2%)
	50–70	Bt1	ND	26.0 (8.68%)	ND	273 (91.3%)
	70–90	Bt2	ND	17.4 (7.62%)	ND	211 (92.4%)
	>120	C	ND	17.5 (9.62%)	0.23 (0.13%)	164 (90.3%)
Plinthaquult	0–24	Ap	0.10 (0.12%)	5.51 (6.57%)	1.82 (2.17%)	76.5 (91.1%)
	24–47	Bt1	0.06 (0.06%)	1.10 (1.00%)	0.24 (0.22%)	109 (98.7%)
	47–69	Bt2	0.06 (0.06%)	0.94 (0.91%)	0.26 (0.25%)	102 (98.8%)
	69–100	Bt3	0.07 (0.06%)	2.18 (1.84%)	0.24 (0.20%)	116 (97.9%)

F1: acid soluble fraction, F2: reducible fraction, F3: oxidizable fraction, and F4: residual fraction; ND: not detectable

Teng, Yang, Sun et al. (2011) evaluated the potential mobility of vanadium in soils from contaminated regions (mining and smelting areas) and uncontaminated regions (agricultural areas and urban parks) by modified BCR sequential extraction. Table 4.3 shows the resemblance in the fractionation of soil vanadium among four land-use districts, with the sequence as: residual fraction > oxidizable fraction > reducible fraction > acid soluble fraction by the BCR scheme. In addition, no significant difference was found between the proportion of PMF in polluted regions (mining area 6.5–24.3%; smelting area 6.9–24.5%) and that in unpolluted regions (agricultural area 8.4–21.2%; urban park 8.2–23.7%). However, the PMF vanadium was clearly higher in soils from the contaminated regions (mining area 18.0–83.6 mg/kg; smelting area 41.7–132.1 mg/kg) comparing to that in uncontaminated regions (agricultural area 9.8–26.4 mg/kg; urban park 9.9–25.2 mg/kg), which

TABLE 4.3

Chemical Fractions (mg/kg) of Vanadium and Their Proportions of Total Vanadium in Studied Soils (reorganized from Teng, Yang, Sun et al., 2011)

	F1	F2	F3	F4
Mining area				
1	0.20 (0.03%)	16.4 (2.83%)	32.7 (5.65%)	530 (91.5%)
2	0.20 (0.05%)	11.2 (2.70%)	28.1 (6.77%)	376 (90.5%)
3	0.20 (0.10%)	10.9 (5.57%)	21.9 (11.2%)	163 (83.1%)
4	0.50 (0.15%)	25.6 (7.44%)	57.5 (16.7%)	260 (75.7%)
5	0.20 (0.09%)	9.20 (4.07%)	17.6 (7.79%)	199 (88.0%)
Smelting area				
1	0.70 (0.08%)	39.3 (4.50%)	92.1 (10.5%)	742 (84.9%)
2	0.20 (0.09%)	11.8 (5.34%)	28.0 (12.7%)	181 (81.9%)
3	0.50 (0.20%)	25.9 (10.5%)	34.3 (13.9%)	187 (75.5%)
4	0.60 (0.24%)	25.5 (10.2%)	24.9 (9.99%)	198 (79.5%)
5	0.20 (0.09%)	18.6 (8.78%)	27.9 (13.2%)	165 (78.0%)
Agricultural area				
1	0.10 (0.09%)	6.20 (5.39%)	9.10 (7.91%)	99.6 (86.6%)
2	0.20 (0.23%)	4.70 (5.45%)	7.40 (8.57%)	74.0 (85.8%)
3	0.10 (0.10%)	12.3 (12.0%)	9.30 (9.06%)	80.9 (78.9%)
4	0.10 (0.12%)	7.90 (9.20%)	4.90 (5.70%)	73.0 (85.0%)
5	0.10 (0.11%)	5.40 (6.03%)	7.20 (8.04%)	76.8 (85.8%)
Urban Park				
1	1.10 (0.90%)	3.10 (2.54%)	5.80 (4.75%)	112 (91.8%)
2	1.70 (0.89%)	6.10 (3.21%)	37.3 (19.6%)	145 (76.3%)
3	0.30 (0.24%)	5.80 (4.64%)	14.9 (11.9%)	104 (83.2%)
4	0.30 (0.26%)	5.70 (4.87%)	10.1 (8.63%)	101 (86.3%)
5	0.06 (0.53%)	5.60 (4.96%)	16.5 (14.6%)	90.3 (79.9%)

F1: acid soluble fraction, F2: reducible fraction, F3: oxidizable fraction, and F4: residual fraction highlighted a greater potential environmental risk contributed by industrial activities. A similar result to Teng, Yang, Sun et al. (2011) has been documented by Xiao et al. (2015) that the PMF of vanadium in soils from a stone coal smelting district ranged from 19.2 to 637 mg/kg accounted for approximately 50% of total vanadium.

Vanadium may exist as cationic or anionic species in soils, leading to rather complex incorporation strategies of vanadium into/onto Fe/Al/Mn (hydr)oxides, clay minerals and OM, as described in the last section. However, most of the available SE method presented rather similar fractionation of soil vanadium and low extractability of PMF in literature to show apparent limitations in identifying the differential association preference of vanadium in soil solid phases (Teng, Yang, Sun et al., 2011; Yang et al., 2015; Xiao et al., 2015; Shaheen and Rinklebe, 2018; Wu et al., 2020). Thus, Xu et al. (2017) proposed a newly developed eight-step sequential extraction scheme (SE I) based on the vanadium geochemical dynamic and a combined sequential extraction scheme (SE II) modified from the three most commonly applied SE methods, proposed by Davidson et al. (1998), Tessier et al. (1979), and Wenzel et al. (2001). In SE I, soil vanadium was fractionated into water-soluble, strong-adsorbed, OM-bound, Mn oxides-bound, very poorly crystalline Fe and Al (hydr)oxides-bound, poorly crystalline Fe and Al (hydr)oxides-bound, crystalline Fe and Al (hydr)oxides-bound and residual fractions. Additionally, in the SE II soil vanadium was fractionated into water-soluble, exchangeable, amorphous Fe and Al (hydr)oxides-bound, crystalline Fe and Al (hydr)oxides-bound, OM-bound and residual fractions. As listed in Table 4.4, the proportion of vanadium bound to OM in the forest soil (9.8%) was higher than that in the mining soil (0.3%) based on SE I, corresponding to the higher content of soil organic carbon in the forest soil (8.27%) comparing to the mining soil (2.15%). However, the difference of OM-bound fraction between forest and mining soils became lower by the SE II scheme because of the partial dissolution of Fe/Al (hydr)oxides during HNO_3-H_2O_2 extraction (Table 4.4). The total Fe concentration in the forest soil and mining soil was 1.69% and 11.1%, respectively. Wisawapipat and Kretzschmar (2017) and Wu et al. (2020) identified that Fe oxides greatly retained vanadium in soils; thus, a high percentage of Fe/Al (hydr)oxides-bound vanadium can be expected in the mining soil by both sequential extraction schemes in Table 4.4. The Fe/Al (hydr)oxides-bound fractions of vanadium based on the SE I accounted 53.6% and 56.8% in the forest soil and mining soil, respectively; however, only 2.1% of total vanadium was identified in mining soil by SE II as Fe/Al (hydr)oxides-bound vanadium, which was much lower than that in the forest soil (48.2%). Based on the improvement in the capability of SE to fractionate vanadium associating with different soil component sinks, Xu et al. (2017) claimed the eight-step sequential extraction scheme (i.e., SE I) as an optimized strategy to study the geochemical distribution of soil vanadium.

TABLE 4.4

The Proportion of Vanadium in Chemical Fractions of SE I and SE II in the Mining and Forest Soil (reorganized from Xu et al., 2017)

Chemical Fraction	Mining Soil	Forest Soil
	Proportion (%)	
SE I		
Water soluble	0.1	0.8
Strongly adsorbed	0.1	3.9
OM bound	0.3	9.8
Mn oxides bound	0.3	9.3
(Very poorly/poorly) crystalline Fe and Al (hydr)oxides bound	57	54
Residual	42	23
Recovery (%)	116	84
SE II		
Water soluble	0.1	1.5
Exchangeable	0.1	2.2
Amorphous/crystalline Fe and Al (hydr)oxides bound	2.1	48
OM bound	5.9	8.7
Residual	92	39
Recovery (%)	109	47

4.2.2 Single Extraction

To assess the bioavailable pool of vanadium in soils, the authors in this chapter applied four single extraction methods including $CaCl_2$, HCl, EDTA and $NaHCO_3$ as available proxies. The 0.01 M $CaCl_2$ is a neutral salt-based extractant that is able to exchange weakly sorbed cations on the soil surfaces (Meers et al., 2007), and 0.1 N HCl is an aggressive acid that can lower the pH and promote the release of metals from soils (Baker and Amacher, 1983). However, both extraction methods presented relatively low extractability, that is, 0.01–0.48% and ND-2.43% (Table 4.5), respectively, which might be due to the fact that the readily mobile vanadium in soils occurs mostly in anionic V(V) species (Wisawapipat and Kretzschmar, 2017). On the other hand, the 0.5 M $NaHCO_3$ extraction method was designed as an extractant for anions such as arsenic (Woolson et al., 1971), which extracted ND-3.11% of total soil vanadium (Table 4.5). As a non-specific chelating agent, EDTA is able to promote the dissolution of poorly crystalline Fe (oxyhydr)oxides as well as organically bound vanadium (Meers et al., 2007), and thus the EDTA presented the highest extraction efficiency (0.08–3.65%) amount among all extractants. The greater extractability of EDTA for soil vanadium has been confirmed by Teng, Yang, Wang et al. (2011), comparing with HCl, $NaNO_3$ and HOAc and by Wisawapipat and Kretzschmar (2017), comparing with $NaHCO_3$ and $BaCl_2$.

TABLE 4.5
Extractable Vanadium Content (mg/kg) and Their Proportions of Total Vanadium by Different Single Extractions

Soil	Depth	Horizon	$CaCl_2$	HCl	EDTA	$NaHCO_3$
	cm					
Plinthaquult	0–24	Ap	0.02 (0.02%)	2.34 (2.43%)	0.81 (0.84%)	1.24 (1.29%)
	24–47	Bt1	0.01 (0.01%)	0.45 (0.37%)	0.99 (0.80%)	0.35 (0.28%)
	47–69	Bt2	0.02 (0.02%)	0.47 (0.44%)	0.45 (0.42%)	0.39 (0.36%)
	69–100	Bt3	0.01 (0.01%)	1.19 (0.91%)	0.43 (0.33%)	0.70 (0.53%)
Endoaquept	0–20	Ap	0.02 (0.02%)	2.90 (3.58%)	0.40 (0.49%)	0.78 (0.96%)
	20–35	BA	0.03 (0.04%)	1.25 (1.54%)	1.08 (1.33%)	0.42 (0.52%)
	35–55	Bt1	0.02 (0.02%)	ND	1.15 (1.05%)	0.15 (0.14%)
	55–75	Bt2	0.03 (0.03%)	ND	0.42 (0.44%)	0.09 (0.09%)
	75–100	BC	0.04 (0.04%)	0.28 (0.31%)	0.18 (0.20%)	0.17 (0.19%)
Haplorthod	0–20	A	0.17 (0.48%)	0.87 (2.46%)	0.42 (1.19%)	1.10 (3.11%)
	20–25	E	0.05 (0.09%)	0.37 (0.68%)	1.22 (2.23%)	0.71 (1.30%)
	25–35	Bhs/Bt1	0.05 (0.04%)	0.33 (0.26%)	0.80 (0.62%)	0.55 (0.43%)
	35–40	Bhs/Bt2	0.05 (0.04%)	0.20 (0.15%)	0.85 (0.64%)	0.31 (0.23%)
	40–45	C	0.05 (0.05%)	0.12 (0.12%)	0.58 (0.56%)	0.17 (0.17%)
Haplustalf	0–10	A	0.07 (0.02%)	ND	0.47 (0.16%)	2.35 (0.82%)
	10–20	BA	0.06 (0.01%)	ND	0.38 (0.08%)	2.14 (0.45%)
	20–40	Bt1	0.06 (0.02%)	0.40 (0.12%)	5.67 (1.73%)	1.30 (0.40%)
	40–63	Bt2	0.09 (0.04%)	3.64 (1.49%)	8.81 (3.61%)	1.36 (0.56%)
	63–80	Bt3	0.07 (0.02%)	2.68 (0.94%)	5.23 (1.84%)	1.21 (0.43%)
	80–100	Bt4	0.06 (0.02%)	3.24 (0.91%)	13.1 (3.65%)	2.93 (0.82%)
Humudept	0–17	A1	0.02 (0.02%)	0.06 (0.07%)	0.34 (0.38%)	ND
	17–30	A2	0.02 (0.02%)	0.03 (0.02%)	0.39 (0.32%)	ND
	30–40	AB	0.02 (0.02%)	0.03 (0.02%)	0.45 (0.37%)	ND
	40–60	Bw1	0.02 (0.01%)	0.03 (0.02%)	0.70 (0.46%)	ND
	60–75	Bw2	0.02 (0.01%)	0.05 (0.04%)	0.99 (0.71%)	0.04 (0.03%)
	>75	C	0.02 (0.01%)	0.06 (0.04%)	1.05 (0.78%)	0.06 (0.04%)

ND: not detectable

Conventionally, the available pool of soil vanadium can be determined via SE methods and single extraction methods. SE methods aim to segregate soil vanadium into operationally defined phases in order to reveal its solubility and potential mobility; meanwhile, single extraction methods use chelating agents or a weak acid/base, simulating a plant's uptake mechanisms, which can be utilized to estimate the bioavailability of vanadium. Besides chemical extraction strategies, diffusive gradients in thin films (DGT) is an in-situ sampling technique that is able to provide the information of time-weighted concentrate of heavy metals with high spatial resolution in soils (Agbenin and Welp, 2012; Wang et al., 2016), in sediments (Gao et al., 2017) and in water

(Luo et al., 2010; Shi et al., 2016). Currently, the evaluation of labile vanadium concentration using DGT has been performed mostly in water, for example, Luo et al. (2010), Shi et al. (2016) and Luko et al. (2017); however, the binding layers their applied in DGT, that is, ferrihydrite and anionic exchange resin, embrace great potential in determining the availability of vanadium soils basing on the superiority of the Fe oxides-bound fraction in PMF of vanadium and the existence of vanadium oxyanions in soils, respectively.

4.3 Bioavailability of Vanadium to Plants

4.3.1 Vanadium Content in Plants

The transportation of vanadium in soil-plant systems has been emphasized by many studies because of the importance of human health risk through the food chain (Teng, Yang, Wang et al., 2011; Yangv et al., 2017; Yu and Yang, 2019; Chen et al., 2020). The uptake of vanadium largely relies on the plant species; as listed in Table 4.6, the vanadium level in the tissues of Fabaceae were commonly higher than those in other botanical families. As reported by Imtiaz, Ashraf et al. (2018), vanadium contents in the shoot and root of chickpeas were 602 mg/kg and 3753 mg/kg, respectively, based on a soil pot experiment. Considerable uptake of vanadium by Fabaceae was also found by Yang, Wang et al. (2017); the vanadium content was 5.2–67.1 mg/kg and 6.3–788.5 mg/kg in shoot and root tissues of soybean, respectively, based on a soil pot experiment. Nevertheless, in most plant species, vanadium content in plant tissues generally decreases in the order of roots > shoots (Chen et al., 2020); for example, the vanadium concentration in root tissues of alfalfa was 22.3 times higher than that in shoot tissues (Yang et al., 2011), reflecting the sequestration of vanadium in roots. Regardless of plant species, any tested plant with hydroponic cultivation demonstrated a higher capacity for vanadium accumulation in both shoot and root tissues compared to plants collected in fields and cultivated in pots (Table 4.6). For further information, please see Chapter 10 in this book.

4.3.2 Soil Factors Controlling Availability of Vanadium to Plants

The immobilization mechanism of vanadium by soil components mainly depends on its speciation, which further affects the biological accessibility and toxicity of vanadium in soils. V(IV) tends to form hydrous oxides and strong covalent bonding with OM, lowering its bioavailability (Tracey et al., 2007). Additionally, the preferential retention of V(IV) onto aluminum (hydr) oxides and smectite may reduce its mobility as well (Wehrli and Stumm, 1990). Nevertheless, as a redox sensitive element, the hydrolyzed V(IV)

TABLE 4.6
Vanadium Uptake (mg/kg) by Different Plant Species (reorganized from Chen et al., 2020)

Family Name	Plant Species	Shoot	Root	Experiment Condition	References
Fabaceae	Alfalfa	154.32	3440.14	Soil pot	Yang et al. (2011)
	Chickpea	601.6	3752.8	Soil pot	Imtiaz, Ashraf et al. (2018)
	Chickpea	2500	6000	Hydroponics	Imtiaz et al. (2016)
	Chickpea	850	4500	Hydroponics	Imtiaz et al. (2015)
	Common bean	78.45	3428.55	Hydroponics	Abeywardane (2019)
	Soybean	5.20–67.08	6.25–788.46	Soil pot	Yang, Wang et al. (2017)
Brassicaceae	Chinese cabbage	2.08–2.71	14.4–24.9	Soil pot	Tian et al. (2014)
	Chinese cabbage	4.92	39.75	Soil pot	Liao and Yang (2020)
	Chinese green mustard	668	9666	Hydroponics	Vachirapatama et al. (2011)
	Mustard	3000	5000	Hydroponics	Imtiaz, Mushtaq et al. (2018)
	Rape	1.56	6.47	Soil pot	Tian et al. (2015)
	Rape	18.1	73.3	Soil pot	Gokul et al. (2018)
Poaceae	Corn	200	198	Field	Ameh et al. (2019)
	Corn	80	61	Field	Ameh et al. (2019)
	Dog tail grass	1208.3	276.8	Field	Aihemaiti et al. (2017)
	Lemon grass	22	-	Field	Owolabi et al. (2016)
	Rice	346.5	2008	Hydroponics	Chongkid et al. (2007)
	Smile grass	0.11	34.2	Field	Guarino et al. (2019)
	Wheat	648.45	6620.66	Hydroponics	Abeywardane (2019)

Solanaceae	Box-thorn	0.66	2.73	Soil pot	Hou et al. (2014)
	Chilies	146	-	Field	Owolabi et al. (2016)
	Pepper	27	25	Hydroponics	García-Jiménez et al. (2018)
	Tomato	4.04	4010	Hydroponics	Vachirapatama et al. (2011)
	Tomato	51	-	Field	Owolabi et al. (2016)
Asteraceae	Blackjack	89	-	Field	Owolabi et al. (2016)
	Coneflower	44	-	Field	Owolabi et al. (2016)
	Common wormwood	0.01	1.69	Field	Guarino et al. (2019)
	Golden rods	35	-	Field	Owolabi et al. (2016)
	Groundsel	2.1	6.45	Field	Guarino et al. (2019)
	Mugwort	12	89.4	Field	Qian et al. (2014)
	Wild wormwood	66	-	Field	Owolabi et al. (2016)
Amaranthaceae	Beetroot	-	46	Field	Owolabi et al. (2016)
	Chenopodium album	13.0	384.3	Field	Aihemaiti et al. (2017)
	Slim amaranth	92	75	Field	Ameh et al. (2019)
Lamiaceae	Catmint	104	-	Field	Owolabi et al. (2016)
	Garden sage	439	-	Field	Owolabi et al. (2016)
	Pennyroyal	12	26	Soil pot	Barouchas and Kyramariou (2015)

(*Continued*)

TABLE 4.6 (Continued)

Family Name	Plant Species	Shoot	Root	Experiment Condition	References
	Rosemary	33	-	Field	Owolabi et al. (2016)
	Sweet basil	1.34	11.4	Soil pot	Akoumianaki-Ioannidou et al. (2016)
	Thyme	44	-	Field	Owolabi et al. (2016)
Anacardiaceae	Karee	84	-	Field	Owolabi et al. (2016)
Asphodelaceae	Aloe vera	61	-	Field	Owolabi et al. (2016)
Betulaceae	Gray birch	9.38	79.2	Field	Qian et al. (2014)
Cactaceae	Bunny cactus	2.36–2586.3	7.41–5908.5	Hydroponics	Yang and Tang (2015)
Canellaceae	Pepper bark	25	-	Field	Owolabi et al. (2016)
Chenopodiaceae	Kochia scoparia	66.4	1454.7	Field	Aihemaiti et al. (2017)
Cucurbitaceae	Watermelon	0.3	1.3	Hydroponics	Nawaz et al. (2018)
Hypoxidaceae	African potato	112	-	Field	Owolabi et al. (2016)
Polygonaceae	Asian knotweed	12.2	225	Field	Qian et al. (2014)
Rhamnaceae	Buffalo thorn	122	-	Field	Owolabi et al. (2016)
Typhaceae	Bulrush	5.1	218	Field	Qian et al. (2014)
Urticaceae	Stinging nettle	18	-	Field	Owolabi et al. (2016)
Verbenaceae	Lemon verbena	63	-	Field	Owolabi et al. (2016)

species can be easily oxidized to V(V) species, which tends to form weaker absorption onto oxides, clay minerals and OM (Wehrli and Stumm, 1989). Therefore, V(V) is considered the most mobile species of vanadium in soils and is verified as a possible carcinogen to humans as well as be listed in the same hazard pollutant class as Hg, Pb, As and Cd (WHO, 2006; Shaheen et al., 2019). Inorganic V(IV) exists mainly as cationic species of vanadyl (VO^{2+}) in soils and as the further hydrolyzed species of $VO(OH)^+$ when pH value is above 4 (Baes and Mesmer, 1976; Crans et al., 1998). On the other hand, at pH 3 to 7, inorganic V(V) occurs mostly as anionic species of vanadate ($H_2VO_4^-$) (Tracey et al., 2007), while the further deprotonated vanadate species dominate merely in alkaline environments base on their pK_a values (i.e., $H_2VO_4^-$, $pK_{a1} = 7.1$ and $pK_{a2} = 12$) (Heath and Howarth, 1981). In addition, V(IV) oxycationic species are able to form stronger complexes with OM compared to V(V) oxyanionic species (Wehrli, 1987); thus, the reduction from V(V) to V(IV) has frequently be observed in OM-rich systems even under aerobic conditions reducing the bioavailability of vanadium (Larsson et al., 2015). However, the hydrophilicity of OM increases with the increase of pH due to the deprotonation of organic functional groups at higher pH, leading to the enhancement of the mobility of organic vanadium complexes and the decline of the reduction of V(V) to V(IV), which is in agreement of Reijonen et al. (2016) that the amount of PMF of vanadium by SE method increased as the pH elevated from 4.0 to 6.9. Wu et al. (2020) recommended that $NaHCO_3$ is a potential bioavailability index for soil vanadium compared to the other applied single extractants ($CaCl_2$, HCl, and EDTA), because a better linear correlation was found between $NaHCO_3$-extractable vanadium in soils and plant shoot uptake vanadium ($r = 0.71$, $p < 0.01$), which further affirmed the greater bioavailability of vanadium at higher pH.

The availability of vanadium in soils is controlled by various soil solid phases (e.g., exchangeable, Fe/Al/Mn-oxides bound, OM bound), speciation and pH that require a considerable cost of work and time consumption to conduct a comprehensive determination on. Wu et al. (2020) found significant correlations between $NaHCO_3$-extracted vanadium and total soil vanadium (V_{total}), pH and cation exchange capacity (CEC) (Table 4.7). Thus, Wu

TABLE 4.7

Linear Correlation Matrix between Soil Properties and Vanadium in Soils ($n = 94$) (reorganized from Wu et al., 2020)

	Total V	Clay	pH	OC	CEC
0.01 M $CaCl_2$-extractable V	0.25*	−0.30*	NS	0.57***	0.22*
0.1 N HCl-extractable V	NS	0.24*	NS	NS	NS
0.05 M EDTA-extractable V	0.45***	NS	0.27**	−0.21*	0.60***
0.5 M $NaHCO_3$-extractable V	0.69***	NS	0.40***	NS	0.31**

*, **, and *** significant at $p < 0.05$, 0.01, and 0.001; NS: not significant.

et al. (2020) proposed a multivariate regression to evaluate the extractable vanadium as a function of V_{total}, pH, and CEC with the following equation ($r = 0.71, n = 94, p < 0.001$):

$$\log\left(V_{extractable}\right) = 2.0 \times \log\left(V_{total}\right) - 1.0 \times \log\left(CEC\right) + 0.12 \times pH - 4.4 \quad (4.1)$$

This equation is able to provide the information on relevant soil factors for predicting bioavailability of vanadium efficiently. However, these results required further examination using different soils and plant species in the future.

References

Abeywardane, M. 2019. Vanadium phytotoxicity: Vanadium uptake, translocation and interactions with nutrients in wheat and common bean (Doctoral dissertation). Royal Melbourne Institute of Technology University.

Agbenin, J. O., and G. Welp. 2012. Bioavailability of copper, cadmium, zinc, and lead in tropical savanna soils assessed by diffusive gradient in thin films (DGT) and ion exchange resin membranes. *Environmental Monitoring and Assessment* 184:2275–2284.

Aide, M. 2005. Geochemical assessment of iron and vanadium relationships in oxic soil environments. *Soil & Sediment Contamination* 14:403–416.

Aihemaiti, A., J. Jiang, D. Li, T. Li, W. Zhang, and X. Ding. 2017. Toxic metal tolerance in native plant species grown in a vanadium mining area. *Environmental Science and Pollution Research* 24:26839–26850.

Akoumianaki-Ioannidou, A., P. E. Barouchas, E. Ilia, A. Kyramariou, and N. K. Moustakas. 2016. Effect of vanadium on dry matter and nutrient concentration in sweet basil (*Ocimum basilicum* L.). *Australian Journal of Crop Science* 10:199–206.

Ameh, E. G., O. D. Omatola, and S. B. Akinde. 2019. Phytoremediation of toxic metal polluted soil: Screening for new indigenous accumulator and translocator plant species, northern Anambra Basin, Nigeria. *Environmental Earth Sciences* 78:345.

Baes, C. F., and R. E. Mesmer. 1976. *The Hydrolysis of Cations*. Wiley-Interscience: New York.

Baken, S., M. A. Larsson, J. P. Gustafsson, F. Cubadda, and E. Smolders. 2012. Ageing of V in soils and consequences for bioavailability. European *Journal of Soil Science* 63:839–847.

Baker, D. E., and M. C. Amacher. 1983. Nickel, copper, zinc and cadmium. In *Methods of Soil Analysis*, Part 2, 2nd ed., eds. A. L. Page, R. H. Miller, and D. R. Keeney, 323–336. ASA and SSSA: Madison, WI.

Barouchas, P. E., and A. Kyramariou. 2015. Effect of vanadium on dry matter and nutrient concentration in pennyroyal (*Mentha pulegium* L.). Bulletin of university of agricultural sciences and veterinary medicine Cluj-Napoca. *Horticulture* 72:295–298.

Bosque-Sendra, J.M., M.C. Valencia, and S. Boudra. 1998. Speciation of V (IV) and V (V) with Eriochrome cyanine R in natural waters by solid phase spectrophotometry. Fresenius. *Journal of Analytical Chemistry* 360:31–37.

Burke, I. T., W. M. Mayes, C. L. Peacock, A. P. Brown, A. P. Jarvis, and K. Gruiz. 2012. Speciation of arsenic, chromium, and vanadium in red mud samples from the Ajka spill site, Hungary. *Environmental Science & Technology* 46:3085–3092.

Cappuyns, V., and E. Slabbinck. 2012. Occurrence of vanadium in Belgian and European alluvial soils. *Applied and Environmental Soil Science* 2012:1–12.

Chandrajith, R., C. B. Dissanayake, and H. J. Tobschall. 2005. The abundances of rarer trace elements in paddy (rice) soils of Sri Lanka. *Chemosphere* 58:1415–1420.

Chen, L., J. R. Liu, J. Gao, W. F. Hu, and J. Y. Yang. 2020. Vanadium in soil-plant system: Source, fate, toxicity, and bioremediation. *Journal of Hazardous Materials* 405:124200.

Chongkid, B., N. Vachirapattama, and Y. Jirakiattikul. 2007. Effects of vanadium on rice growth and vanadium accumulation in rice tissues. *Kasetsart Journal Natural Science* 41:28–33.

Cicchella, D., L. Giaccio, E. Dinelli, S. Albanese, A. Lima, D. Zuzolo, P. Valera, and B. de Vivo. 2015. GEMAS: Spatial distribution of chemical elements in agricultural and grazing land soil of Italy. *Journal of Geochemical Exploration* 154:129–142.

Crans, D., S. Amin, and A. Keramidas. 1998. Chemistry of relevance to V in the environment. In *V in the Environment. Part 1: Chemistry and Biochemistry*, eds. J.O. Nriagu, 73–96. New York: John Wiley and Sons.

Davidson, C. M., A. L. Duncan, D. Littlejohn, A. M. Ure, and L. M. Garden. 1998. A critical evaluation of the three-stage BCR sequential extraction procedure to assess the potential mobility and toxicity of heavy metals in industrially-contaminated land. *Analytica Chimica Acta* 363:45–55.

Dudka, S., and B. Market. 1992. Baseline contents of As, Ba, Be, Li, Nb, Sr and V in surface soils of Poland. *Science of the Total Environment* 122:279–290.

Ferreira, R. S. G., P. G. P. de Oliveira, and F. B. Noronha. 2001. The effect of the nature of vanadium species on benzene total oxidation. *Applied Catalysis B: Environmental* 29:275–283.

Gao, B., L. Gao, Y. Zhou, D. Xu, and X. Zhao. 2017. Evaluation of the dynamic mobilization of vanadium in tributary sediments of the three gorges reservoir after water impoundment. *Journal of Hydrology* 551:92–99.

García-Jiménez, A., L. I. Trejo-Téllez, D. Guillén-Sánchez, and F. C. Gómez-Merino. 2018. Vanadium stimulates pepper plant growth and flowering, increases concentrations of amino acids, sugars and chlorophylls, and modifies nutrient concentrations. *PLoS One* 13:e0201908.

Gehring, A. U., I. V. Fry, J. Luster, and G. Sposito. 1994. V in sepiolite—a redox—indicator for an ancient closed brine system in the Madrid basin, central Spain. *Geochimica et Cosmochimica Acta* 58:3345–3351.

Gokul, A., L. F. Cyster, and M. Keyster. 2018. Efficient superoxide scavenging and metal immobilization in roots determines the level of tolerance to Vanadium stress in two contrasting Brassica napus genotypes. *South African Journal of Botany* 119:17–27.

Govindaraju, K. 1994. Compilation of working values and sample description for 383 geostandards. *Geostandards Newsletter Special Issue* 18:1–158.

Granero, S., and J. L. Domingo. 2002. Levels of metals in soils of Alcala de Henares, Spain: Human health risks. *Environment International* 28:159–164.

Guagliardi, I., D. Cicchella, R. de Rosa, N. Ricca, and G. Buttafuoco. 2018. Geochemical sources of vanadium in soils: Evidences in a southern Italy area. *Journal of Geochemical Exploration* 184:358–364.

Guarino, C., D. Zuzolo, and M. Marziano. 2019. Identification of native-metal tolerant plant species in situ: Environmental implications and functional traits. *Science of the Total Environment* 650:3156–3167.

Gustafsson, J. P. 2019. Vanadium geochemistry in the biogeosphere—speciation, solid-solution interactions, and ecotoxicity. *Applied Geochemistry* 102:1–25.

Heath, E., and O. W. Howarth. 1981. Vanadium-51 and oxygen-17 nuclear magnetic resonance study of vanadate (V) equilibria and kinetics. *Journal of the Chemical Society, Dalton Transactions* 5:1105–1110.

Hernandez, H., and R. Rodriguez. 2012. Geochemical evidence for the origin of V in an urban environment. *Environmental Monitoring and Assessment* 184:5327–5342.

Hope, B. K. 1997. An assessment of the global impact of anthropogenic V. *Biogeochemistry* 37:1–13.

Hou, M., C. Lu, and K. Wei. 2014. Accumulation and speciation of vanadium in Lycium seedling. *Biological Trace Element Research* 159:373–378.

Huang, J. H., F. Huang, L. Evans, and S. Glasauer. 2015. V: Global (bio)geochemistry. *Chemical Geology* 417:68–89.

Hurlbut, C. S., and C. Klein. 1977. *Manual of Mineralogy* (after James D. Dana). Wiley: Hoboken, NJ.

Imtiaz, M., M. Ashraf, M. S. Rizwan, M. A. Nawaz, M. Rizwan, S. Mehmood, B. Yousaf, Y. Yuan, A. Ditta, M. A. Mumtaz, M. Ali, S. Mahmood, and S. Tu. 2018. Vanadium toxicity in chickpea (*Cicer arietinum* L.) grown in red soil: Effects on cell death, ROS and antioxidative systems. *Ecotoxicology and Environmental Safety* 158:139–144.

Imtiaz, M., M. A. Mushtaq, M. A Nawaz, M. Ashraf, M. S. Rizwan, S. Mehmood, O. Aziz, M. Rizwan, M. S. Virk, Q. Shakeel, R. Ijaz, V. Androutsopoulos, A. Tsatsakis, and M. D. Coleman. 2018. Physiological and anthocyanin biosynthesis genes response induced by vanadium stress in mustard genotypes with distinct photosynthetic activity. *Environmental Toxicology Pharmacology* 62:20–29.

Imtiaz, M., M. A. Mushtaq, M. S. Rizwan, M. S. Arif, B. Yousaf, M. Ashraf, X. Shuanglian, M. Rizwan, S. Mehmood, and S. Tu. 2016. Comparison of antioxidant enzyme activities and DNA damage in chickpea (*Cicer arietinum* L.) genotypes exposed to V. *Environmental Science and Pollution Research* 23:19787–19796.

Imtiaz, M., M. S. Rizwan, S. Xiong, H. Li, M. Ashraf, S. M. Shahzad, M. Shahzad, S. Rizwan, and S. Tu. 2015. Vanadium, recent advancements and research prospects: A review. *Environment International* 80:79–88.

Jayawardana, D. T., H. M. T. G. A. Pitawala, and H. Ishiga. 2015. Geochemical evidence for the accumulation of vanadium in soils of chronic kidney disease areas in Sri Lanka. *Environmental Earth Sciences* 73:5415–5424.

Jiao, X., and Y. Teng. 2008. Techniques on soil remediation and disposal of V pollution. *Chinese Journal of Soil Science* 39:448–452.

Kabata-Pendias, A. 2011. Elements of group 5 (previously group Vb). In *Trace Elements in Soils and Plants*, ed. A. Kabata-Pendias, 173–177. Boca Raton: CRC Press.

Largo Resources. 2019. About vanadium. www.largoresources.com/company/about—vanadium/default.aspx.

Larsson, M.A., M. D'Amato, F. Cubadda, A. Raggi, I. Öbornc, D. B. Kleja, and J. P. Gustafsson. 2015. Long-term fate and transformations of Vin a pine forest soil with added converter lime. *Geoderma* 259:271–278.

Larsson, M. A., G. Hadialhejazi, and J. P. Gustafsson. 2017. Vanadium sorption by mineral soils: Development of a predictive model. *Chemosphere* 168:925–932.

Larsson, M. A., I. Persson, C. Sjöstedt, and J. P. Gustafsson. 2017. Vanadate complexation to ferrihydrite: X-ray absorption spectroscopy and CD-MUSIC modelling. *Environmental Chemistry* 14:141–150.

Liao, Y., and J. Yang. 2020. Remediation of vanadium contaminated soil by nanohydroxyapatite. *Journal of Soils and Sediments* 20:1534–1544.

Luko, K. S., A. A. Menegário, C. A. Suárez, M. Tafurt-Cardona, J. H. Pedrobom, A. M. C. M. Rolisola, E. T. Sulato, and C. H. Kiang. 2017. In situ determination of V (V) by diffusive gradients in thin films and inductively coupled plasma mass spectrometry techniques using amberlite IRA-410 resin as a binding layer. *Analytica Chimica Acta* 950:32–40.

Luo, J., H. Zhang, J. Santner, W. Davison. 2010. Performance characteristics of diffusive gradients in thin films equipped with a binding gel layer containing precipitated ferrihydrite for measuring Arsenic(V), Selenium(VI), Vanadium(V), and Antimony(V). *Analytical Chemistry* 82:8903–8909.

McDonough, W. F., and S. S. Sun. 1995. The composition of the Earth. *Chemical Geology* 120:223–253.

Manta, D. S., M. Angelone, A. Bellance, R. Neri, and M. Sprovieri. 2001. Heavy metals in urban soils: A case study from the city of Palermo (Sicily), Italy. *Science of the Total Environment* 300:229–243.

Meers, E., G. Du Laing, V. Unamuno, A. Ruttens, J. Vangronsveld, F. M. G. Tack, and M. G. Verloo. 2007. Comparison of cadmium extractability from soils by commonly used single extraction protocols. *Geoderma* 141:247–259.

Mermut, A. R., J. C. Jain, L. Song, R. Kerrich, L. Kozak, and S. Jana. 1996. Trace element concentrations of selected soils and fertilizers in Saskatchewan, Canada. *Journal of Environmental Quality* 25:845–853.

Nawaz, M. A., Y. Jiao, and C. Chen. 2018. Melatonin pretreatment improves vanadium stress tolerance of watermelon seedlings by reducing vanadium concentration in the leaves and regulating melatonin biosynthesis and antioxidant-related gene expression. *Journal of Plant Physiology* 220:115–127.

Nriagu, J.O. 1998. History, occurrence, and use of V. In *V in the Environment. Part 1: Chemistry and Biochemistry*, ed. J. O. Nriagu, 1–24. John Wiley & Sons: New York.

Ovari, M., M. Csukas, and G. Y. Zaray. 2001. Speciation of beryllium, nickel, and V in soil samples from Csepel Island, Hungary. *Fresenius Journal of Analytical Chemistry* 370:768–775.

Owolabi, I. A., K. L. Mandiwana, and N. Panichev. 2016. Speciation of chromium and vanadium in medicinal plants. *South African Journal of Chemistry* 69:67–71.

Panichev, N., K. Mandiwana, D. Moema, R. Molatlhegi, and P. Ngobeni. 2006. Distribution of vanadium(V) species between soil and plants in the vicinity of vanadium mine. *Journal of Hazardous Materials* 137:649–653.

Park, E.J., G. H. Lee, C. Yoon, and D. W. Kim. 2016. Comparison of distribution and toxicity following repeated oral dosing of different V oxide nanoparticles in mice. *Environmental Research* 150:154–165.

Poggioli, R., R. Arletti, A. Bertolini, G. Frigeri, and A. Benelli. 2001. Behavioral and developmental outcomes of prenatal and postnatal V exposure in the rat. *Pharmacological Research* 43:341–347.

Protasova, N.A., and M. T. Kopayeva. 1985. Trace and dispersed elements in soils of Russian Plateau. *Pochvovedeniye* 1:29–37.

Qian, Y., F. J. Gallagher, H. Feng, M. Wu, and Q. Zhu. 2014. Vanadium uptake and translocation in dominant plant species on an urban coastal brownfield site. *Science of the Total Environment* 476:696–704.

Rawlins, B. G., T. R. Lister, and A. Mackenzie. 2002. Trace metal pollution of soils in northern England. *Environmental Geology* 42:612–620.

Reijonen, I., M. Metzler, and H. Hartikainen. 2016. Impact of soil pH and organic matter on the chemical bioavailability of vanadium species: The underlying basis for risk assessment. *Environmental Pollution* 210:371–379.

Reimann, C., U. Siewers, T. Tarvainen, L. Bityukova, J. Eriksson, A. Giucis, V. Gregorauskiene, V. K. Lukashev, N. N. Matinian, and A. Pasieczna. 2003. *Agricultural Soils in Northern Europe: A Geochemical Atlas*. Geologisches Jahrbuch, Sonderhefte, Reihe D. Heft SD 5.

Salminen, R., and V. Gregorauskiene. 2000. Considerations regarding the definition of a geochemical baseline of elements in the surficial materials in areas differing in basic geology. *Applied Geochemistry* 15:647–653.

Schwertmann, U., and G. Pfab. 1996. Structural vanadium and chromium in lateritic iron oxides: Genetic implications. *Geochimica et Cosmochimica Acta* 60:4279–4283.

Shaheen, S. M., D. S. Alessi, F. M. G. Tack, Y. S. Ok, K. H. Kim, J. P. Gustafsson, D. L. Sparks, and J. Rinklebe. 2019. Redox chemistry of vanadium in soils and sediments: Interactions with colloidal materials, mobilization, speciation, and relevant environmental implications—A review. *Advances in Colloid and Interface Science* 265:1–13.

Shaheen, S. M., and J. Rinklebe. 2018. Vanadium in thirteen different soil profiles originating from Germany and Egypt: Geochemical fractionation and potential mobilization. *Applied Geochemistry* 88:288–301.

Shi, Y. X., V. Mangal, and C. Guéguen. 2016. Influence of dissolved organic matter on dissolved vanadium speciation in the Churchill River estuary (Manitoba, Canada). *Chemosphere* 154:367–374.

Su, T., S. Shu, H. Shi, J. Wang, C. Adams, and E. C. Witt. 2008. Distribution of toxic trace elements in soil/sediment in post—Katrina New Orleans and the Louisiana Delta. *Environmental Pollution* 156:944–950.

Takeda, A., K. Kimurab, and S. Yamasaki. 2004. Analysis of 57 elements in Japanese soils, with special reference to soil group and agricultural use. *Geoderma* 119:291–307.

Taylor, R., and J. Giles. 1970. The association of V and molybdenum with iron oxides in soils. *Soil Science* 21:203–215.

Teng, Y., J. Yang, Z. Sun, J. Wang, R. Zuo, and J. Zheng. 2011. Environmental vanadium distribution, mobility and bioaccumulation in different land-use Districts in Panzhihua Region, SW China. *Environmental Monitoring and Assessment* 176:605–620.

Teng, Y., J. Yang, J. Wang, and L. Song. 2011. Bioavailability of V extracted by EDTA, HCl, HOAC, and NaNO3 in topsoil in the Panzhihua Urban Park, located in Southwest China. *Biological Trace Element Research* 144:1394–1404.

Tessier, A., P. G. Campbell, and M. Bisson. 1979. Sequential extraction procedure for the speciation of particulate trace metals. *Analytical Chemistry* 51:844–851.

Tian, L. Y., J. Y. Yang, C. Alewell, and J. H. Huang. 2014. Speciation of vanadium in Chinese cabbage (Brassica rapa L.) and soils in response to different levels of vanadium in soils and cabbage growth. *Chemosphere* 111:89–95.

Tian, L. Y., J. Y. Yang, and J. H. Huang. 2015. Uptake and speciation of vanadium in the rhizosphere soils of rape (*Brassica juncea* L.). *Environmental Science and Pollution Research* 22:9215–9223.

Tracey, A. S., G. R. Willsky, and E. S. Takeuchi. 2007. *Vanadium: Chemistry, Biochemistry, Pharmacology and Practical Applications.* CRC Press: Boca Raton, FL.

Tsadilas, C. D., and S. M. Shaheen. 2010. Distribution of total and ammonium bicarbonate-DTPA-extractable soil vanadium from Greece and Egypt and their correlation to soil properties. *Soil Science* 175:535–543.

Tume, P., J. Bech, and L. Longan. 2006. Trace elements in natural surface soils in Sant Climent (Catalonia, Spain). *Ecological Engineering* 27:145–152.

Vachirapatama, N., Y. Jirakiattiku, G. W. Dicinoski, A. T. Townsend, and P. R. Haddad. 2011. Effect of vanadium on plant growth and its accumulation in plant tissues. *Songklanakarin Journal of Science and Technology* 33:255–261.

van Zinderen Bakker, E. M., and J. F. Jaworski. 1980. *Effects of Vanadium in the Canadian Environment.* Jaworski J. Ottawa, National Research Council of Canada [i.e. National Research Council Canada], Associate Committee on Scientific Criteria for Environmental Quality: Ottawa, ON.

Wang, P., T. Wang, Y. Yao, C. Wang, C. Liu, and Y. Yuan. 2016. A diffusive gradient-in-thin-film technique for evaluation of the bioavailability of Cd in soil contaminated with Cd and Pb. *International Journal of Environmental Research and Public Health* 13:556.

Wehrli, B. 1987. Vanadium in der Hydrosphäre: Oberflächenkomplexe und Oxidationskinetik (Doctoral dissertation). ETH Zurich.

Wehrli, B., S. Ibric, and W. Stumm. 1990. Adsorption kinetics of vanadyl (IV) and chromium (III) to aluminum oxide: Evidence for a two-step mechanism. *Colloids Surfaces* 51:77–88.

Wehrli, B., and W. Stumm. 1989. Vanadyl in natural waters: Adsorption and hydrolysis promote oxygenation. *Geochimica et Cosmochimica Acta* 53:69–77.

Wenzel, W. W., N. Kirchbaumer, T. Prohaska, G. Stingeder, E. Lombi, and D. C. Adriano. 2001. Arsenic fractionation in soils using an improved sequential extraction procedure. *Analytica Chimica Acta* 436:309–323.

Wisawapipat, W., and R. Kretzschmar. 2017. Solid phase speciation and solubility of V in highly weathered soils. *Environmental Science & Technology* 51:8254–8262.

Woolson, E. A., J. H. Axley, and P. C. Kearney. 1971. Correlation between available soil arsenic, estimated by six methods, and response of corn (*Zea mays* L.). *Soil Science Society of America Journal* 35:101–105.

World health organization. 2006. *IARC Monographs on the Evaluation of Carcinogenic Risks to Humans. (vol. 86) Cobalt in Hard Metals and Cobalt Sulfate, Gallium Arsenide, Indium Phosphide and Vanadium Pentoxide.* Lyon, France: 2003 IARC Working Group on the Evaluation of Carcinogenic Risks to Humans.

Wu, C. Y., M. Asano, Y. Hashimoto, J. Rinklebe, S. M. Shaheen, S. L. Wang, and Z. Y. Hseu. 2020. Evaluating vanadium bioavailability to cabbage in rural soils using geochemical and micro-spectroscopic techniques. *Environmental Pollution* 258:113699.

Xiao, X. Y., Y. A. N. G. Miao, Z. H. Guo, Z. C. Jiang, Y. N. Liu, and C. A. O. Xia. 2015. Soil vanadium pollution and microbial response characteristics from stone coal smelting district. *Transactions of Nonferrous Metals Society of China* 25:1271–1278.

Xu, Y.H., J. H. Huang, and H. Brandl. 2017. An optimized sequential extraction scheme for evaluation of V mobility in soils. *Journal of Environmental Sciences* 53:173–183.

Yang, J., and Y. Tang. 2015. Accumulation and biotransformation of vanadium in Opuntia microdasys. *Bulletin of Environmental Contamination and Toxicology* 94:448–452.

Yang, J., Y. Teng, J. Wang, and J. Li. 2011. Vanadium uptake by Alfalfa grown in V-Cd-contaminated soil by pot experiment. *Biological Trace Element Research* 142:787–795.

Yang, J., Y. Teng, J. Wu, H. Chen, G. Wang, L. Song, W. Yue, R. Zuo, and Y. Zhai. 2017. Current status and associated human health risk of V in soil in China. *Chemosphere* 171:635–643.

Yang, J., M. Wang, Y. Jia, M. Gou, and J. Zeyer. 2017. Toxicity of vanadium in soil on soybean at different growth stages. *Environmental Pollution* 231:48–58.

Yay, O. D., O. Alagha, and G. Tuncel. 2008. Multivariate statistics to investigate metal contamination in surface soil. *Journal of Environmental Management* 86:581–594.

Yu, Y., and J. Yang. 2019. Oral bioaccessibility and health risk assessment of vanadium (IV) and vanadium(V) in a vanadium titanomagnetite mining region by a whole digestive system in-vitro method (WDSM). *Chemosphere* 215:294–304.

Zhang, C., D. Fay, D. McGrath, E. Grennan, and O. T. Carton. 2008. Statistical analyses of geochemical variables in soils of Ireland. *Geoderma* 146:378–390.

5

Redox Chemistry of Vanadium in Soils and Sediments: Biogeochemical Factors Governing the Redox-Induced Mobilization of Vanadium in Soils

Jörg Rinklebe[1], Vasileios Antoniadis[2] and Sabry M. Shaheen[1,3,4]

[1] *University of Wuppertal, School of Architecture and Civil Engineering, Institute of Foundation Engineering, Water- and Waste-Management, Laboratory of Soil- and Groundwater-Management, Wuppertal, Germany*

[2] *University of Thessaly, Department of Agriculture Crop Production and Rural Environment, Volos, Greece*

[3] *King Abdulaziz University, Faculty of Meteorology, Environment, and Arid Land Agriculture, Department of Arid Land Agriculture, Jeddah, Saudi Arabia*

[4] *University of Kafrelsheikh, Faculty of Agriculture, Department of Soil and Water Sciences, Kafr El-Sheikh, Egypt*

CONTENTS

5.1 Introduction 96
5.2 Impact of E_H Changes on Vanadium Release and Mobilization 97
5.3 Biogeochemical Factors Govern the Redox-Induced Mobilization of Vanadium in Soils 101
5.3.1 Soil pH 101
5.3.2 Fe and Mn (Hydr)Oxides 103
5.3.3 Dissolved Organic Carbon 104
5.3.4 Sulfur 104
5.4 Impact of Redox Changes on Vanadium Speciation 105
5.5 Conclusions and Necessary Future Research 107
5.5 References 108

DOI: 10.1201/9781003173274-5

5.1 Introduction

Vanadium (V) is a non-essential metal for living organisms, with many known valences dependent on redox potential (E_H) (Shaheen et al., 2019; Bing et al., 2020). Among them, pentavalent V is the most mobile and toxic; thus, high concentrations may cause adverse effects to the environment, including plants and animals (Shaheen et al., 2019). The known valencies of V are six: −1, 0, +2, +3, +4, and +5; the latter three have significance in conditions found in the environment (Crans et al., 1998; Takeno, 2005), with V(V) and V(IV) being the dominant (Figure 5.1). As for V(III), it is usually a short-lived species, existing only in highly reducing conditions, such as in prolonged saturated soils or in peat deposits; it is also insoluble and readily oxidized to higher valency species even when O_2 is sparingly present (Imtiaz et al., 2015).

Vanadium soil dynamics is highly influenced by redox conditions, as indexed by E_H. Such effects may be direct (i.e., affecting the V chemical behavior itself) or indirect (i.e., E_H changes cause changes to intermediate factors, which in turn influence V behavior). Such intermediate factors may be pH, oxides of Fe and Mn, sulfur and organic C (Shaheen et al., 2019; Bing et al., 2020).

In this chapter we discuss the most up-to-date research on the behavior and fate of V concerning its geochemical dynamics, redox-influenced

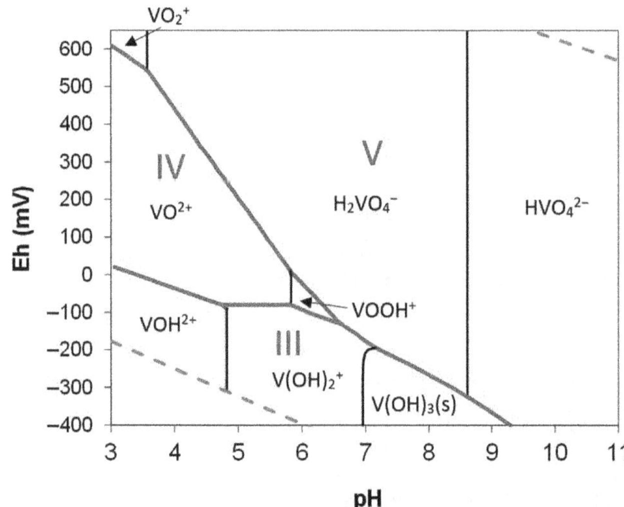

FIGURE 5.1
Vanadium E_H-pH diagram indicates V species in water at 1 μmol L^{-1}. III = V(+3); IV = V(+4); V= V(+5).

Source: Redrawn from Gustafsson and Johnsson (2004) and Shaheen et al. (2019) with permission from the publisher.

speciation and its analysis and measurement with spectroscopic techniques; this helps clarify aspects linked to the metal's mobilization dynamics in the solution-colloid interface of soils and sediments, especially those affected by oxic/anoxic conditions, that is, in wetlands or in waterlogged areas, so that potential risks may be identified, particularly in soils. We duly recognize the necessity for more in-depth subsequent research on the redox-induced release dynamics and mobilization of vanadium in soils. Therefore, we include in this chapter relevant suggestions about the impact of soil factors governing the redox-induced mobilization of vanadium in soils, including the redox-dependent changes on soil pH, Fe-Mn oxides, dissolved organic carbon and sulfate.

5.2 Impact of E_H Changes on Vanadium Release and Mobilization

Soil V, similarly to all redox-sensitive elements, is dependent upon redox conditions (Shaheen et al., 2019; Bing et al., 2020). In a microcosm apparatus, a series of investigations on the role of E_H were conducted with samples from places across the globe: the United States (Shaheen et al., 2016), Egypt (Shaheen, Rinklebe, Frohne et al., 2014) and Germany (Elbe river—Shaheen, Rinklebe, Rupp et al., 2014; Wupper river—Frohne et al., 2015). Soluble V from Egypt (Nile River; Figure 5.2) ranged from 15.0 to 71.0 µg L^{-1}, with the increased values found as expected at higher E_H values. As a result, dissolved V was correlated positively with E_H (Table 5.1). The possible explanation is the relatively fast oxidation of trivalent V to its tetra- and pentavalent species, both much more soluble than V(III) (Shaheen, Rinklebe, Rupp et al., 2014, Shaheen et al., 2016, 2019; Frohne et al., 2015). In contrast, soluble V in the US samples (Mississippi river sediments; Figure 5.2) ranged between 0.9 and 48.6 µg L^{-1}, the Wupper samples (Figure 5.2) ranged from 1.7 to 12.8 µg L^{-1}, and those from the Elbe river (Figure 5.3) from 0.3 to 4.35 µg L^{-1}, all with the opposite tendency—a decrease when reducing conditions prevailed (i.e., with lower E_H).

As redox is affected by soil moisture content, wet-dry cycles are expected to cause redox fluctuations. Shaheen, Rinklebe, Rupp et al. (2014), in a lysimeter study (Figure 5.3), established long-term (LT, for 94 days) and short-term (ST, for 21 days) wet-dry cycles. They found that soluble V in the ST cycles lysimeter was higher than that at LT, an outcome likely influenced by pH fluctuations and Fe oxides solubility.

As already mentioned, soil redox changes have a significant impact on soil V speciation, with pentavalent V being the dominant species in oxic conditions and tetravalent V in anoxic (and with trivalent V being present under

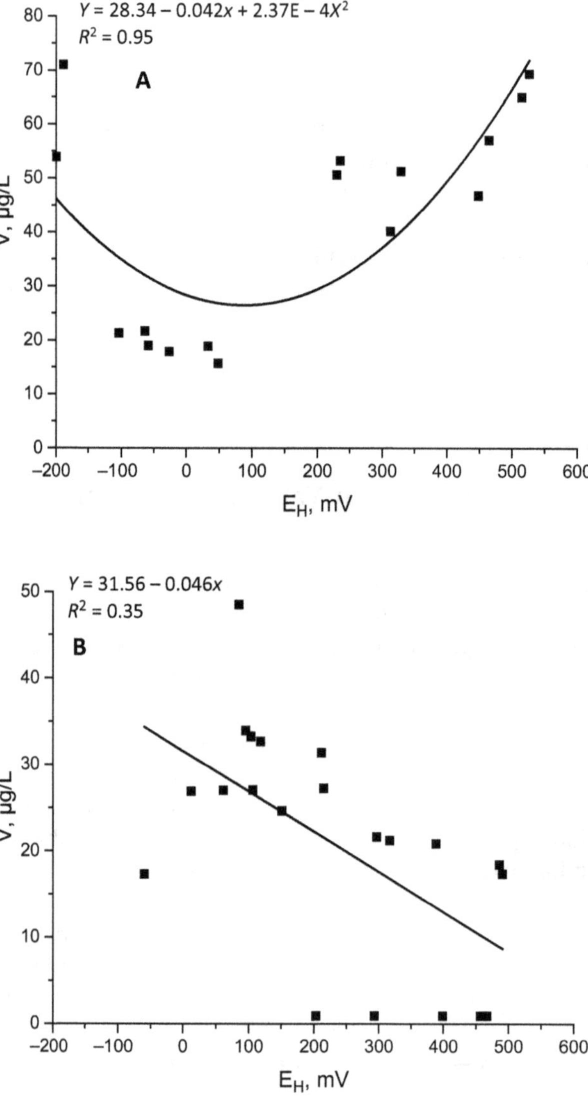

FIGURE 5.2
Water soluble vanadium vs E_H in different soils. *(Continued)*
Source: Data reproduced from Shaheen et al. (2019), with permission from the publisher.

certain conditions, and usually only short-lived under real soil conditions). It is interesting that when oxic conditions prevail in strongly acidic soils, cation V(IV) is readily oxidized to V(V) (anionic); if tetravalent V is dissolved (a likely state in a strongly acidic soil solution), it is possible that it will subsequently be sorbed as pentavalent onto positively charged surfaces, for

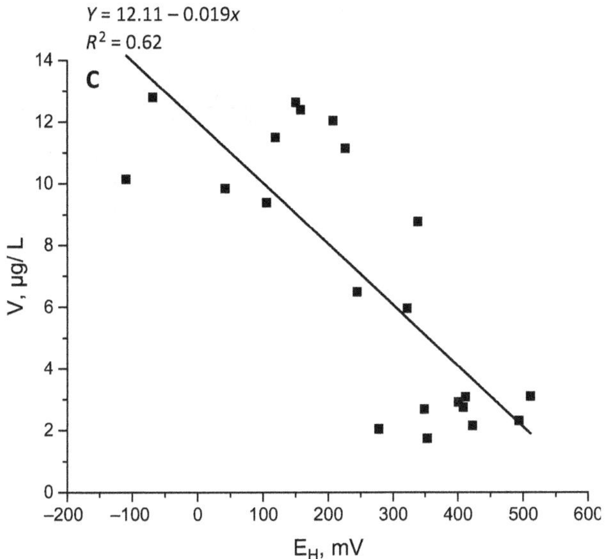

FIGURE 5.2 (Continued)

example, Fe/Al oxides. Such solubility decrease has also been hypothesized by Shaheen, Rinklebe, Rupp et al. (2014), Shaheen et al. (2016) and Frohne et al. (2015). The case is similar in highly weathered soils: Wisawapipat and Kretzschmar (2017) reported that re-sorption of newly oxidized V(V) from previously soluble V(IV) can occur on abundant kaolinite.

On the other hand, it must be noted that V(IV) is not a species of lithogenic origin, but it rather derives from the reduction of V(V) when there is abundance of electron donors: such phases may be organic carbon being decomposed to $C^{IV}O_2$ (Wanty and Goldhaber, 1992) (e.g., in lavishly organic matter-rich soils, such as organic planes) or O^{-II} in liquid water ($H^{I}_2O^{-II}$) being oxidized to gaseous O^0_2.

The reduction half reaction (electron reception process) is expressed as follows:

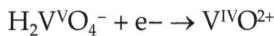

$$H_2V^VO_4^- + e- \rightarrow V^{IV}O^{2+}$$

The electron donation in the presence of organic C (oxidation half reaction) is as follows:

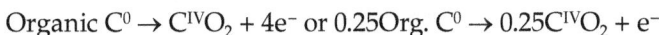

$$\text{Organic } C^0 \rightarrow C^{IV}O_2 + 4e^- \text{ or } 0.25\text{Org. } C^0 \rightarrow 0.25C^{IV}O_2 + e^-$$

The resultant redox reaction is a proton consuming process:

$$H_2V^VO_4^- + 0.25\text{Org. } C^0 + 3H^+ \rightarrow V^{IV}O^{2+} + 0.25C^{IV}O_2 + 2.5H_2O$$

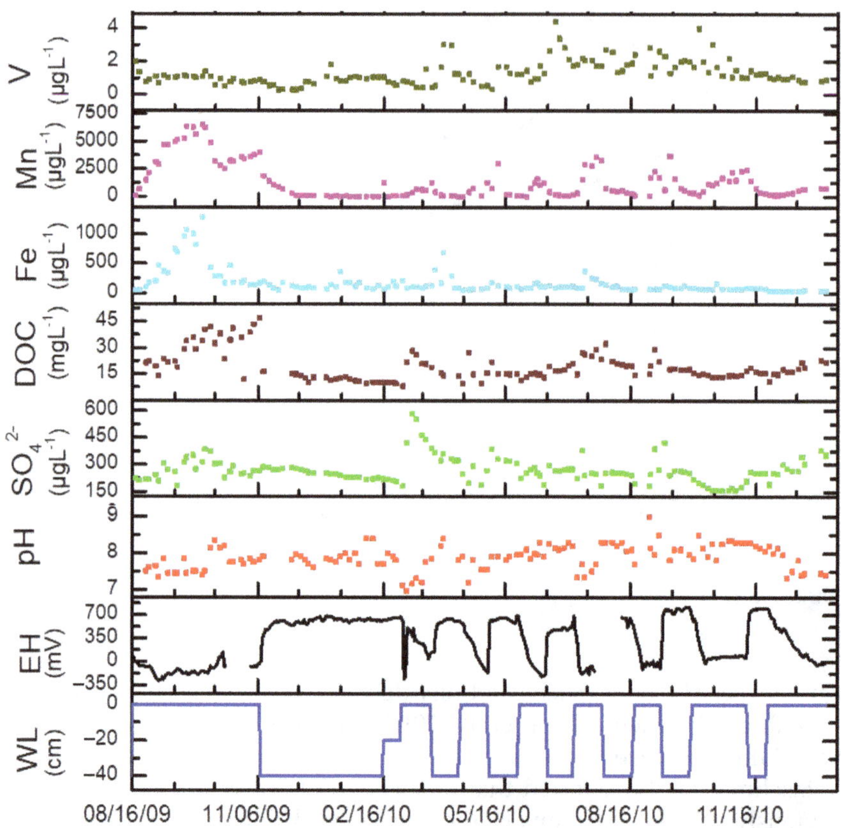

FIGURE 5.3
Pore water concentrations of vanadium and governing factors as affected by the changes of water level (WL) in the lysimeters.
Source: Reproduced and adapted from Shaheen, Rinklebe, Rupp et al., 2014 with permission from the publisher.

Indeed, it has been reported in the literature that dissolved organic C (DOC) can act as a readily available reductant agent (Frohne et al., 2015; Shaheen, Rinklebe, Rupp et al., 2014, 2016). The further reduction to trivalent V is also possible, although the process under normal conditions is much slower (waterlogged conditions must be prolonged over months), and it also requires the presence of robust reductant species, most notorious of which is S^{-II} (Wanty and Goldhaber, 1992). Sulfides, in turn, are readily oxidized to S^0 (as in elemental sulfur) and S^{II} (as in the thiosulfate oxyanion; $S^{II}_2O_3^{2-}$), and in the terminal and most stable S^{VI} sulfate species ($S^{VI}O_4^{2-}$). The abundance of sulfides was provided as an explanation for the fact that V(IV) was reduced to V(III) at E_H=–180 mV (a combination of strongly anoxic environment and the ample presence of a highly efficient electron donor), and thus total V was found to reduce its solution solubility, as reported by Frohne et al. (2015) and Shaheen et al. (2016). On the other hand, an increase in V solubility may also be evident: this could occur in aerated (i.e., oxic) soils that happen to be alkaline, as per Shaheen, Rinklebe, Frohne et al. (2014). Such experimental findings are also supported by some earlier studies (e.g., Aide, 2005; Tsadilas and Shaheen, 2010; Wright and Belitz, 2010).

5.3 Biogeochemical Factors Govern the Redox-Induced Mobilization of Vanadium in Soils

5.3.1 Soil pH

The extent to which E_H will affect pH depends on soil pH buffering capacity, which in turn depends on the presence of soil constituents, such as clay (especially when comprising of highly reactive secondary minerals), organic matter and oxides (Rinklebe et al., 2017; Shaheen et al., 2019; Bing et al., 2020). The expected effect would be that the two factors are adversely proportional (an increase of the one causes the decrease of the other). In a well-established effect, a decrease in E_H would cause the reduction of redox-sensitive elements in soil from their previous oxidation state: for example, $N^{V}O_3^-$ to $N^{-III}H_2$ in amines (when immobilized in micro-organisms), Mn(III/IV) to Mn^{2+}, and oxides-constructing Fe^{III} to free ion Fe^{2+}. Such reduction mechanisms require the consumption of solution H^+, and this in turn increases solution pH (Frohne et al., 2011; Rinklebe, Shaheen, and Frohne, 2016). For example, NO_3^- reduction consumes 2 mol of H^+ per mol of N reduced, and the same mol of H^+ are taken from solution per mol of reduced Mn^{VI} and per mol of reduced Fe_2O_3. However, in a complex system such as soil, this chemically viable option is not always materialized: there is a number of works (e.g., Shaheen, Rinklebe, Frohne et al., 2014; Rinklebe, Shaheen, and Yu, 2016) that has reported decreased pH with decreased E_H as a likely result of concurrent production of CO_2 (a readily soluble gas that produces carbonic acid and thus generates acidity, as per the reaction $CO_2 + H_2O \rightarrow H_2CO_3 \rightarrow HCO_3^- + H^+$), eluted from soil microbially-driven processes, that is, organic matter decomposition and biota respiration (including roots of higher plants and micro-organisms) (Wang et al., 2016). Likewise, Shaheen, Rinklebe, Frohne et al. (2014) and Shaheen, Rinklebe, Rupp et al. (2014), found a concurrent decrease of E_H and pH, an unexpected effect that nevertheless caused a decrease in soluble V, as predicted due to the formation of the less-soluble reduced species (Table 5.1). This decreased solubility, apart from being caused by the decreased E_H, may also be caused by decreased pH itself. Inversely, if this is a valid assumption, increased pH would cause increased V solubility. Indeed, it has repeatedly been reported that V solubility in alkaline soils is

TABLE 5.1

Pearson Correlation Coefficients (r) between Concentrations of Water-Soluble V and Governing Factors

Parameter	r	Soils	Reference
Fe	ns	Nile soil	Shaheen, Rinklebe, Frohne et al. (2014)
	+0.54*	Mississippi soil	Shaheen et al. (2016)
	+0.22*	Elbe soil	Shaheen, Rinklebe, Rupp et al. (2014)
	+0.76**	Wupper soil	Frohne et al. (2015)
Mn	−0.56*	Nile soil	Shaheen, Rinklebe, Frohne et al. (2014)
	ns	Mississippi soil	Shaheen et al. (2016)
	ns	Elbe soil	Shaheen, Rinklebe, Rupp et al. (2014)
	−0.68**	Wupper soil	Frohne et al. (2015)
DOC	−0.56*	Nile soil	Shaheen, Rinklebe, Frohne et al. (2014)
	+0.49*	Mississippi soil	Shaheen et al. (2016)
	ns	Elbe soil	Shaheen, Rinklebe, Rupp et al. (2014)
	+0.61**	Wupper soil	Frohne et al. (2015)
SUVA	+0.69*	Nile soil	Shaheen, Rinklebe, Frohne et al. (2014)
	+0.53*	Mississippi soil	Shaheen et al. (2016)
	+0.89*	Wupper soil	Shaheen, Rinklebe, Rupp et al. (2014)
Sulfate	ns	Nile soil	Frohne et al. (2015)
	ns	Mississippi soil	Shaheen, Rinklebe, Frohne et al. (2014)
	ns	Elbe soil	Shaheen et al. (2016)
	ns	Wupper soil	Shaheen, Rinklebe, Rupp et al. (2014)
pH	+0.56*	Nile soil	Shaheen, Rinklebe, Frohne et al. (2014)
	ns	Mississippi soil	Shaheen et al. (2016)
	+0.51*	Elbe soil	Shaheen, Rinklebe, Rupp et al. (2014)
	ns	Wupper soil	Frohne et al. (2015)
E_H	+0.51*	Nile soil	Shaheen, Rinklebe, Frohne et al. (2014)
	−0.59**	Mississippi soil	Shaheen et al. (2016)
	−0.30*	Elbe soil	Shaheen, Rinklebe, Rupp et al. (2014)
	−0.78**	Wupper soil	Frohne et al. (2015)

*, **significant at the 0.05 and 0.01 level, respectively

E_H = Redox potential; DOC = Dissolved organic carbon; SUVA = Specific UV absorbance: calculated as the absorbance (measured at 254 nm) normalized to the DOC.

ns: not significant

higher than that in acidic (Panichev et al., 2006; Tsadilas and Shaheen, 2010; Shaheen and Rinklebe, 2018). This is driven primarily by V(V), as reported by Panichev et al. (2006), and Reijonen et al. (2016). More specifically, the latter study found that both V(IV) and V(V) are responsible for such solubility increase due to (a) V(V) being protected from reduction by SOM and (b) V(IV) being readily oxidized, with O_2 acting as an electron receptor. This same work also identified another indirect connection between pH and V solubility: the fact that in alkaline soils (hydr)oxides activity is dramatically diminished, which in turn causes a decrease in their retention ability concerning V. Likewise, in an early work, V sorption was found to be decreased (which would indicate enhanced solubility), similar to the solubility of organic V (Blackmore et al., 1996).

5.3.2 Fe and Mn (Hydr)Oxides

Iron and Mn (hydr)oxides reactivity is greatly influenced by soil redox conditions, and thus they govern metals' (including vanadium's) retention and release depending on soil E_H (Rinklebe et al., 2017; Shaheen et al., 2019; Bing et al., 2020). Indeed, oxides have a strong tendency to retain V, a net result of a number of mechanisms such as adsorption (either specific electrostatic inner-sphere chemisorption or less direct outer-sphere sorption via surface-connected reactive groups) and occlusion (physical entrapment and subsequent removal from solution). On the other hand, oxides activity is reported to decrease with decreased E_H due to the reduction of structural Fe^{III} to the much more soluble divalent Fe species (Frohne et al., 2015; Shaheen, Rinklebe, Rupp et al., 2014; Shaheen et al., 2016), and the case is similar with Mn; thus, increased V solubility with decreasing E_H could well be the result of the indirect effect of diminished Fe/Mn oxides reactivity.

Conversely, the expectedly diminished V solubility at oxic conditions could be the indirect effect of the enhanced reactivity of Fe/Mn oxides, which in turn would tend to bind V and remove it from solution, thus decreasing V solubility. Indeed, this chain-resembling association of "changed E_H-affecting oxides reactivity-affecting V retention/solubility" was also identified by Martin (2005) and Wright et al. (2014). The mechanism of oxides redox-driven dissolution may be entirely chemical or biological. Concerning the former, apart from the previously stated reduction of structural Fe(III) to soluble Fe(II) in the presence of O_2 as an electron donor, this could also be driven by the much more robust reductant sulfide species, H_2S, which readily donates electrons to its own oxidation to S^0, S^{II} and S^{VI} (Wright et al., 2014). As for the latter (the biological mechanism), this is linked to the involvement of organic matter decomposer microbes, which turn C^0 (organic) to the stable gaseous inorganic $C^{IV}O_2$: this reaction proceeds with electrons being donated from organic C to oxide-Fe(III) and results in oxides dissolution. Thus, it would be expected that if the indirect connection between E_H and V solubility via the oxides was correct, Fe and Mn solubility (induced at low E_H due to the dissolution of oxides crystals) would be correlated positively with V solubility (also induced in low E_H due to the fact that they are released to solution from dissolving oxides). However, this effect is not always observed in the literature. For example, Shaheen, Rinklebe, Frohne et al. (2014), Frohne et al. (2015), and Shaheen et al. (2016) found a negative correlation between V solubility and Mn solubility (Table 5.1). In the first of the three cited works, this occurred due to the fact that V solubility increased at oxic E_H values. Concerning the latter two works, V solubility was indeed found to have increased under reducing conditions, but Mn solubility increased, instead of being decreased as expected, under oxic conditions.

5.3.3 Dissolved Organic Carbon

Organic matter affects V dynamics, as it does with a host of other elements: during decomposition, it liberates previously borne V in soil (Frohne et al., 2015; Shaheen et al., 2019). The decomposition itself differs depending on soil redox conditions. The soluble organic fraction, known as DOC, comprises aliphatic and aromatic compounds, conveniently measured through indirect spectrophotometer measurements of soil extractions at the UV wavelength range: specific UV absorbance (SUVA) can be an index of the presence of aromatic compounds referred to as DAC (dissolved aromatic C) (Leenheer and Croue, 2003; Frohne et al., 2015), with low SUVA measurements indicating low aromatic compounds content (Weishaar et al., 2003; Grybos et al., 2009; Shaheen, Rinklebe, Frohne et al., 2014). Aliphatic C is then calculated as the difference between total C (as in the DOC) and DAC. There is evidence that V solubility decreases with enhanced aliphatic C content and with decreased aromatic C. Indeed, conducting this indirect DOC measurement, Shaheen, Rinklebe, Frohne et al. (2014) confirmed these trends in a Nile river soil (Table 5.1). However, this is not always the case: Frohne et al. (2015) and Shaheen et al. (2016) reported a positive correlation between V versus both DOC and SUVA (Table 5.1).

As for redox, both DOC and DAC are known to increase in reducing conditions. On the other hand, both aliphatic and aromatic C compounds act as electron donors that function as driving forces for V(V) reduction to V(IV) (Wanty and Goldhaber 1992; Frohne et al., 2015; Shaheen et al., 2016). This is a reaction that would lead to the decreased V solubility—the produced V(IV) at low E_H can be sorbed by the highly reactive DOC (Reijonen et al., 2016). The same study also reported that at high E_H DOC does affect V (in its pentavalent species) by increasing its solubility due to the formation of low molecular weight organo-vanadium soluble complexes. The interesting findings reported in that work are summarized in Figure 5.4.

5.3.4 Sulfur

Sulfur is an element with many oxidation states, all relatively stable and present under varying natural environmental conditions; sulfur oxidation is a fast and proton releasing (i.e., acidifying) process that readily occurs in the presence of electron receptors (among which the most frequently available is gaseous O_2); other receptor elements are also affected (Du Laing et al., 2009; Rinklebe et al., 2017), and one of these elements could be V(V), which is reduced to the less soluble V(IV). Although S oxidation is a robust driving

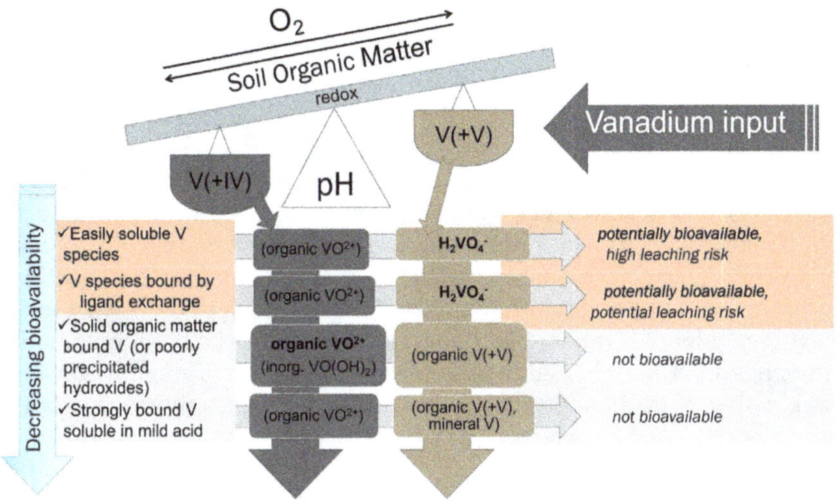

FIGURE 5.4
Scheme to show the impact of E_H, pH, and soil organic matter on the bioavailability of vanadium species. V(+4) and V(+5) indicate the dominant species at pH 4–7.
Source: Reproduced from Reijonen et al., 2016 with permission from the publisher.

force for V reduction, the opposite process (S reduction) is rather slow, generated over long periods of time, and thus experiments concluded within days or weeks cannot possibly record its effect. Indeed, a number of redox studies (e.g., Frohne et al., 2015; Shaheen, Rinklebe, Frohne et al., 2014; Shaheen, Rinklebe, Rupp et al., 2014; Shaheen et al., 2016) have found no connection between E_H and $S^{VI}O_4$ (as also presented in Table 5.1). Another explanation is provided by Du Laing et al. (2009): rapid formation of various S species in wetland soils. Further understanding of the role of S geochemistry is highly necessary, as there are aspects of its behavior that are yet to be elucidated.

5.4 Impact of Redox Changes on Vanadium Speciation

Vanadium species are shown in Figure 5.1, in its E_H-pH predominance diagram. The pentavalent V (vanadate; $H_2V^VO_4^-$) is predominant in usual pH and E_H values found under normal soil conditions. In strongly acidic and aerated soils, tetravalent V (vanadyl, also referred to as oxovanadium; $V^{IV}O^{2+}$, with a characteristic blue aqueous solution) is rather predominant. Both these species can be found soluble in the soil solution, with V(V) being more soluble than V(IV). However, as for V found in mineral lattices, its form is that of its tri- and tetravalent species (Gehring et al., 1993; Schwertmann and Pfab, 1994; Shaheen et al., 2019).

Analytical measurements target the V species with a likelihood of being found soluble, that is, V(V) and V(IV) (Huang et al., 2015; Imtiaz et al., 2015; Shaheen et al., 2019). Vanadium (IV) can be extracted with strongly acidic solutions, but at pH > 5 V is readily oxidized to V(V) (Pyrzyńska, 2006). Conversely, when at low pH, V(V) is reduced to V(IV), but such reduction is very slow and thus does not jeopardize laboratory extractions, typically conducted within 30 minutes (Gamage et al., 2010). As for the extraction of V(III), although it is of limited solubility, it is possible to be detected, but only at extremely low E_H values, and only if such low E_H is being maintained for a prolonged period (Huang et al., 2015); if this strict condition fails, V(III) is quickly oxidized (Li and Le, 2007). Concerning the analysis of the prevalent V species, one must first fathom the complexity of the factors resulting in the metal's valency. For example, apart from H_2S (Wanty and Goldhaber, 1992) and organic matter (Lu et al., 1998), microbes may even have a role in generating V(IV) from V(V), that is, *Geobacter metallireducens* and *Shewanella* spp. (Zhang et al., 2015). When V(IV) is eventually generated, due to its low solubility, its newly formed complexes are readily precipitated; such precipitates may include vanadyl phosphate sincosite [$CaV_2(PO)_4(OH)_4{\cdot}3H_2O$]. Also, associations of V(IV) with colloidal phases (that also drive the newly formed V out of the solution) are also possible, both inorganic (typically V(IV) retained onto (hydr)oxides) and organic. Moreover, V(V), being the most highly soluble and thus toxic (Ma and Fu, 2009), and also the most stable even in groundwater (Wright et al., 2014), is controlled mainly by various soil solution factors, that is, pH, anionic soluble pairs and DOC, as well as by the metal's association with colloidal surfaces (that is, (hydr)oxides; Wällstedt et al. (2010), Larsson, Hadialhejazi et al. (2017; Larsson, Persson et al., 2017); soil organic matter; Reijonen et al. (2016); clay particles; Mikkonen and Tummavuori (1993); Wang et al. (2016).

The fact that V speciation is dependent on many complex and interconnected soil variables and factors makes its analysis a formidable task. Pyrzyńska (2004), in his early work, and subsequently Chen and Owens (2008) reviewed the analytical techniques for V. Usually samples require delicate handling and high experience, as pretreatment for concentration and avoidance of matrix interference are necessary.

One category of V determination is its extraction from solid matrices, including soil. The geochemical fractionation of the metal to various soil "pools" of significance related to its mobility is conducted with well-established sequential extraction protocols, as suggested and explained by many works (e.g., Połedniok and Buhl, 2003; von Gunten et al., 2017; Xu et al., 2017); such extraction does not identify V in any specific oxidation state, but rather V(T) is measured (total vanadium, i.e., the summation of all valencies of V in any given geochemical fraction; Shaheen and Rinklebe (2018)). By this it may be realized that vital information is missing from such extractions; however,

Połedniok and Buhl (2003) suggested a method for identifying to some extent soil V(IV) and V(V).

Beyond such an approach, V speciation can be addressed with advanced methods, that is, XAS (synchrotron-powered X-ray absorption spectroscopy), a technique that has the major advantage of using whole soil (Kelly et al., 2009). This is achieved at core 1s electron (K-edge), with excitation energy of 5,465 eV. Such a low energy, however, may exhibit problems if soils are of low V content (Shahen et al., 2019). X-ray absorption near edge structure (XANES) is also a powerful advanced technique often employed for V analysis (Wisawapipat and Kretzschmar, 2017). This analysis requires the fitting of the produced spectra of the tested sample to known reference spectra of given "clean" constituents. An example of XANES is presented in Figure 5.2. One can observe (a) a pre-edge feature at ca. 5,470 eV being increased and (b) a 1–2 eV increase in the absorption edge position ($E_{1/2}$) for the step-wise V oxidation state from V(III) to V(V) (Chaurand et al., 2007).

Also, another similarly advanced spectroscopic technique is the extended X-ray absorption fine structure (EXAFS); this may even go to the point of assessing the V molecular coordination. Such analysis has been used in sorption tests with colloidal suspensions of (hydr)oxides (Terzano et al., 2007; Chaurand et al., 2007; Peacock and Sherman, 2004). Further description and explanations of the analytical techniques can be found in Chapter 6.

5.5 Conclusions and Necessary Future Research

Vanadium speciation – and thus solubility and mobility – is highly influenced by redox conditions; pentavalent V (an anionic species) is by far the most stable and soluble and predominant species under normal aerated soil conditions with circum-neutral pH. Tetravalent V, a cation, is produced from V(V) when prolonged anoxic conditions prevail or in strongly acidic soils, even under aerated conditions. However, it is less soluble compared to V(V), and thus when soil E_H is decreased, V solubility is expected to decrease. There are instances where this effect does not materialize due to indirect influence of other soil factors and constituents affecting V solubility in manners that may vary. For example, Fe/Mn oxides are dissociated at low E_H and previously held V is liberated, and thus its solubility increases. Similarly, indirect effects on V solubility do exist in relation to DOC (either aliphatic or aromatic), pH and sulfides. Vanadium solubility is also greatly influenced by biological soil factors such as certain reducing micro-organisms. Its transformations and changes among oxidation states make the metal's extraction and analysis a dire task requiring experience in laboratory handling and treatment. Vanadium can be analyzed and quantified with both conventional techniques (e.g., sequential fractionation) and advanced spectroscopic techniques, such as XAS, XANES, and EXAFS.

There are still many aspects concerning V behavior to be elucidated by future research. Vanadium (III) is rare, as it requires strictly anoxic conditions and the presence of S^{-II} ("euxinia" conditions). Sample handling must be delicate, as V(III) is easily oxidized; thus, its investigation and examination is still in want for the scientific community. Along these lines, the role of volatile sulfides on V dynamics is a topic that needs to be further elucidated. Also, V speciation and its apportionment among water soluble and sorbed species onto a variety of colloidal phases, especially in soils with enriched V levels, are areas that need to be further investigated. Lastly, the V redox-induced relationships with soil amendments, both organic and inorganic, are not fully understood and must be further explored with the use of spectroscopic and microscopic techniques.

References

Aide, M., 2005. Geochemical assessment of iron and vanadium-relationships in Oxic soil environments. *Soil Sediment Contam* 14, 403–416.

Bing, H., Zhong, Z., Wang, X., Zhu, H., Yanhong Wu, Y., 2020. Spatiotemporal distribution of vanadium in the flooding soils mediated by entrained-sediment flow and altitude in the Three Gorges Reservoir. *Science of the Total Environment* 724, 138246.

Blackmore, D. P. T., Ellis, J., Riley, P. J., 1996. Treatment of a vanadium-containing effluent by adsorption/coprecipitation with iron oxyhydroxide. *Water Research* 30, 2512–2516.

Chaurand, P., Rose, J., Briois, V., Salome, M., Proux, O., Nassif, V., Olivi, L., Susini, J., Hazemann, J.-L., Bottero, J.-Y., 2007. New methodological approach for the vanadium K-edge X-ray absorption near-edge structure interpretation: Application to the speciation of vanadium in oxide phases from steel slag. *Journal of Physical Chemistry B* 111, 5101–5110.

Chen, Z. L., Owens, G., 2008. Trends in speciation analysis of vanadium in environmental samples and biological fluids-A review. *Analytica Chimica Acta* 607, 1–14.

Crans, D. C., Amin, S. S., Keramidas, A. D., 1998. Chemistry of relevance to vanadium in the environment. In: *Vanadium in the Environment*, J. O. Nriagu, Ed. John Wiley & Sons, Inc., New York, NY, pp. 73–95.

Du Laing, G., Rinklebe, J., Vandecasteele, B., Meers, E., Tack, F. M. G., 2009. Trace metal behavior in estuarine and riverine floodplain soils and sediments: A review. *Science of the Total Environment* 407, 3972–3985.

Frohne, T., Diaz-Bone, R. A., Du Laing, G., Rinklebe, J., 2015. Impact of systematic change of redox potential on the leaching of Ba, Cr, Sr, and V from a riverine soil into water. *Journal of Soils and Sediments* 15, 623–633.

Frohne, T., Rinklebe, J., Diaz-Bone, R. A., Du Laing, G., 2011. Controlled variation of redox conditions in a floodplain soil: Impact on metal mobilization and biomethylation of arsenic and antimony. *Geoderma* 160, 414–424.

Gamage, S. V., Hodge, V. F., Cizdziel, J. V., Lindley, K., 2010. Determination of vanadium (IV) and (V) in Southern Nevada groundwater by ion chromatography-inductively coupled plasma mass spectrometry. *Open Chemical and Biomedical Methods Journal* 3, 10–17.

Gehring, A. U., Fry, V., Luster, J., Sposito, G., 1993. The chemical form of vanadium (IV)in kaolinite. *Clays Clay Miner* 41, 167–662.

Grybos, M., Davranche, G., Gruau, P., Petitjean, Pedrot, M., 2009. Increasing pH drives organic matter solubilization from wetland soils under reducing conditions. *Geoderma* 154, 13–19.

Gustafsson, J. P., Johnsson, L., 2004. *Vanadin i svensk miljö—förekomst och toxicitet.* TRITA-LWR Report 3009. Department of Land and Water Resources Engineering, KTH Royal Institute of Technology, Stockholm.

Huang, J. H., Huang, F., Evans, L., Glasauer, S., 2015. Vanadium: Global (bio)geochemistry. *Chemical Geology* 417, 68–89.

Imtiaz, M., Rizwan, M. S., Xiong, S., Li, H., Ashraf, M., Shahzad, S. M., Shahzad, M., Rizwan, M., Tu, S., 2015. Vanadium, recent advancements and research prospects: A review. *Environment International* 80, 79–88.

Kelly, S. D., Hesterberg, D., Ravel, B., 2009. Analysis of soils and minerals using X-ray absorption spectroscopy. In: *Methods of Soil Analysis, Part 5—Mineralogical Methods,* A. L. Ulery and L. R. Drees, Eds. Soil Science Society of America, Madison, WI, pp. 387–463.

Larsson, M. A., D'Amato, M., Cubadda, F., Raggi, A., Öborn, I., Kleja, D. B., Gustafsson, J. P., 2015. Long-term fate and transformations of vanadium in a pine forest soil with added converter lime. *Geoderma* 259–260, 271–278.

Larsson, M. A., Hadialhejazi, G., Gustafsson, J. P., 2017. Vanadium sorption by mineral soils: Development of a predictive model. *Chemosphere* 168, 925–932.

Larsson, M. A., Persson, I., Sjöstedt, C., Gustafsson, J. P., 2017. Vanadate complexation to ferrihydrite: X-ray absorption spectroscopy and CD-MUSIC modelling. *Environmental Chemistry* 14(3), 141–150.

Leenheer, J. A., Croue, J.-P., 2003. Characterized aquatic dissolved organic matter. *Environmental Science & Technology* 37, 2323–2331.

Li, X. S., Le, X. C., 2007. Speciation of vanadium in oilsand coke and bacterial culture by high performance liquid chromatography inductively coupled plasma mass spectrometry. *Analytica Chimica Acta* 602, 17–22.

Lu, X., Johnson, W. D., Hook, J., 1998. Reaction of vanadate with aquatic humic substances: An ESR and 51V NMR study. *Environmental Science & Technology* 32, 2257–2263.

Ma, Z., Fu, Q., 2009. Comparison of hypoglycemic activity and toxicity of vanadium (IV) and vanadium (V) adsorbed in fermented mushroom of Coprinus comatus. *Biological Trace Element Research* 132(1–3), 278–284.

Martin, S. T., 2005. Precipitation and dissolution of iron and manganese oxides. In: *Environmental Catalysis,* V. H. Grassian, Ed. CRC Press, Boca Raton, FL, pp. 61–81.

Mikkonen, A., Tummavuori, J., 1993. Retention of vanadium (V), molybdenum (VI) and tungsten (VI) by kaolin. *Acta Agriculturae Scandinavica, Section B, Soil and Plant Science* 43, 11–15.

Panichev, N., Mandiwana, K., Moema, D., Molatlhegi, R., Ngobeni, P., 2006. Distribution of vanadium (V) species between soil and plants in the vicinity of vanadium mine. *Journal of Hazardous Materials* 137, 649–653.

Peacock, C. L., Sherman, D. M., 2004. Vanadium (V) adsorption onto goethite (α-FeOOH) at pH 1.5 to 12: A surface complexation model based on ab initio molecular geometries and EXAFS spectroscopy. *Geochimica et Cosmochimica Acta* 68(8), 1723–1733.

Połedniok, J., Buhl, F., 2003. Speciation of vanadium in soil. *Talanta* 59, 1–8.

Pyrzyńska, K., 2004. Determination of vanadium species in environmental samples. *Talanta* 64, 823–829.

Pyrzyńska, K., 2006. Selected problems in speciation analysis of vanadium in water samples. *Chem. Anal. Warsaw* 51, 339.

Reijonen, I., Metzler, M., Hartikainen, H., 2016. Impact of soil pH and organic matter on the chemical bioavailability of vanadium species: The underlying basis for risk assessment. *Environmental Pollution* 210, 371–379.

Rinklebe, J., Knox, A. S., Paller, M., 2017. *Trace Elements in Waterlogged Soils and Sediments.* CRC Press; Taylor & Francis Group, New York.

Rinklebe, J., Shaheen, S. M., Frohne, T., 2016. Amendment of biochar reduces the release of toxic elements under dynamic redox conditions in a contaminated floodplain soil. *Chemosphere* 142, 41–47.

Rinklebe, J., Shaheen, S. M., Yu, K., 2016. Release of As, Ba, Cd, Cu, Pb, and Sr under pre-definite redox conditions in different rice paddy soils originating from the U.S.A. and Asia. *Geoderma* 270, 21–32.

Schwertmann, U., Pfab, G., 1994. Structural vanadium in synthetic goethites. *Geochimica et Cosmochimica Acta* 58, 4349–4352.

Shaheen, S. M., Alessi, D. S., Tack, F. M. G., Ok, Y. S., Kim, K.-H., Gustafsson, J. P., Sparks, D. L., Rinklebe, J., 2019. Redox chemistry of vanadium in soils and sediments: Interactions with colloidal materials, mobilization, speciation, and relevant environmental implications- A review. *Advances in Colloid and Interface Science* 265, 1–13.

Shaheen, S. M., Rinklebe, J., 2018. Vanadium in thirteen different soil profiles originating from Germany and Egypt: Geochemical fractionation and potential mobilization. *Applied Geochemistry* 88, 288–301.

Shaheen, S. M., Rinklebe, J., Frohne, T., White, J., DeLaune, R. D., 2014. Biogeochemical factors governing cobalt, nickel, selenium, and vanadium dynamics in periodically flooded Egyptian North Nile Delta rice soils. *Soil Science Society of America Journal* 78, 1065–1078.

Shaheen, S. M., Rinklebe, J., Frohne, T., White, J., DeLaune, R. D., 2016. Redox effects on release kinetics of arsenic, cadmium, cobalt, and vanadium in Wax Lake Deltaic soils. *Chemosphere* 150, 740–748.

Shaheen, S. M., Rinklebe, J., Rupp, H., Meissner, R., 2014. Lysimeter trials to assess the impact of different flood-dry-cycles on the dynamics of pore water concentrations of As, Cr, Mo and V in a contaminated floodplain soil. *Geoderma* 228–229, 5–13.

Takeno, N., 2005. *National Institute of Advanced Industrial Science and Technology. Atlas of Eh-pH Diagrams.* Geological Survey of Japan Open File Report. Takeno, Naoto.

Terzano, R., Spagnuolo, M., Vekemans, B., De Nolf, W., Janssens, K., Falkenberg, G., Fiore, S., Ruggiero, P., 2007. Assessing the origin and fate of Cr, Ni, Cu, Zn, Pb, and V in industrial polluted soil by combined microspectroscopic techniques and bulk extraction methods. *Environmental Science & Technology* 41(19), 6762–6769.

Tsadilas, C. D., Shaheen, S. M., 2010. Distribution of total and AB-DTPA-extractable soil vanadium from Greece and Egypt and their correlation with soil properties. *Soil Science* 175, 535–543.

von Gunten, K., Alam, M. S., Hubmann, M., Ok, Y. S., Konhauser, K. O., Alessi, D. S., 2017. Modified sequential extraction method for biochar and petroleum coke: Metal release potential and its environmental implications. *Bioresource Technology* 236, 106–110.

Wällstedt, T., Björkvald, L., Gustafsson, J. P., 2010. Increasing concentrations of arsenic and vanadium in (southern) Swedish streams. *Applied Geochemistry* 25, 1162–1175.

Wang, Y., Yin, X., Sun, H., Wang, C., 2016. Transport of vanadium (V) in saturated porous media: Effects of pH, ionic-strength and clay mineral. *Chemical Speciation & Bioavailability* 28(1–4), 7–12.

Wanty, R. B., Goldhaber, M. B., 1992. Thermodynamics and kinetics of reactions involving vanadium in natural systems: Accumulation of vanadium in sedimentary rocks. *Geochimica et Cosmochimica Acta* 56, 1471–1483.

Weishaar, J. L., Aiken, G. R., Bergamaschi, B. A., Fram, M. S. Fujii, R., Mopper, K., 2003. Evaluation of specific ultraviolet absorbance as an indicator of the chemical composition and reactivity of dissolved organic carbon. *Environmental Science & Technology* 37, 4702–4708.

Wisawapipat, W., Kretzschmar, R., 2017. Solid-phase speciation and solubility of vanadium in highly weathered soils. *Environmental Science & Technology* 51, 8254–8262.

Wright, M. T., Belitz, K., 2010. Factors controlling the regional distribution of vanadium in groundwater. *Ground Water* 48, 515–525.

Wright, M. T., Stollenwerk, K. G., Belitz, K., 2014. Assessing the solubility controls on vanadium in groundwater, northeastern San Joaquin Valley, CA. *Applied Geochemistry* 48, 41–52.

Xu, Y-. H., Huang, J-. H., Brandl, H., 2017. An optimised sequential extraction scheme for the evaluation of vanadium mobility in soils. *Journal of Environmental Sciences* 53, 173–183.

Zhang, B., Hao, L., Tian, C., Yuan, S., Feng, C., Ni, J., Borthwick, A. G. L., 2015. Microbial reduction and precipitation of vanadium (V) in groundwater by immobilized mixed anaerobic culture. *Bioresource Technology* 192, 410–417.

6

Vanadium Speciation in Soil Aqueous and Solid Phases

Worachart Wisawapipat[a], Yohey Hashimoto[b] and Shan-Li Wang[c]

[a] *Department of Soil Science, Faculty of Agriculture, Kasetsart University, Bangkok, Thailand*

[b] *Department of Bio-Applications and Systems Engineering, Tokyo University of Agriculture and Technology, Tokyo, Japan*

[c] *Department of Agricultural Chemistry, National Taiwan University, Taipei, Taiwan*

CONTENTS

6.1 The Overview of Chemical Speciation 113
6.2 Aqueous Species of Vanadium 115
6.3 X-Ray Absorption Spectroscopy for Vanadium Speciation 118
6.3.1 Principles of X-Ray Absorption Spectroscopy 118
6.3.2 Vanadium *K*-Edge XANES Spectral Features 123
6.3.3 Speciation of Vanadium in Minerals, Soils, and Sediments 125
6.3.4 Conclusions 129
References 130

6.1 The Overview of Chemical Speciation

Chemical speciation in soil chemistry and its related fields refers to the identification and quantitation of different chemical forms (i.e., species) of an element in soil solid, pore water, and gaseous phases. The distribution of the chemical species of an element in soil solids is determined by soil reactions, which can be conceptually categorized into five principal mechanisms: (i) outer-sphere surface complexation, (ii) inner-sphere surface complexation, (iii) multinuclear surface complexation, (iv) homogeneous precipitation, and (v) lattice diffusion (Manceau et al., 2002). These reactions also determine the distributions of chemical species between soil aqueous and solid phases. In aqueous solution, the chemical speciation of elements is a critical

DOI: 10.1201/9781003173274-6

parameter that controls solubility, (bio)availability, solid-solution phase partition and transport in soils. Most hazardous and nutritional elements are mainly absorbed by biota as free ion forms in soil solution. Various inorganic and organic ligands to form metal complexes also occur in the soil solution, including inorganic anions, dissolved organic matter and low-molecular organic acids. Because of strong interactions with biological ligands and colloidal surfaces, particular attention is given to information on the free ion activity and its distribution of the target element. Elucidating chemical speciation in the soil aqueous and solid phases can bring insight into the key roles of different reactions in determining the mobility and bioavailability of elements in soils.

Macroscale and molecular-scale research techniques have been used to qualitatively and quantitatively characterize chemical species of nutrients, and toxic and potentially toxic elements in soil systems. Sequential extraction is a representative method of the macroscopic techniques, which fractionate metal(loid)s in the soil based on their solubility in reagents with different chemical natures. This method can operationally determine the fractionation of metal(loid)s and therefore does not represent actual chemical species in the soils. Differing extracting solutions can also change intrinsic chemical species in the samples by causing redistribution and formation of freshly precipitated species, thereby resulting in overestimations or underestimations of some species reported for phosphorus (Gu et al., 2020). Although this macroscopic technique inherits some limitations, it is simple, inexpensive, and able to perform in most laboratories and provides operationally defined fractions and solubility in samples with low concentrations.

In contrast, molecular spectroscopic techniques target intact species of soil metal(loid)s without drastic physical and chemical alterations to the original field conditions. Spectroscopic techniques such as synchrotron-based X-ray absorption spectroscopy (XAS) have been used to determine chemical speciation. Much current understanding of the chemical speciation of elements in soil has been obtained by XAS since it is an element-specific technique that can be applied to determine the local structure of an absorbing element in naturally complex materials, such as soil. One primary challenge is to combine the information and principles developed from microscale mechanisms to macroscale analyses (e.g., single and sequential extractions) to assess the reactions of target nutrient, and toxic and potentially toxic elements in soil.

This chapter reviewed studies on aqueous chemistry and X-ray absorption near edge structure (XANES) based speciation of vanadium (V) in soil systems, as it is documented as an essential or toxic element for humans, depending on its concentration and speciation (Crans et al., 2013; WHO, 2000). The chapter starts with the aqueous speciation of V in a simulated soil solution to demonstrate the importance of aqueous V speciation in the presence of inorganic and organic ligands. In the following section, the state of the synchrotron XANES techniques that have been used to identify and quantify V speciation in minerals, soils, and sediments are described, with an emphasis on the differences of V-containing minerals from natural and anthropogenic sources in the soil and sediment systems.

6.2 Aqueous Species of Vanadium

Vanadium is a redox-sensitive element and exists mainly in three oxidation states, including V(III), V(IV) and V(V), in the natural environment. A recent paper by Gustafsson (2019) has reported an E_h-pH diagram of V for a total V concentration of 1 μM at 25°C (Figure 6.1). The aqueous V is dominated by vanadate (i.e., V(V)) at alkaline pH ranges and at high E_h ranges at acidic pH ranges. Vanadyl (i.e., V(IV)) is predominant at acidic pH ranges and at intermediate E_h ranges. Vanadium(III) is stable at low E_h ranges. The vanadium(II) ion is thermodynamically unstable in water, but it might occur in specific conditions at high temperatures or by forming complexes with organic ligands (Gustafsson, 2019).

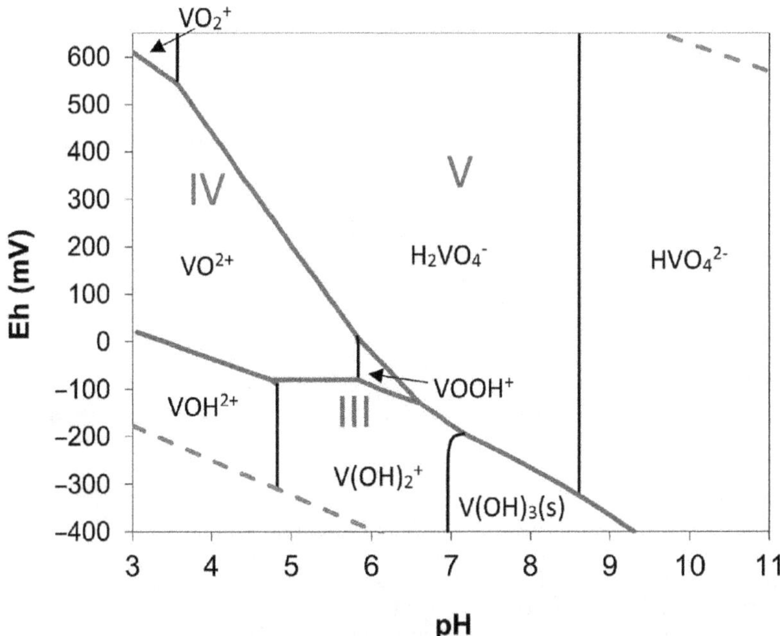

FIGURE 6.1
Predominance E_h-pH diagram showing V speciation at 25°C.
Source: Gustafsson, J.P. 2019. Vanadium geochemistry in the biogeosphere—speciation, solid-solution interactions and ecotoxicity. Applied Geochemistry 102:1–25.

The aqueous species of V(IV) is dominated by the vanadyl ion (VO^{2+}) and VO^{2+}-associated with inorganic and organic ligands. The predicted aqueous speciation of V(IV) in a simulated soil solution shows the predominance of VO^{2+} with a minor extent of $V(OH)_3^+$ and $VOSO_4^0$ in acidic solutions (Figure 6.2a). This result suggests that VO^{2+} may be a predominant V species in strongly acidic soils. The proportion of $VOSO_4^0$ to the total V(IV) species may increase in soil solutions with higher SO_4 concentrations. The vanadyl ion complexed with the carbonate ion (i.e., $VOOHCO_3^-$) is predominant in alkaline conditions. In neutral pH conditions, the large quantity of V^{4+} in solution precipitates as $V_2O_{4(s)}$ when oversaturated solids in solution are allowed to precipitate during the computation process. Vanadium oxides, including $V_2O_{4(s)}$ and $V_2O_{3(s)}$, are commonly prepared at high temperatures, and therefore these V species may be unlikely to occur in ordinary soil conditions (Wang et al., 2007). In the presence of dissolved organic carbon (DOC) in solution, however, nearly 100% of V(IV) complexes with DOC in the solution at pH > 4 (Figure 6.2b). The calculated aqueous V(IV) speciation shows the occurrence of DOC in solution enhances V(IV) complexation and prevents V(IV) from precipitation in the entire pH range. In ordinary soil conditions, therefore, V(IV) may rarely form precipitates and likely exists as complexes with organic ligands.

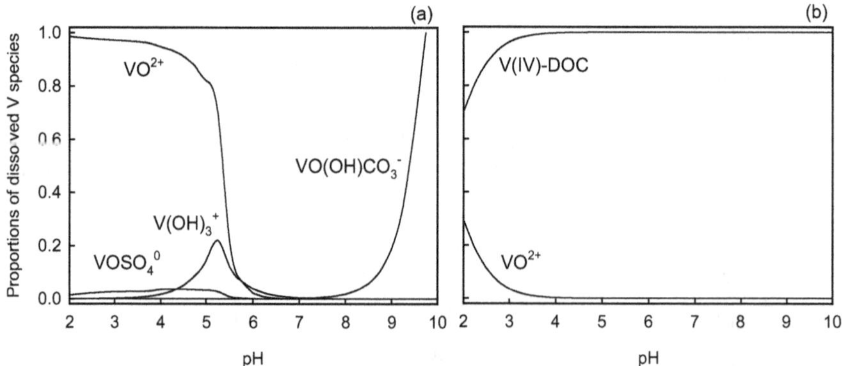

FIGURE 6.2
Distribution of aqueous V^{4+} species as a function of solution pH at 25 °C, predicted by (a) no DOC in solution, and (b) 155 mg L^{-1} of DOC in a simulated soil solution. The distribution is the concentration of a specified species divided by the total concentration of V^{4+} at 10^{-6} M in the solution. The diagram was constructed using a simulated soil solution containing cations (e.g., Al^{3+} and Ca^{2+}), anions (e.g., $SO_4^{2-} = 2.13 \times 10^{-6}$ M) and DOC (155 mg L^{-1}) at $pCO_{2(g)} = 4 \times 10^{-4}$ atm. The soil solution data were referred to the Norfolk soil in North Carolina (Hashimoto, 2007). The chemical reactions and pK_a values used to construct the diagrams were derived from the data compiled in Visual MINTEQ ver. 3.1. The NICA-Donnan model was used to compute V complexation with dissolved organic matter.

In solutions with low V concentrations in the absence of organic ligands, V(V) occurs mainly as the pervanadyl ion (VO_2^+) in strongly acidic conditions and vanadate oxyanions ($H_2VO_4^-$ and HVO_4^{2-}) in neutral and alkaline conditions, respectively (Figure 6.3a). Vanadic acid ($H_3VO_4^0$) is a minor species in the pH around 3–4, and VO_4^{3-} is predominant under extremely alkaline conditions (pH > 12). Vanadium(V) forms complexes with organic ligands at acidic pH conditions in which VO_2^+ is predominant. Figure 6.3b shows that V^{5+} complexes with oxalate [VO_2(oxalate)$^-$] occur in acidic conditions at an oxalate concentration of 100 µM. Formations of V^{5+}-organic complexes may be enhanced in the rhizosphere at higher organic acid concentrations, depending on the abundance of major cations (e.g., Al^{3+}) that preferentially form complexes with organic ligands. The aqueous chemistry of V(V) is extremely complex at higher V concentrations due to formations of poly-nuclear V(V) species. Various poly-nuclear V(V) species have been reported, which contain up to 10 vanadium atoms (e.g., $V_{10}O_{28}^{6-}$). The poly-nuclear V(V) complexes are minor species at low V concentrations (10^{-6} M, Figure 6.1), whereas these species become predominant at high V concentrations (>10^{-4} M) in acidic solutions (Pettersson et al., 2003).

The aqueous species of V(III) is dominated by the vanadic ion (V^{3+}) in strongly acidic conditions (pH ~2) and the hydrolyzed form VOH^{2+} in weak acidic conditions (pH ~4) (Figure 6.4). The other hydrolyzed form, $V(OH)_2^+$, occurs in solutions with pH values between 3 and 6. A trace amount of V^{3+} complex with SO_4^{2-}(VSO_4^+) occurs in strongly acidic conditions. The calculated aqueous speciation shows that V^{3+} ions start precipitating as $V_2O_{3(s)}$ at pH 4.5 and above and hardly form complexes with organic ligands in the entire pH range. However, precipitation of V(III) and the other oxidation

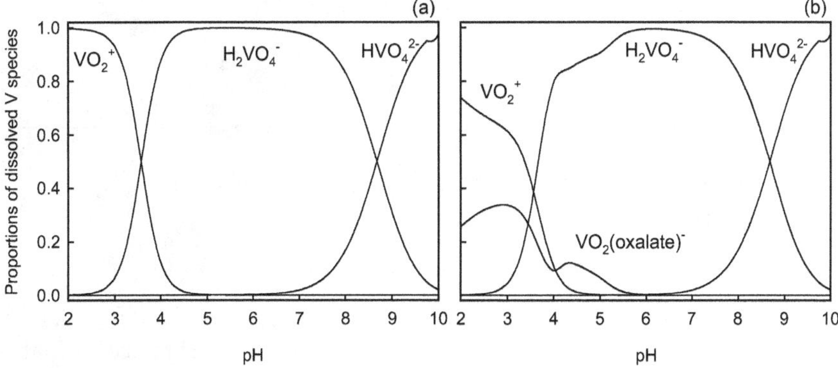

FIGURE 6.3
Distribution of aqueous V^{5+} species as a function of solution pH at 25 °C predicted by (a) no organic ligands in solution and (b) 0.1 mM of oxalate in a simulated soil solution. The distribution is the concentration of a specified species divided by the total concentration of V^{5+} at 10^{-6} M in the solution.

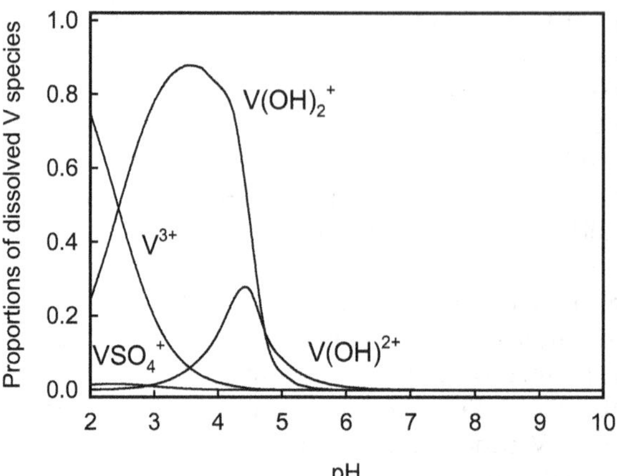

FIGURE 6.4
Distribution of aqueous V^{3+} species as a function of solution pH at 25 °C predicted in a simulated soil solution. The distribution is the concentration of a specified species divided by the total concentration of V^{3+} at 10^{-6} M in the solution.

states of V may be of minor importance in ordinary soil conditions. A V(III) species of $V(OH)_{3(s)}$, a possible precursor of $V_2O_{3(s)}$, occurs in strongly reducing conditions. In addition, V^{3+} can be substituted for octahedral Fe^{3+} in goethite (Schwertmann and Cornell, 2000; Schwertmann and Pfab, 1996). These suggest that V(III) precipitates may rarely be found in ordinary soils. A large part of the aqueous chemistry of V(III), such as complexation with inorganic and organic ligands, remains unknown due to incomplete information of thermodynamic chemical equilibria (Gustafsson, 2019).

6.3 X-Ray Absorption Spectroscopy for Vanadium Speciation

6.3.1 Principles of X-Ray Absorption Spectroscopy

X-ray absorption spectroscopy (XAS) is a powerful technique for identifying and quantifying chemical speciation (e.g., oxidation state, coordination geometry, and bond distance) of elements, including V in soil and environmental samples. The chemical species are of great importance for deliberating information about elemental (bio)availability, mobility, and solubility in terrestrial and aquatic systems.

The principle of XAS for application to soils was meticulously described by Fendorf et al. (1994) and Kelly et al. (2008). Another excellent textbook about the XAS principle is *XAFS for Everyone*, written by Calvin (2013), with engaging cartoon understudies of people engaged in XAS research. This spectroscopic technique is fundamentally involved in understanding the interaction of X-ray photons with the atoms of a target element in examined samples. Several processes occur when X-ray photons react with the atoms. In general, two of the primary processes occur during the interaction: (1) X-ray absorption and (2) X-ray fluorescence emission. These facilitate the investigation of elemental species in samples (Figure 6.5).

The XAS technique involves the absorption of X-ray photon energy to atoms in the adsorbing materials. The incident energy of the X-ray photons is required to sufficiently excite the core-shell electron of atoms, consisting of an electron from the *K*-shell to higher electron shells (i.e., *L*, *M* and *N* shells) of the atoms, or even into the continuum if the electron is freely liberated from the atoms. The electron-binding energies are specific to each element, determined as the absorption coefficient (μ)—the probability of an X-ray being absorbed by a sample—for the investigated element in samples at a specific energy range. The binding energy for the excitation of 1s electrons (*K* shell) in V is 5,465 eV, referred to as the vanadium *K* edge energy. As aforementioned, the absorption of X-rays is the first step when the incident X-ray photons react with the atom of an element. Therefore, the absorption coefficient can be estimated by transmission X-rays (I_t) as:

$$\mu x = \ln\left(\frac{I_0}{I_t}\right)$$

where μ is the absorption coefficient, x is the sample thickness, I_0 is the incident X-rays and I_t is the X-ray transmitted through the sample. The μ

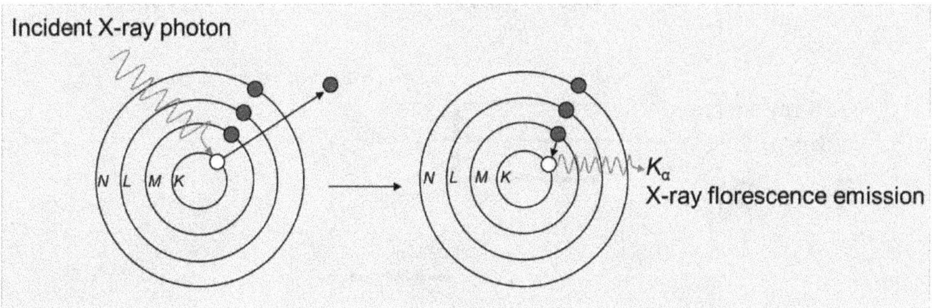

FIGURE 6.5
Simplified action of absorption of an X-ray photon (left) and subsequent emission of X-ray fluorescence in atom (right).

coefficient is inversely associated with the sample thickness and the transmission X-ray, meaning that the X-ray photons are more absorbed when the sample is thicker, resulting in lower amounts of transmitted X-ray.

X-ray absorption of an atom results in subsequent X-ray fluorescence emission from the excited atom, which consists of characteristic X-ray energies corresponding to the *K* or *L* transitions of electrons in the excited atom. Therefore, the μ coefficient can be estimated from the emitted fluorescence X-rays (I_f) as:

$$\mu x \propto \left(\frac{I_F}{I_0}\right)$$

This relationship means that the numbers of emitted fluorescence X-rays are directly associated with the numbers of X-ray photons absorbed by the atoms of the examined element. It also indicates that the higher concentration of the element of interest, the greater the amount of X-ray fluorescence emitted.

There are several essential steps for XAS experiments on elemental speciation in mineral samples, including V in soil samples. Measuring total V concentration in the samples is the first essential procedure for determining the experimental setups (i.e., transmission and fluorescence modes) for XAS data acquisition. Natural soil samples, which typically contain low V concentrations, are suited to the fluorescent experimental setup (Figure 6.6). Comparatively, the transmission mode is well suitable for reference standard compounds containing high V concentrations. Using too low of a concentration of the target element in a sample could make it challenging to acquire usable spectra for the sample. Different synchrotron facilities with differing

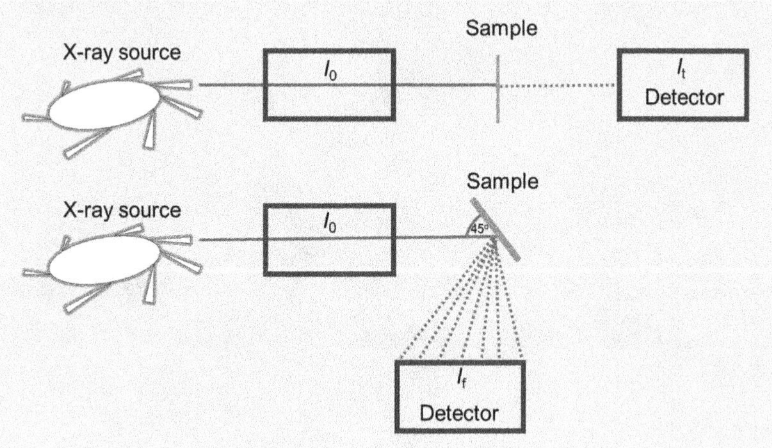

FIGURE 6.6
Schematic picture of fluorescent and transmission XAS experimental setup.

photon fluxes and numbers of X-ray fluorescence detectors can result in different spectral quality. Therefore, the synchrotron facilities (e.g., The Stanford Synchrotron Radiation Lightsource and the Australian Synchrotron) with higher numbers of fluorescence detectors offer a greater possibility of acquiring usable XAS spectra from a very dilute sample. The characterization of the primary components of the minerals and compounds in the samples is the second step, which provides information on the potential host phases that could be sources and sinks for V. X-ray diffraction (XRD) is a complementary technique to gain information about mineral types in samples. When obtaining the potential host minerals of V in samples, the following process is a compilation of reference standards for V-containing solids. Some reference standards may need to be prepared fresh before XAS spectral acquisition. The final step involves performing spectral fitting of the samples with selected reference standards. Before fitting, all the spectra need to be processed for energy calibration, background removal and normalization (Kelly et al., 2008). The sample spectra are reconstructed with a combination of selected reference spectra referred to as linear combination fitting (LCF). The relative percent of each reference spectrum in the samples (i.e., chemical speciation) is reported along with the goodness of fit.

Two regions of the XAS spectra can be observed after the spectral acquisition, consisting of X-ray absorption near edge structure (XANES) and extended X-ray absorption fine structure (EXAFS). Figure 6.7 presents the

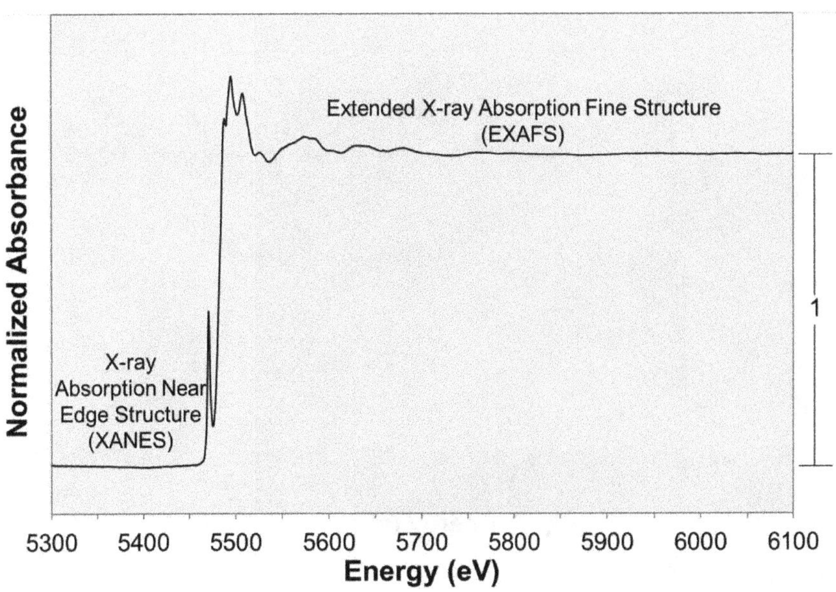

FIGURE 6.7
Vanadium *K*-edge XAFS of vanadium (V) oxide (V_2O_5), showing regions of XANES and EXAFS.

V *K*-edge XAFS of vanadium (V) oxide (V_2O_5), demonstrating the XANES and EXAFS regions. The XANES region provides information about the elemental oxidation state and binding geometry. In contrast, the EXAFS region offers information about the molecular bonding environments (e.g., coordination numbers and atomic distances) of the elements. The V *K*-edge XANES data can be collected from −50 to +200 eV relative to its absorption edge (i.e., 5,465 eV), but 30 eV above the edge is typically used. The EXAFS can be extended beyond +1,000 eV above the edge. To the best of our knowledge, relatively low V concentrations in soil and sediment samples are the primary challenge in achieving usable EXAFS spectra. Another key challenge in the investigation of V *K*-edge EXAFS in soil and other natural samples is caused by the spectral interferences from some naturally inherent elements such as cerium (Ce) and lanthanum (La), as the Ce L_3 absorption edge at 5,723 eV and the La L_2 absorption edge at 5,891 eV overlap with the V *K*-edge EXAFS region (Figure 6.8). Specifically, a study on V speciation in an urban atmospheric particulate matter showed that the presence of barium (Ba) L_2 at 5623 eV could hinder the V *K*-edge EXAFS study, which might also be the case for soil materials (Huggins et al., 2000). An investigation of red mud revealed that the V *K*-edge XANES was affected by the interference from the La L_3 edge at 5,483 eV (Burke et al., 2012). Practically, this spectral interference is well resolved using a 100 mm^2 13-element Ge detector (Canberra Ultra-LEGe) with a high energy resolution, which is nominally less than 150 eV. With this excellent energy resolution, the emission lines of the La L_α (~4,646 eV) and La

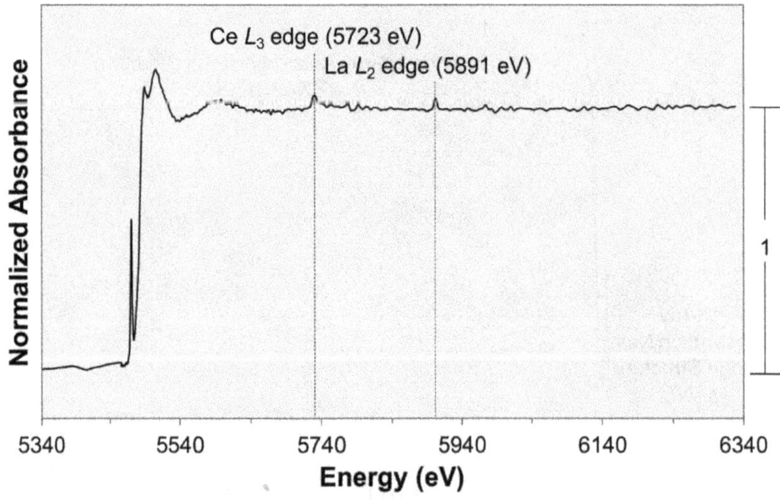

FIGURE 6.8
Normalized V *K*-edge EXAFS spectra of highly weathered soil containing 131 mg Ce kg^{-1} and 62 mg La kg^{-1} and spiked with sodium vanadate ($NaVO_3$) at 448 mg V kg^{-1} soil, showing presence of Ce L_3 absorption edge at 5,723 eV and La L_2 absorption edge at 5,891 eV.

L_β/V K_β (~5,373 eV) can be well separated from V K_α (~4,949 eV), so the measurement of V K_α fluorescence can be achieved with minimal interference by the La L_3 absorption edge (L_α and L_β) (Wisawapipat and Kretzschmar, 2017).

This chapter is focused on V *K*-edge XANES spectroscopy, as it is rich in several types of useful information (e.g., oxidation state and coordination geometry) of V in soils. Furthermore, it can be used as fingerprints to determine the chemical speciation of V via LCF. The details of V *K*-edge XANES features are provided in the following section.

6.3.2 Vanadium *K*-Edge XANES Spectral Features

XANES spectra consist of various spectral features in the pre-edge, edge and post-edge regions of the spectra. The features include the threshold, centroid, height, full width at half maximum, and intensity of pre-edge peak, as well as the edge energy and white-line energy of the main-edge peak. Details of such spectral features of vanadium oxides, V-containing minerals and V-adsorbed solids have been reported by Wong et al. (1984) and Wisawapipat and Kretzschmar (2017). Some essential features such as the threshold, pre-edge peak intensity, pre-edge energy, main-edge energy and white-line energy can be used for obtaining information about the oxidation state and coordination symmetry of V compounds and other natural samples.

The threshold value is defined as the energy position of the first maximum in the first derivate spectrum (Figure 6.9). The pre-edge centroid denotes the energy position at the maximum absorbance of the pre-edge peak. The pre-edge peak intensity is obtained by multiplying the pre-edge peak height by its full width at half maximum. The energy value of the absorption edge is determined using the energy position of the second maximum in the first derivate spectrum. The white-line energy is defined as the maximum absorbance energy above the absorption edge.

Several parameters (e.g., height, intensity, and threshold) are valuable indicators of V oxidation states. The main edge-energy of V reference compounds (both oxides and salts) is associated with a nominal V oxidation state (Figure 6.10a). The threshold parameter is also linearly related to the nominal V oxidation state (Figure 6.10b). The pre-edge peak intensity was also reported to be related to the oxidation state with a third-order polynomial function (r^2 = 0.99) for basaltic glasses (Sutton et al., 2005). Therefore, these parameters are frequently used to determine the oxidation state of V in soil and environmental samples.

In addition to various V oxides, V^{3+}/V^{4+} substituted into mineral structures and V(V) adsorbed to soil solids are important compounds when considering V species in soil and environmental samples. As mentioned earlier, the V oxidation state can be determined by the threshold and the main edge-energy in XANES spectra. However, the oxyanions of V(V) (i.e., vanadate) are thought to be the most toxic form in biota (Costa et al., 2006). Thus, information

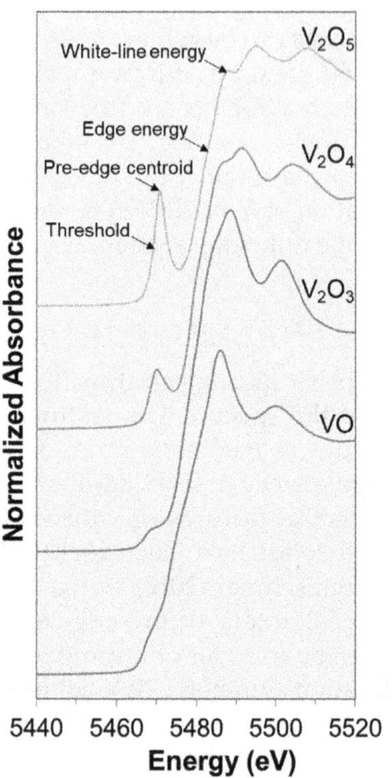

FIGURE 6.9
Vanadium *K*-edge XANES spectra of selected V oxide compounds.

about the interactions of V(V) with soil solids should be useful in assessing V mobility and toxicity in soil systems. Only minor spectral features have been observed for V(V) adsorbed to solid phases, such as clay minerals (e.g., kaolinite), Al oxyhydroxides (e.g., gibbsite), and Fe oxyhydroxides (e.g., goethite and ferrihydrite) (Figure 6.11). Information from complementary techniques (such as XRD) is essential for describing the potential host phases adsorbing V(V) species.

The absence and the presence of pre-edge peaks are related to the respective symmetry and asymmetry of the V coordination geometry. The VO oxide compound without the presence of the pre-edge peak designates ideal octahedral symmetry. However, the pre-edge peaks for V_2O_3, V_2O_4 and V_2O_5 demonstrate asymmetry of the V bonding environment. The respective elevated pre-edge peak intensity for V_2O_3 and V_2O_4 indicates the higher distorted degrees of the octahedral molecular geometry. V_2O_5, with the highest pre-edge peak intensity, is conformed with the square pyramidal molecular geometry.

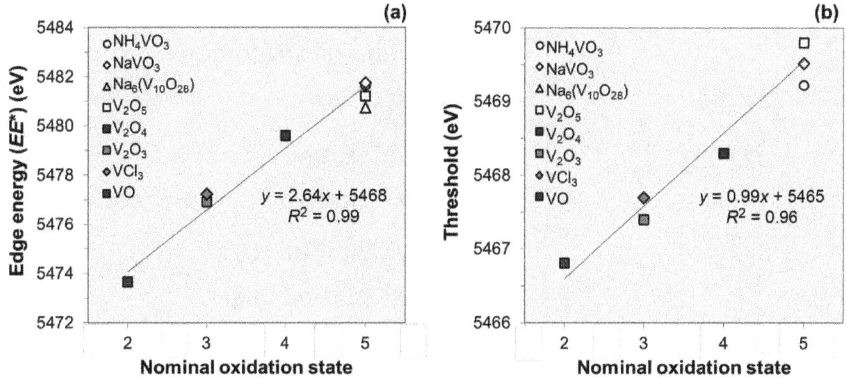

FIGURE 6.10
Linear relationships for V standard components (VO, V_2O_3, V_2O_4 and V_2O_5): (a) main edge-energy (EE*) vs. nominal oxidation state and (b) pre-edge peak threshold of V oxides vs. nominal oxidation state for V standard compounds.

Source: Reprinted (adapted) with permission from (Wisawapipat, W. and Kretzschmar, R., 2017. Solid phase speciation and solubility of vanadium in highly weathered soils. Environmental Science & Technology, 51(15), pp. 8254–8262). Copyright (2017) American Chemical Society.

6.3.3 Speciation of Vanadium in Minerals, Soils, and Sediments

Speciation of V (e.g., oxidation state, chemical forms) in solid materials (minerals, soils, and sediments) is essential information for assessing the solubility, mobility, and (bio)availability of V in terrestrial and aquatic ecosystems. Primary sources of V in natural systems are typically related to V mines, slags, and vehicle-emitted particulate matter (Burke et al., 2012; Gilligan and Nikoloski, 2020; Shafer et al., 2012), which can contaminate soils and sediments.

Ores related to titanomagnetite $Fe^{2+}(Fe^{3+}, Ti)_2O_4$ or magnetite ($Fe^{2+}(Fe^{3+})_2O_4$) are the primary sources of V production (Gilligan and Nikoloski, 2020). The concurrence of V with Fe and Ti is attributed to relatively identical radii of V^{3+} (64.0 pm) with Fe^{3+} (64.5 pm) and six-coordinated Ti^{4+} (60.0 pm) (Ahrens, 1952), resulting in the typically isomorphic substitution of Fe^{3+} by V^{3+} in the octahedral location of the spinel structure. The V *K*-edge XANES spectroscopy also elaborated that V^{3+} substituted for Fe^{3+} in the octahedral sites with a minor amount of V^{4+}, while the most V-concentrated sample contained V^{4+} up to 10% of the total V substitution (Balan et al., 2006; Bordage et al., 2011). In the synthesis of V-Ti doped magnetite for contaminant removal, it was found that V^{3+} and Ti^{4+} were substituted in the octahedral site (Liang et al., 2012). In addition, V(IV) was shown to be the dominant oxidation state for V-doped titanium oxides, titania, which is widely used for photocatalysts and semiconductors (El Koura et al., 2018; Rossi et al., 2018; Wu and Chen,

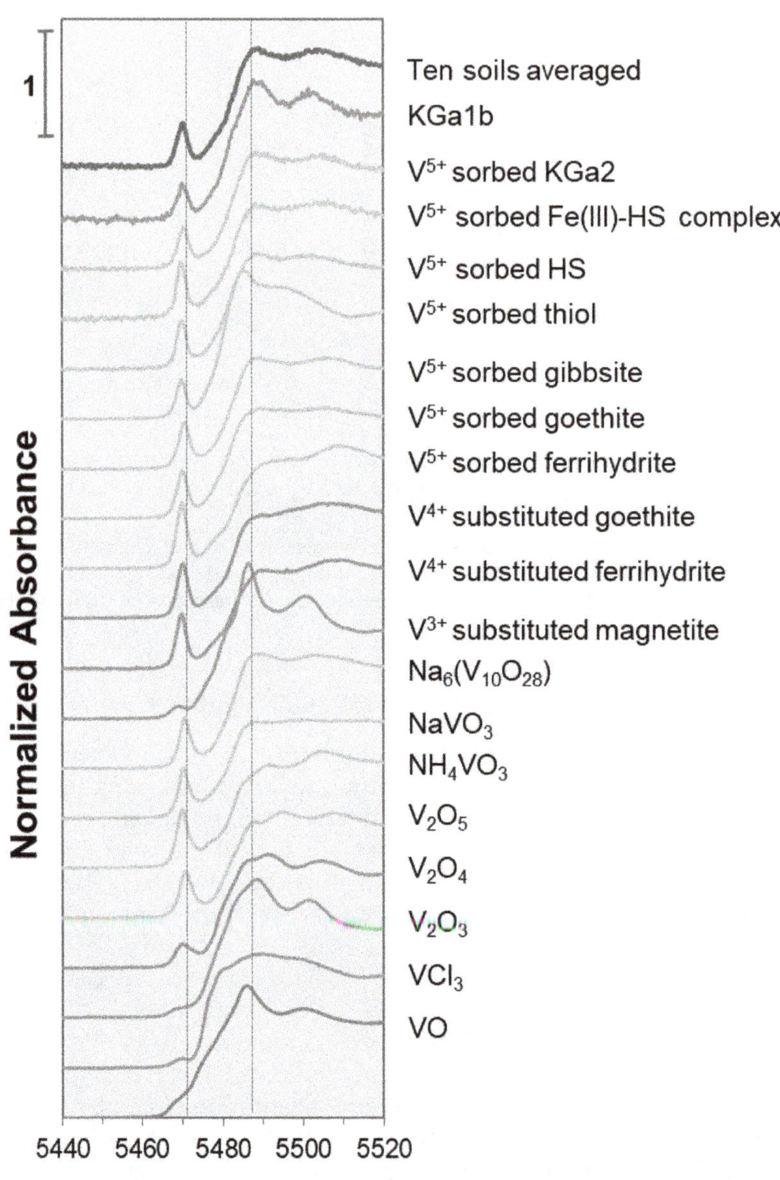

FIGURE 6.11
Vanadium *K*-edge XANES spectra of selected V reference compounds, $V^{3+/4+}$ substituted Fe minerals, KGa1b kaolinite and V^{5+}-adsorbed minerals, and the averaged spectrum of all 10 soil samples for comparison.
Source: Reprinted (adapted) with permission from (Wisawapipat, W. and Kretzschmar, R., 2017. Solid phase speciation and solubility of vanadium in highly weathered soils. Environmental Science & Technology, 51(15), pp. 8254–8262). Copyright (2017) American Chemical Society.

2004). Nevertheless, magnetite in soil and environmental systems could be oxidized and transformed into maghemite (γ-Fe_2O_3) and hematite (α-Fe_2O_3) under oxic conditions or into other Fe oxyhydroxides such as goethite (α-FeOOH) under naturally complex conditions (Cornell and Schwertmann, 2003). The V-substituted goethite could be present as variable oxidation states from V(III) to V(IV)/V(V) with increasing synthetic temperatures, as confirmed by the threshold energy of the XANES spectra. The goethite substituted with higher degrees of V(IV)/V(V) oxidation states was susceptible to dissolving by proton-promoted dissolution (Kaur et al., 2009). In addition to such V-containing Fe oxyhydroxides, other reported V-containing minerals were coulsonite (FeV_2O_4; V^{3+}), pascoite (($Ca_3(V_{10}O_{28})\cdot 17H_2O$; V^{4+}), cavansite ($Ca(VO)(Si_4O_{10})\cdot 4H_2O$; V^{4+}), and vanadinite ($Pb_5(VO_4)_3Cl$; V^{5+}) (Chaurand et al., 2007). Kaolinite (well-ordered, KGa1b) was also documented to contain V^{4+} within its octahedral structure (Wisawapipat and Kretzschmar, 2017).

Slags are byproducts of titanomagnetite smelting and are the primary V contaminant source (Chen et al., 2013). Much research has used XANES spectroscopy to investigate V oxidation in the slags and their related materials. An average V oxidation state of basic oxygen furnace (BOF) steel slag was 4.1, whereas the naturally aged BOF slag had a higher oxidation state of 4.3 (Chaurand et al., 2006). The analysis of pre-edge features (centroid position and normalized area) of micro-XANES spectra allows the description of the spatial distribution of the V oxidation state within different weathered zones of the leached steel slags. The intact steel slags contained V(III) as the main oxidation state, and a slightly altered zone showed mixed oxidation states of V(III), V(IV) and V(V), whereas V(V) was the main species in the altered zone (Chaurand et al., 2007). A study showed consistent data, demonstrating that V(III) was the most dominant V oxidation state in two blast furnace slags as confirmed by the pre-edge peak intensity and the edge energy (Larsson, Baken et al., 2015). However, another study documented that the primary V oxidation state in the BOF slag samples mainly occurred as V(V), presumably occurred as the dicalcium silicate phase ($Ca_2Si(V^{5+})O_4$) in the tetrahedral coordination (Hobson et al., 2017). An investigation of alkaline slag leachate from a field site and in a laboratory-based experiment concluded that V(V) was the dominant oxidation state in the leachate, which could be retained by kaolinite and goethite (Hobson et al., 2018). This study also observed that goethite had greater V(V) retention capability than kaolinite, which neutral solutions had higher V retention of solid phases than alkaline solutions. Red mud is a bauxite rock byproduct from the Bayer processes, that is, the Al extraction using sodium hydroxide solution under heated (150–200 °C) pressure conditions, with being abundant in V. A study of the V speciation in red mud revealed that V(V) was the primary oxidation state, presumably related to Ca-aluminosilicate phases, based on the pre-edge feature and the edge energy (Burke et al., 2012).

To date, information has been paltry on V speciation in soil samples. V speciation was investigated in pine forest soils from organic and mineral soil sampled at two depths of 0–10 and 10–20 cm (Larsson, D'Amato et al., 2015). Their results indicated that the V oxidation state was in the range 3.7–4.6, depending on the samples. The predominant V species in the organic layer was V(IV) associated with the organic matter, whereas Al/Fe (hydr)oxides were the main sorbents for V(V) in the mineral soil. Another study on two Swedish sandy soils with acidic conditions revealed that the native oxidation state in mineral soils was dominated by V(IV) (4.0–4.2) (Larsson, Baken et al., 2015). Further investigation of vanadate adsorption by these soil samples showed that V(V) adsorbed to ferrihydrite and $Al(OH)_3$ were the main V species in the vanadate-treated soils (Larsson et al., 2017). The V oxidation state was also examined in different soils from various land uses (i.e., paddy soil, forest soils and mining soils) from diverse countries (Germany, Bangladesh and China) (Xu, 2016). This study reported that V(IV) and V(III) were the main V species for the studied soils, with a predominance of V(III) in the mining soils. In addition, after the soils had been extracted using oxalate and 4M HCl solutions, the V(IIII) species was more prevalent relative to the mixed V(IV)/V(III) of the native soils. These data suggested that the V^{3+} ions could substitute in the mineral lattice, making it difficult to be dissolved by the extracting solutions.

Other research on the V speciation of highly weathered soils demonstrated that kaolinite was the dominant V host mineral, occurring as V(IV) in its structure and as V(V) adsorbed to its surface, whereas V(V) adsorbed onto organic/mineral solids also occurred in the studied soils (Wisawapipat and Kretzschmar, 2017). However, the spectral features of the V(V)-adsorbed solids (e.g., Al and Fe oxyhydroxide or Fe^{3+}-organic matter complexes) are somewhat similar and can rarely be differentiated using LCF. Therefore, complementary techniques, such as XRD, are needed for the deliberation of V(V)-adsorbed solid species. Data on soil V speciation has been complied with by Gustafsson (2019) and Shaheen et al. (2019). Recently, the V oxidation state was studied in agricultural and forest upland soils from Taiwan. Based on the pre-edge feature of the V spectra, the mixed oxidation states of V(IV)/V(V) were the main V species in the studied soils (Wu et al., 2020).

There is also a serious lack of information on the application of XANES spectroscopy for determining V speciation in sediments. A XANES study of marine sediments from Australia reported a mean V oxidation state of about 3.5, suggesting a mixed oxidation state of V(III)/V(IV). This information was interpreted only from the pre-edge peak because of a very low V content (~50 mg/kg), with the spectral collection using a monolithic 100-channel HP-Ge fluorescence detector. The limitation of a low V concentration meant that it was not possible to obtain accurate quantitative analysis using the LCF approach (Bennett et al., 2018). Another study on riverine sediment from Wilson Creek (USA) demonstrated that V(III) was the primary oxidation state (~75%) with a moderate amount of V(III) (~75%), which was estimated from the weak pre-edge feature of the

XANES spectra (Nedrich et al., 2018). This result suggested that mixed oxidation states of V(III)/V(IV) were relatively immobilized when exposed to oxygen. In addition, V speciation in river sediment was investigated that had been affected by a Cu–Au mine at Mount Polley, Canada (Hudson-Edwards et al., 2019). This study examined magnetite and titanite grains from the spilled tailings of the mine and secondary Fe (oxyhydr)oxide collected from the seep-draining stream. The data showed that V^{3+} substituted magnetite and $V^{3+/4+}$ substituted titanite ($CaTiSiO_5$) were the main V species in the spilled tailings. The secondary Fe (oxyhydr)oxides from the seep-draining stream bank had a strong pre-edge peak, suggesting the adsorption of V(V) species adsorbed onto these Fe (oxyhydr)oxides (Hudson-Edwards et al., 2019).

To date, there has been relatively abundant information on V speciation in V ore minerals (e.g., manganite) and V-doped Ti oxides using XANES spectroscopy. However, knowledge on V speciation (oxidation state and species) in soil and sediment samples remains substantially lacking. A critical challenge that hinders the V speciation measurement using the XANES technique could be the inherently low V concentrations in most natural samples. Collectively, V could be present as $V^{3+/4+}$ in the structure of various minerals or occur as V(V)-adsorbed species on mineral solids. Further investigation is required regarding V chemical species that are subjected to change under diverse environmental conditions (e.g., oxic vs. anoxic conditions) and soil types. Using XANES spectroscopy to model studies on the adsorption of different V oxidation states on the potential host phases of V (e.g., kaolinite and Fe/Al oxyhydroxides) could help to gain a better understanding of V speciation for interpretation of its mobility and availability in soil systems.

6.3.4 Conclusions

The geochemistry of V is complex, with three oxidation states (III, IV and V) that may occur in the natural environment. Knowledge about soil V chemistry is relatively limited, compared to hazardous metal(loid)s including Cd, As, Cu and Pb, which have been recognized as important soil contaminants in soils with serious concerns of public health and environmental conservations. Determination of chemical species of V in soil aqueous and solid phases is a challenging task in soil chemistry and its related studies. One of the challenges of V chemistry in soils includes the completion of the thermodynamic database for V, particularly for V(III), which is the least studied species of V (Gustafsson, 2019). The XAS investigation of soil V can provide information for developing chemical speciation models to predict the bioavailability and mobility of V in the terrestrial environment. Principles developed from microscale investigations using XAS-based chemical speciation give a mechanistic understanding of macroscale phenomena of V in the soil.

References

Ahrens L.H., 1952. The use of ionization potentials Part 1. Ionic radii of the elements. *Geochim. Cosmochim. Acta.* 2, 155–169. https://doi.org/10.1016/0016-7037(52)90004-5.

Balan E., De Villiers J.P.R., Eeckhout S.G., Glatzel P., Toplis M.J., Fritsch E., et al., 2006. The oxidation state of vanadium in titanomagnetite from layered basic intrusions. *Am. Mineral.* 91, 953–956. https://doi.org/10.2138/am.2006.2192.

Bennett W.W., Lombi E., Burton E.D., Johnston S.G., Kappen P., Howard D.L., et al., 2018. Synchrotron X-ray spectroscopy for investigating vanadium speciation in marine sediment: Limitations and opportunities. *J. Anal. At. Spectrom.* 33, 1689–1699. https://doi.org/10.1039/C8JA00231B.

Bordage A., Balan E., De Villiers J.P., Cromarty R., Juhin A., Carvallo C., et al., 2011. V oxidation state in Fe-Ti oxides by high-energy resolution fluorescence-detected X-ray absorption spectroscopy. *Phys. Chem. Miner.* 38, 449–458. https://doi.org/10.1007/s00269-011-0418-3.

Burke I.T., Mayes W.M., Peacock C.L., Brown A.P., Jarvis A.P., Gruiz K., 2012. Speciation of arsenic, chromium, and vanadium in red mud samples from the Ajka spill site, Hungary. *Environ. Sci. Technol.* 46, 3085–3092. https://doi.org/10.1021/es3003475.

Calvin S., 2013. *XAFS for Everyone*. CRC Press.

Chaurand P., Rose J., Briois V., Salome M., Proux O., Nassif V., et al., 2007. New methodological approach for the vanadium *K*-edge X-ray absorption near-edge structure interpretation: Application to the speciation of vanadium in oxide phases from steel slag. *J. Phys. Chem. B.* 111, 5101–5110. https://doi.org/10.1021/jp063186i.

Chaurand P., Rose J., Domas J., Bottero J.-Y., 2006. Speciation of Cr and V within BOF steel slag reused in road constructions. *J. Geochem. Explor.* 88, 10–14. https://doi.org/10.1016/j.gexplo.2005.08.006.

Chen D., Zhao L., Liu Y., Qi T., Wang J., Wang L., 2013. A novel process for recovery of iron, titanium, and vanadium from titanomagnetite concentrates: NaOH molten salt roasting and water leaching processes. *J. Hazard. Mater.* 244–245, 588–595. https://doi.org/10.1016/j.jhazmat.2012.10.052.

Cornell R.M., Schwertmann U., 2003. *The Iron Oxides: Structure, Properties, Reactions, Occurrences and Uses*. John Wiley & Sons.

Costa D.L., Lehmann J.R., Winsett D., Richards J., Ledbetter A.D., Dreher K.L., 2006. Comparative pulmonary toxicological assessment of oil combustion particles following inhalation or instillation exposure. *Toxicol. Sci.* 91, 237–246. https://doi.org/10.1093/toxsci/kfj123.

Crans D.C., Woll K.A., Prusinskas K., Johnson M.D., Norkus E., 2013. Metal speciation in health and medicine represented by iron and vanadium. *Inorg. Chem.* 52, 12262–12275. https://doi.org/10.1021/ic4007873.

El Koura Z., Rossi G., Calizzi M., Amidani L., Pasquini L., Miotello A., et al., 2018. XANES study of vanadium and nitrogen dopants in photocatalytic TiO_2 thin films. *Phys. Chem. Chem. Phys.* 20, 221–231. https://doi.org/10.1039/c7cp06742a.

Fendorf S.E., Sparks D.L., Lamble G.M., Kelley M.J., 1994. Applications of x-ray absorption fine structure spectroscopy to soils. 58, 1583–1595. https://doi.org/10.2136/sssaj1994.03615995005800060001x.

Gilligan R., Nikoloski A.N., 2020. The extraction of vanadium from titanomagnetites and other sources. *Miner. Eng.* 146, 106106. https://doi.org/10.1016/j.mineng.2019.106106.

Gu C., Dam T., Hart S.C., Turner B.L., Chadwick O.A., Berhe A.A., et al., 2020. Quantifying uncertainties in sequential chemical extraction of soil phosphorus using XANES spectroscopy. *Environ. Sci. Technol.* 54, 2257–2267.10.1021/acs.est.9b05278.

Gustafsson J.P., 2019. Vanadium geochemistry in the biogeosphere—speciation, solid-solution interactions, and ecotoxicity. *Appl. Geochem.* 102, 1–25. https://doi.org/10.1016/j.apgeochem.2018.12.027.

Hashimoto Y., 2007. Citrate sorption and biodegradation in acid soils with implications for aluminum rhizotoxicity. *Appl. Geochem.* 22, 2861–2871. https://doi.org/10.1016/j.apgeochem.2007.07.006.

Hobson A.J., Stewart D.I., Bray A.W., Mortimer R.J.G., Mayes W.M., Riley A.L., et al., 2018. Behaviour and fate of vanadium during the aerobic neutralisation of hyperalkaline slag leachate. *Sci. Total Environ.* 643, 1191–1199. https://doi.org/10.1016/j.scitotenv.2018.06.272.

Hobson A.J., Stewart D.I., Bray A.W., Mortimer R.J.G., Mayes W.M., Rogerson M., et al., 2017. Mechanism of vanadium leaching during surface weathering of basic oxygen furnace steel slag blocks: A microfocus X-ray absorption spectroscopy and electron microscopy study. *Environ. Sci. Technol.* 51, 7823–7830. https://doi.org/10.1021/acs.est.7b00874.

Hudson-Edwards K.A., Byrne P., Bird G., Brewer P.A., Burke I.T., Jamieson H.E., et al., 2019. Origin and fate of vanadium in the Hazeltine creek catchment following the 2014 Mount Polley mine tailings spill in British Columbia, Canada. *Environ. Sci. Technol.* 53, 4088–4098. https://doi.org/10.1021/acs.est.8b06391.

Huggins F.E., Huffman G.P., Robertson J.D., 2000. Speciation of elements in NIST particulate matter SRMs 1648 and 1650. *J. Hazard. Mater.* 74, 1–23. https://doi.org/10.1016/S0304-3894(99)00195-8.

Kaur N., Singh B., Kennedy B.J., Gräfe M., 2009. The preparation and characterization of vanadium-substituted goethite: The importance of temperature. *Geochim. Cosmochim. Acta* 73, 582–593. https://doi.org/10.1016/j.gca.2008.10.025.

Kelly S.D., Hesterberg D., Ravel B., 2008. Analysis of soils and minerals using x-ray absorption spectroscopy. In *Methods of Soil Analysis Part 5—Mineralogical Methods*, pp. 387–463. Soil Science Society of America, Inc.

Larsson M.A., Baken S., Smolders E., Cubadda F., Gustafsson J.P., 2015. Vanadium bioavailability in soils amended with blast furnace slag. *J. Hazard. Mater.* 296, 158–165. https://doi.org/10.1016/j.jhazmat.2015.04.034.

Larsson M.A., D'Amato M., Cubadda F., Raggi A., Öborn I., Kleja D.B., et al., 2015. Long-term fate and transformations of vanadium in a pine forest soil with added converter lime. *Geoderma* 259–260, 271–278. https://doi.org/10.1016/j.geoderma.2015.06.012.

Larsson M.A., Hadialhejazi G., Gustafsson J.P., 2017. Vanadium sorption by mineral soils: Development of a predictive model. *Chemosphere* 168, 925–932. https://doi.org/10.1016/j.chemosphere.2016.10.117.

Liang X., Zhong Y., Zhu S., Ma L., Yuan P., Zhu J., et al., 2012. The contribution of vanadium and titanium on improving methylene blue decolorization through heterogeneous UV-Fenton reaction catalyzed by their co-doped magnetite. *J. Hazard. Mater.* 199–200, 247–254. https://doi.org/10.1016/j.jhazmat.2011.11.007.

Manceau A., Marcus M.A., Tamura N., 2002. Quantitative speciation of heavy metals in soils and sediments by synchrotron X-ray techniques. *Rev Mineral Geochem.* 49, 341–428. https://doi.org/10.2138/gsrmg.49.1.341.

Nedrich S.M., Chappaz A., Hudson M.L., Brown S.S., Burton G.A., 2018. Biogeochemical controls on the speciation and aquatic toxicity of vanadium and other metals in sediments from a river reservoir. *Sci. Total Environ.* 612, 313–320. https://doi.org/10.1016/j.scitotenv.2017.08.141.

Pettersson L., Andersson I., Gorzsás A., 2003. Speciation in peroxovanadate systems. *Coord Chem Rev.* 237, 77–87. https://doi.org/10.1016/S0010-8545(02)00223-0.

Rossi G., Pasquini L., Catone D., Piccioni A., Patelli N., Paladini A., et al., 2018. Charge carrier dynamics and visible light photocatalysis in vanadium-doped TiO_2 nanoparticles. *Appl. Catal. B.* 237, 603–612. https://doi.org/10.1016/j.apcatb.2018.06.011.

Schwertmann U., Cornell R.M., 2000. *Iron Oxides in the Laboratory: Preparation and Characterization.* WILEY-VCH.

Schwertmann U., Pfab G., 1996. Structural vanadium and chromium in lateritic iron oxides: Genetic implications. *Geochim. Cosmochim Acta* 60, 4279–4283. https://doi.org/10.1016/S0016-7037(96)00259-1.

Shafer M.M., Toner B.M., Overdier J.T., Schauer J.J., Fakra S.C., Hu S., et al., 2012. Chemical speciation of vanadium in particulate matter emitted from diesel vehicles and urban atmospheric aerosols. *Environ. Sci. Technol.* 46, 189–195. https://doi.org/10.1021/es200463c.

Shaheen S.M., Alessi D.S., Tack F.M.G., Ok Y.S., Kim K.-H., Gustafsson J.P., et al., 2019. Redox chemistry of vanadium in soils and sediments: Interactions with colloidal materials, mobilization, speciation, and relevant environmental implications- A review. *Adv. Colloid Interface Sci.* 265, 1–13. https://doi.org/10.1016/j.cis.2019.01.002.

Sutton S.R., Karner J., Papike J., Delaney J.S., Shearer C., Newville M., et al., 2005. Vanadium *K* edge XANES of synthetic and natural basaltic glasses and application to microscale oxygen barometry. *Geochim. Cosmochim. Acta* 69, 2333–2348. https://doi.org/10.1016/j.gca.2004.10.013.

Wang Y., Li Z., Sheng X., Zhang Z.J.T.J.O.C.P., 2007. Synthesis and optical properties of V_2O_5 nanorods. 126, 164701. https://doi.org/10.1063/1.2722746.

WHO, 2000. *Air Quality Guidelines for Europe—Second Edition Copenhagen: WHO Regional Publications.* European Series, No. 91. WHO.

Wisawapipat W., Kretzschmar R., 2017. Solid phase speciation and solubility of vanadium in highly weathered soils. *Environ. Sci. Technol.* 51, 8254–8262. https://doi.org/10.1021/acs.est.7b01005.

Wong J., Lytle F., Messmer R., Maylotte D., 1984. *K*-edge absorption spectra of selected vanadium compounds. *Phy. Rev. B.* 30, 5596. https://doi.org/10.1103/PhysRevB.30.5596.

Wu C.-Y., Asano M., Hashimoto Y., Rinklebe J., Shaheen S.M., Wang S.-L., et al., 2020. Evaluating vanadium bioavailability to cabbage in rural soils using geochemical and micro-spectroscopic techniques. *Environ. Pollut.* 258, 113699. https://doi.org/10.1016/j.envpol.2019.113699.

Wu J.C.S., Chen C.-H., 2004. A visible-light response vanadium-doped titania nano-catalyst by sol-gel method. *J. Photochem. Photobiol. A: Chem.* 163, 509–515. https://doi.org/10.1016/j.jphotochem.2004.02.007.

Xu Y.-H., 2016. Biogeochemical characterization of Vanadium in soils and its interaction with Fungi. PhD Thesis. University of Zurich.

7

Remediation of Vanadium in the Soil Environment

Sandun Sandanayake[1], S. Keerthanan[1], Ahamed Ashiq[1,2], Arifin Sandhi[3] and Meththika Vithanage[1]

[1] *Ecosphere Resilience Research Center, Faculty of Applied Sciences, University of Sri Jayewardenepura, Nugegoda, Sri Lanka*

[2] *Hydrometallurgy and Environment Laboratory, The Robert M. Buchan Department of Mining, Queen's University, Kingston, Ontario, Canada*

[3] *Department of Biology and Environmental Science, Faculty of Health and Life Sciences, Linnaeus University, Kalmar, Sweden*

CONTENTS

7.1 Soil Vanadium Status 135
7.2 Vanadium Remediation in Soil 137
7.2.1 Physical Techniques 138
7.2.2 Chemical Techniques 138
7.2.3 Using Soil Amendments 140
7.2.4 Phytoremediation of Vanadium 143
7.2.5 Bioremediation of Vanadium in Soils 146
7.3 Factors Affecting Remediation of Vanadium in Soils 148
7.4 Limitations and Prospectives 149
7.5 Conclusions 149
References 150

7.1 Soil Vanadium Status

Among the trace elements, V (Vanadium) is considered the fifth most abundant natural trace element available in the earth crust and is extensively used in many industries, including the aerospace, chemical and steel industries (Teng et al., 2011). Primarily, decomposition of V-containing parental rocks

DOI: 10.1201/9781003173274-7

during the weathering process increases soil availability. Incomplete recovery of V in metal refining industries has discharged nearly 30–40% of the V content to the environment (Shaheen et al., 2019). The presence of a high level of V in the environment has led to a negative impact on plants and animals. The average concentration of V in soil worldwide is roughly calculated as 108 mg kg^{-1} (Chen et al., 2021). The concentration of V (592 mg kg^{-1}) was reported in the soil in China, where discharged industrial wastes, including from coal and polymetallic mines, in the soil play the role for this elevated concentration of V (Long et al., 2021). Similarly, V was also reported in the soil where water spills from a mine dam at up to 3,600 mg kg^{-1} in South Africa (Panichev et al., 2006). Brief information about V status in the soil on a global scale is provided in Table 7.1. An in-depth complete description of global V status is given in Chapter 1 of this book.

In soil, two valence states of V are predominantly present: tetravalent V(IV) and pentavalent V(V), each of which has a different biogeochemical action in the soil. Geochemical fractionation of V is described in Chapter 4 in this book, and hence details are not discussed here. The species V(V) is

TABLE 7.1

Global Vanadium Status in Different Types of Soil

Places	Nature of Soil	Range, mg kg^{-1}	Reference
China	Agricultural soil	V(V): 4.5–21.3 V(IV): 62.6–243.6	Yang et al. (2021)
China	Soil received leachate from industries	V(V): 6–199.3 V(IV): 117.2–555.9	Aihemaiti, Gao, Liu et al. (2020)
Italy	Weathering of rocks	V: 54–239	Guagliardi et al. (2018)
Germany	Grassland and agricultural soil	V: 20.7–133.1	Shaheen and Rinklebe (2018)
Egypt	Agricultural soil	V: 37.4–122.1	Shaheen and Rinklebe (2018)
Korea	Industry, road, and abandoned mine soil	V: 25–41.6	Lee et al. (2018)
Japan	Soil water	V: 0.013–0.04 μM	Kasai et al. (2008)
Belgium	Floodplain soils	V: 45–87	Cappuyns and Slabbinck (2012)
Sri Lanka	Agricultural/ Non-agricultural soil	V: 62–820	Jayawardana et al. (2014)
China	Topsoil close to V smelter	V: 5109.6	Zhang et al. (2020)
China	Farmland soil	V: 115.5	Li et al. (2020)
Ghana	Soil from a mining site	V: 556	Amuah et al. (2021)
Sweden	Mor soil	V: 8	Larsson et al. (2015)

highly toxic; it is linked to bioavailability in plants and soil organisms and has relatively high mobilization in soil (Yang et al., 2021). In contrast, V(IV) shows less toxicity and high immobilization in soil due to the strong interaction with soil particles compared to V(V) (Hanus-Fajerska et al., 2021). Generally, V(V) is stable at pH >5, whereas V(IV) is stable at pH <5 in soil solution (Chen et al., 2021). The V(V) is the most stable species under oxidation circumstances, though it can be reduced to V(IV) by soil organic matter and microorganisms such as V-reducing bacteria (Baken et al., 2012; Shaheen et al., 2019). Several bacterial species such as sulfur-oxidizing bacteria have the capacity to oxidize the V(IV) to V(V) in the soil medium (Aihemaiti et al., 2020a).

Generally, V(V) existing in the soil as oxyanion vanadate such as VO_4^{3-}, HVO_4^{2-}, and $H_2VO_4^-$ exhibited less retention in the soil, whereas V(IV) exists as vanadyl oxo-cation (VO^{2+}), which shows stable and high retention in soil (Chen et al., 2021) due to the strong interaction with soil humic compounds (Baken et al., 2012). Vanadium is not found in its metallic form in the soil, although it can be found as vanadate compounds of Ca, Zn, Pb, Fe and Mn (Jayawardana et al., 2014).

7.2 Vanadium Remediation in Soil

Even though V has been recognized as a trace metal with a significant value in many fields, including industries, the presence of V in the elevated concentration in soils leads to a negative impact on plants, animals and eventually humans (Del Carpio et al., 2018). Compared to other valence states, V(V) has greater mobility in the environment, and it also possesses elevated toxicity towards the biota (Imtiaz et al., 2015). Due to the lack of biodegradability, V tends to be more persistent and accumulated in the environment, and it can be biologically magnified via food chains (Aihemaiti et al., 2020b). A study has found that intake of high levels of V can result in lung tumors and renal masses (Wang et al., 2020). The people residing in regions such as the United States, China, South Africa and Russia have a great potential to get affected by this emerging environmental hazard from V pollution (Gummow et al., 2006; Wu, Yang, and Zhang, 2021). Therefore, remediation of V in contaminated soil has become a major global concern in recent times. The treatment of toxic metal-contaminated soil can be done in several ways, such as removal of the contaminant, removal of contaminated soil fraction, immobilization of the contaminant within the bulk soil and so on (Kumpiene et al., 2019). The major soil V-remediation strategies proposed and commonly practiced in the different affected regions can be classified under the chemical, physical and biological processes and graphically illustrated in Figure 7.1 (Jiang et al., 2017).

7.2.1 Physical Techniques

A number of remediation techniques have been developed to reduce/remove or regulate contaminated soils with extended content of different heavy metals. The most commonly used physical methods include soil replacement, electrokinetic remediation and thermal desorption (Yang et al., 2020). Compared to other heavy metal remediation methods, only a few previous studies have suggested the remediation of V-contaminated soils by using physical techniques. These techniques include V-contaminated soil excavation, soil replacement, thermal treatment and electrolysis (Dong et al., 2021; He et al., 2019). One of the conventional physical remediation methods for contaminated soil is excavating polluted soil and disposing of it in landfills. This technique reduces the threat of spreading of V pollution from the contaminated area to the surrounding environment, although the threat remains at the landfill site (Vocciante et al., 2016). Removal of contaminated soil followed by replacing clean soil is another V-remediation technique. According to (Kumpiene et al., 2019), this practice is well suited for smaller areas rather than more significantly contaminated areas where metal ore mining, smelting and refining activities are performed on a large scale. Thermal treatment of soils contaminated with metals is another physical method to remediate toxic metals. This method often requires high energy because the treatment usually happens in temperatures above 600°C, and another factor is that the treated soil cannot be used for agricultural purposes (Ma et al., 2015). Compared to the physical-based techniques, environmentally sustainable approaches have also been investigated for remediation of V-contaminated soil. For example, electrokinetic remediation technology has been proposed by Vocciante et al. (2016), which could remediate contaminated soils from different kinds of toxic trace metals. This technique is a kind of soil washing method with the assistance of electrochemistry, and the simulated results were more advantageous than conventional soil washing techniques for heavy metal removal.

7.2.2 Chemical Techniques

One of the major reasons for using the chemical approach is to use chemical remediation, which could either help to remove or immobilize the contaminants (Zou et al., 2019b). A wide range of chemicals have been investigated and used to remediate soils contaminated with trace metals, including V-contaminated soil; processes include adsorption of contaminants, reduction, and leaching from the soil matrix, which ultimately results in a reduced contaminant hazard potential. Meanwhile, in the soil stabilization technique, the mobility, solubility and toxicity of the metal contaminants in soil are reduced to limit their hazard potential (Zou, Li et al., 2019). According to He et al. (2020), the major chemical processes those could be involved in the remediation of contaminant soils are ion exchange, adsorption-desorption, precipitation, complexation, oxidation-reduction and so on.

Among chemical approaches to removing soil contaminants, the soil washing method can be considered as a practical and effective long-term heavy metal treatment in the soil medium. V-contaminated soil remediation can be improved by using appropriate soil washing chemicals under ideal conditions. To obtain effective soil washing performance, the selection of an appropriate chemical reagent is crucial. A wide range of chemical reagents for soil remediation have been implicated: inorganic acids (HCl) (Udovic and Lestan, 2012), organic acids (tartaric acid, citric acid, oxalic acid) (Di Palma and Mecozzi, 2007), non-ionic surfactants (Yang et al., 2006), alkalis (Chen et al., 2016) and chelates (ethylenediaminetetraacetic acid (EDTA)) (Zeng et al., 2005). Besides these, biodegradable washing substances also take part in the soil remediation process. For example, bio-surfactants (humic acid) (Conte et al., 2005), biodegradable chelates ([S,S]-ethylenediaminedisuccinic acid (EDDS)) (Zou, Xiang et al., 2019), dissolved organic carbon solutions and organic metabolites produced by fungi are also used as soil washing reagents (Arwidsson et al., 2010).

Although chemical reagent-based remediation of soils is considered to be an effective method, a large-scale chemical application has many disadvantages, including the high expenditure for those reagents (Dermont et al., 2008). The usage of alkaline solutions and inorganic acids as soil washing reagents contribute high and low pH conditions to the soil, respectively, promoting the dissolution of many metals together with non-contaminants (Jiang et al., 2017). For toxic metal extraction from contaminated soils, one of the most frequently used effective chelating reagents is EDTA (Jez and Lestan, 2016). This chemical is more effective than other anionic surfactants and has less impact on soil properties than acids. However, EDTA may form strong complexes with cations such as Mg, Al, Ca and Fe in addition to the target toxic metals and may affect the structure and physical properties of the soil (Zeng et al., 2005). Besides these, due to low biodegradability and high persistence in the soil environment, EDTA affects soil microbes, plants and groundwater sources (Zou, Xiang et al., 2019b). As mentioned in Aihemaiti et al. (2020a), the soil washing method can be a vital reason for water pollution, as wastewater with concentrated toxic metals is also produced during this process.

Meanwhile, for substitutions for inorganic soil washing reagents, organic acids have been used for the soils contaminated with toxic metals, including V. Due to a lack of effects on soil properties and higher biodegradability, organic acids are extensively used in the soil washing process. Zou et al. (2019b) have researched using volatile fatty acids (VFAs) extracted from food waste as a soil washing substance for the remediation of soils contaminated with Cr and V. The optimum conditions for a single washing run were cited as 30 g L^{-1} of VFA concentration, 10:1 of liquid-to-solid ratio, and four hours of contact time (Zou, Xiang et al., 2019). Besides that, organic matter such as humic acid can fix heavy metals in the soil by forming specific ligand complexes with heavy metals. The heavy metal morphology is affected by complexation and adsorption reactions happening with the organic matter (Dong et al., 2021). Moreover, the addition of organic matter with suitable nutrition is an effective remediation approach for V-contaminated soil and reduces food contamination. This approach is effective since it is an economic option and required less time (Imtiaz et al., 2015).

7.2.3 Using Soil Amendments

Generally, mobilization of V in the soil-plant system greatly depends on a number of factors including soil pH level, cation exchange capacity, organic matter content and microbial activity in the plant-soil system (El-Naggar et al., 2021). Vanadium, when present in excess amount in acidic soil (>2 mg kg^{-1}), strongly influences the chlorophyll activity, potassium consumption and nitrogen assimilation, thereby proving to be a toxin that can cause chlorosis (El-Naggar et al., 2021). On the other hand, trace concentrations of V are critical for growth and nitrogen fixation in certain plants. A number of studies have indicated that V can endanger livestock, as vanadate ($H_2VO_4^-$) structurally mimics phosphate ($H_2PO_4^-$) ions in plants, which ultimately leads to osteoporosis among grass-eating mammals because of lack of phosphates (Panichev et al., 2006; Wu et al., 2020; Yang et al., 2017). The mobile V in the soil can exchange with salts in dissolved Fe or Mn oxides, clay minerals and other organic matters present in the soil, potentially harming the ecosystems (El-Naggar et al., 2021; Shaheen and Rinklebe, 2018). Several morphological and physiological processes in the plants are affected by V content in the soil. The optimal amounts of V in the soil could improve chlorophyll synthesis, germination rates and shoot and root growth, to an extent of a threshold concentration that is needed to remediate V-contaminated soils effectively and further minimize the repercussions involved (Dong et al., 2018; Wu et al., 2020). Soil amendments are widely considered as a solution for remediation and restoration for V-contaminated soils because they do not potentially harm the environment in the long run after their applications (Wu et al., 2020).

Numerous studies have been well documented on remediation of heavy metal contaminated in soils, such as utilizing zero-valent iron, humic acid, biochar, phosphogypsum, activated carbon and many other carbon-based adsorbents to reduce the mobility of heavy metals into the plant system (Wu et al., 2020; Yang et al., 2011). As far as V-contaminated soils are concerned, very little has been investigated on their remediation and amendment techniques. A number of studies have found that application of soil conditioners such as polyacrylamide (PAM), carbon-based peat, sludge compost and bentonite have proven to stabilize certain heavy metals through chemical bonds with the contaminant and soil, thereby providing more nutrients to the soil for the continuation of the plant growth process (Mamedov et al., 2009; Mamedov et al., 2017; Zhang, Jiang et al., 2018). Zhang, Jiang et al. (2018) have found that PAM, peat, humic acid and sludge compost as soil conditioners gave optimum growth for canola plants in the soil, whereas application of bentonite and lime caused negative growth on the plant. The biomass increased by 1.2-fold when PAM, humic acid and peat were added in combination, which significantly improved the translocation and accumulation of V into the canola plant, and the height of the plant increased by 59.49%, as compared with the control having only V-contaminated soil.

According to the speciation perspective, V exists in trivalent [V(III)], V(IV) and V(V) forms in soil and aqueous media, among which V(III) is the least formed in the environmental matrices. The V(V) is known to be less static than V(IV) and hence more potential to pose a risk in biological systems. Yang et al. (2017) has found that the availability of V(V) in soil and its translocation varied with different growth stages, particularly to leguminous plant types, for example, soybeans. Studies have found that 30–50 mg kg^{-1} of V in the soil inhibits biomass production for both root and shoot systems and the aboveground parts of the soybean plants (Wang and Liu, 1999; Yang et al., 2017). Nevertheless, Chinese cabbage (*Brassica rapa L.*) showed an increase in the biomass production within the range of 120–350 mg kg^{-1}; it is speculated this is due to the spike in the availability of nitrogen to the soil concerning the presence of other cations present in the soil, thereby making a low uptake environment for V present in the soil. This result is echoed in the other studies, where the toxicity of high V in soils is associated with the cations in soils and also the plants exposed (Tian et al., 2014; Tian et al., 2015; Yang et al., 2017). It has been found that retention of V is most probable in their root system, as studied in the soybean plant, wherein levels up to 2,500 mg kg^{-1} are retained. These studies have suggested that the soybeans adsorb more V as their growth stage changes from the development to the reproductive stage, wherein the greatest accumulation of V is adsorbed in the roots to a higher extent (Yang et al., 2011, 2017). In Zhang, Jiang et al. (2018), where they have studied the effect of soil conditioners on *Brassica campestris L.*, fly ash amended in soils showed a decrease in the height and biomass in cases where a 2 wt% of fly ash was added to soil. The decrease was noted in root length, leaf area, pigment content and biomass in which both the amendments using fly ash and bentonite clays showed the same trend (Sikka and Kansal, 1995; Yang et al., 2017).

Because of the agronomic and environmental potentials it possesses, when applied to environmental systems, biochar, a carbon-rich pyrolyzed biomass, has been consistently proved to act as a comprehensive soil conditioner. Its activity in soils is greatly dependent on the feedstock types and pyrolyzing conditions. The amorphous and crystalline structure largely affects the adsorption being carried out in the systems, especially in acidic conditions with the presence of V in soil matrices (Ginebra et al., 2022). Nevertheless, a pH condition is very conducive for studying the performances of V-uptake from soil-plant systems as it pertains to the buffering capacity of the system (Atugoda et al., 2021; Garbuz et al., 2020; Vithanage et al., 2017). Generally, biochar studies have been relevant for the adsorption of cationic heavy metals and are known to have a lesser significant adsorption capacity for anionic forms. However, they are further improved for the pollutant at an anionic moiety through physical and chemical treatments of biochar surface. Multiple studies have shown that in the pretreatment of biomass before pyrolysis, the higher specific surface area is a common obtainable feature (Vithanage et al., 2015). These studies mainly emphasized aqueous media for the removal of V; thus, the potential of biochar applied on soils with different feedstocks and changes in the surface functionalities have not been investigated for the immobilization of V in soil matrices. Even the most predominant feedstocks – that is, wood and rice husk – utilized to synthesize biochar to be used as the soil conditioners for other contaminated soils have not been studied for changes in the V fractionation in soils and uptake by plants (Vithanage et al., 2016; Bandara et al., 2017).

Yu et al. (2020) has shown that soil samples influenced by extensive mining operations were collected with 92.43 mg kg^{-1} of V in topsoil. These soils were utilized for soil remediation and proven to effectively immobilize V through an agricultural by-product, corn cobs and rice straw to prepare biochars. In the case of biochar-based studies, different particle sizes of the biochar in soil were implanted in soils along with varied dosages to recognize the optimal conditions. From the study, it was concluded that an increased pH, higher organic matter content and a reduced bioavailable V in the soils were observed for rice straw biochar (pyrolytic temperature 650°C and particle size 0.075 mm), in which the reduced bioavailability of V is associated with the increased organic matter that can bind with V(V) to form insoluble complexes. Besides this, studies have found that other complexes formed and tend to be soluble with the organics and further permeate through other media being available for translocation to plants (Reijonen et al., 2016; Yu et al., 2020).

The potential of biochar for removal of V has been further investigated in El-Naggar et al. (2021), where the idea of different surface functionalities has competed with induced charges present in the soil to mobilize or immobilize bioavailable V. Different biogeochemical instances where different cations released in the soil can counteract the release dynamics of V from the soil have been thoroughly studied. Three different biochars are produced from rice husk (500°C), wood chip and forestry waste as feedstock (250 and 500°C). Up to a 5% dosage with soil samples were utilized where V (3,750 mg kg^{-1}) was synthetically induced at varied concentrations to investigate optimal conditions. As the study has mentioned, organic matter content arising from aliphatic and aromatic organic carbon in soils can affect the solubility of V in soil. Vanadium is preferentially bound with the aliphatic organic content that is actively present in the rice husk biochar (500°C) and wood biochar (500°C) compared with the lower pyrolyzing biochar (250°C). This was further supported in Frohne et al. (2015), where the correlation between organic content in soils was investigated with the V uptake in plants. The reduction of V(V) to V(IV) was established on the dissolved organic content in the soil media with their different treatment strategy (El-Naggar et al., 2021; Frohne et al., 2015; Shaheen et al., 2016).

7.2.4 Phytoremediation of Vanadium

Phytoremediation can be considered an eco-friendly, economic, in-situ and sustainable environmental management approach to remediating environmental contaminants (Vithanage et al., 2011). The phytoremediation management could also be applicable for the mitigation of V from the contaminated soil without affecting the soil functions. Generally, phytoremediation can be classified into various categories, including phytoextraction, rhizofiltration or phytofiltration, phytostabilization, phytovolatilization and phytodegradation (Gunarathne et al., 2020). Among them, phytoextraction and phytostabilization are the most applicable techniques for V mitigating in the soil, since heavy metals are non-degradable (Hanus-Fajerska et al., 2021).

Phytoextraction means the accumulation of contaminants from the contaminated soil are removed by the plant uptake and translocated to the aboveground/aerial parts of the plant (Oladoye et al., 2021). Meanwhile, the plant used for the phytoextraction process should have a few characteristics to be a hyperaccumulator plant species. Those are tolerance to the high contamination of specific heavy metals, high uptake, translocation and accumulation towards and high growth in a short time (Chen et al., 2021). Currently, few plants have been identified as a hyperaccumulator toward V. For instance, Aihemaiti et al. (2017) identified two native plants (*Setaria viridis* and *Eleusine indica*) as hyperaccumulators for V (Table 7.2). Similarly, Ray et al. (2020) have found that several mangrove plant species, including *Aegiceras corniculatum* and *Avicennia alba*, showed hyperaccumulation capacity towards V.

Phytostabilization is a strategy for accumulating large amounts of heavy metals stored in the plant root systems, thus restricting the transfer of the heavy metals to the aboveground parts and entry into the food chain through the consumption of aerial parts. Generally, the V in the soil is retained by the phenomena of adsorption onto the root cell wall and sequestered inside the root tissue. Therefore, there are only a few factors to consider when choosing plants for phytostabilization, especially for the removal of V from the contaminated soil: the plant should be resistant to the V state in the soil and have rapid growth, a dense and effective root system and a long life span. (Marques et al., 2009). A wide range of plant species has been identified as a phytostabilizer for V in soil – for instance, *Nicotiana tabacum* (Wu et al., 2021b), *Artemisia vulgaris, Polygonum cuspidatum, Betula populifolia, Populus deltoides* (Qian et al., 2014), *Festuca arundinacea* and *Lotus corniculatus* (Guarino et al., 2019). More plants and their efficiencies are presented in Table 7.2. Generally, the effectiveness and success of phytoremediation management depend on the bioavailability of heavy metals on the soil, as the uptake and translocation of heavy metals in the plants greatly depends on the species and metals. It could be predicted that the pH, soil organic matter and redox potential

TABLE 7.2

This Table Summarizes the Uptake and Phytoremediation Investigation of V by Different Plant Species Reported in the Literature

Plant Name	V_{root}, mg kg^{-1}	V_{shoot}, mg kg^{-1}	TF	Phytoremediation Classification		Reference
				Phytostabilization	Phytoextraction	
Kochia scoparia	1474.7	95.5	0.06	✓		Aihemaiti et al. (2017)
Eleusine indica	335	495.7	1.48	✓	✓	
Setaria viridis	320.7	1245.4	3.88		✓	
Medicago sativa	6.9–43.8	1.6–13.9	0.23–0.32	✓		Gan et al. (2021)
Astragalus membranaceus	7.4–22.9	1.5–7.7	0.2–0.34	✓		
Ipomoea aquatica	4.1–36.1	0.4–16	0.1–0.44	✓		
Nicotiana tabacum	14.1–272.5	1.3–65.2	0.09–0.23	✓		Wu, Yang, Zhang, Wang et al. (2021)
Medicago sativa	6	1.2	0.2	✓		Gan, Chen et al. (2020)
Plantago lanceolata	7.71	0.01	0.001	✓		Guarino et al. (2019)
Piptatherum miliaceum	34.2	0.11	0.003	✓		
Senecio vulgaris	6.45	2.1	0.33	✓		

Artemisia vulgaris	25.7–113	6.98–11.96	0.27–0.11	✓		Qian et al. (2014)
Polygonum cuspidatum	225	9.4	0.04	✓		
Phragmites australis	218	2.06	0.009	✓		
Rhus copallinum	32.5–116	4.28–8.71	0.13–0.08	✓		
Betula populifolia	33.1–280	5.96–12.1	0.18–0.04	✓		
Populus deltoides	67.3–119	4.91–8.24	0.07	✓		
Setaria viridis	6–145.8	8.32–159.5	1.39–1.09		✓	Aihemaiti et al. (2018)
Kochia scoparia	5.3–245.9	3.8–9.5	0.72–0.04	✓		
Chenopodium album	1.2–383.6	7.3–14.7	6.08–0.04	✓		
Rape plant	12.6	Leaf: 26.6	2.1		✓	Teng et al. (2011)
Alfalfa	3.46–3440	2.16–154.3	0.62–0.04	✓		Yang et al. (2011)
Setaria viridis	15.8–1733.6	2.1–340	0.13–0.2	✓		Aihemaiti, Gao, Liu et al. (2020)
Brassica chinensis	13.95–39.75	2.8–4.9	0.2–0.1	✓		Liao and Yang (2019)

V_{root} = V concentration in root parts

V_{shoot} = V concentration in shoot parts

TF = Translocation factor

enhance the bioavailable fraction of V in the soil medium and thus can help to increase the phytoremediation process. Besides these, the bioavailability of V in the soil can be increased by applying chemical agents and inoculating the soil with microorganisms. Gan, Chen et al. (2020) have found that inoculated with *Arthrobacter sp.* strain increased the phytoremediation of V by *Medicago sativa* in V-contaminated soil.

As a sustainable heavy metal remediation management, phytoremediation is a well-established environment-friendly technology for remediation of different toxic trace elements from the contaminated soil. Although phytoremediation is a cost-effective and ecologically friendly technique, it has a few limitations, including a delayed process, a reliance on the metal's bioavailability fraction and post-remediation maintenance.

7.2.5 Bioremediation of Vanadium in Soils

Bioremediation strategies by plants either by phytoextraction and/or phytostabilization have proven to be promising remediation strategies, as detailed in the previous section. Microorganisms in soil have strong reducing and immobilizing effects on heavy metal-contaminated soils that can be taken as a strategy for remediation. The other bioremediation strategy is directly utilizing microorganisms through biosorption and precipitation, which utilizes the reduction process for conversion to less toxic V(IV), both processes are an in-situ mode of remediation (van Marwijk et al., 2009; Zhang et al., 2014). They can catalytically convert soluble V(V) to insoluble V(IV), which can curb the migration towards the plant phloem. Microbial catalyzation using autotrophic organisms has been extensively studied for bio-transforming into insoluble V in varied simulated soil conditions (Shi et al., 2020; Zhang, Qiu et al., 2018). Besides this, elemental sulfur is used by these organisms as an inorganic electron donor for the reduction process. Therefore, microorganism-based bioremediation could be an excellent economic and environmentally friendly process compared to other treatment methods for V-contaminated soil remediation (Chen et al., 2021; Shi et al., 2020).

Most metal-tolerant microorganisms such as Bacteroidetes, Proteobacteria, and Firmicutes predominately reduce the V(V) in soil biota. Besides that, Zhang, Wang et al. (2019) have found that other microorganism genera, namely *Geobacter*, *Pseudomonas* and *Comamonas* also co-exist in different soil samples. Despite higher concentrations present in soil matrices, V played a vital role in the growth of microbial communities. Enriched *Bacillus megaterium* grown in a simulated soil-inoculated bioreactor enhanced the reduction of V(V), as studied in Rivas-Castillo et al. (2017). Autotropic microbes, upon oxidation of elemental sulfur, used bicarbonate present in the growth media to produce volatile fatty acids as by-products. These fatty acids provide for the substrate for reducing V(V) to V(IV) through two intracellular membrane-based mechanisms. Vanadyl that is, oxovanadium (IV) cation, a further by-product after reduction, is pushed out of the cell through bio-reduction of the vanadate oxyanion, in which a small amount of V gets bioaccumulated within the cell. In the next pathway, vanadate reductase primarily present in the cell membrane reduces the V(V) to V(IV) for energy utilization, and no V enters into the cell, unlike the latter pathway.

Therefore, for in-situ soil remediation of V-contaminated soils, these applications alongside soil conditioners play a vital role in the growth of microorganisms. As studied in van Marwijk et al. (2009), the use of fulvic and humic acids in the soil also largely influences the V bioavailability, toxicity and vanadate reduction in the soil biota. Considering the sensitivity of V to redox potential, pH of the growth media and the presence of organic matter and cations, the use of aquatic plants, namely *Phragmites australis* and *Acorus calamus*, have proven to reduce 60% of V-contaminated from the soil matrices (Jiang et al., 2018; Lin et al., 2019).

Microbial reduction performed by *Shewanella loihica* reduced the V(V) to about 87% in the soil collected from the mining site. *Shewanella oneidensis* and *Geobacter metallireducens* are metal-reducing in the eukaryotic domain, capable of reducing V(V) to vanadyl. *Saccharomyces cerevisiae* has also been studied as the sole electron acceptor in soil matrices and grown even at lower concentrations of V (Carpentier et al., 2003; Carpentier et al., 2005; Zhang et al., 2014). However, most of these studies have been performed in pure cultures rather than mixed cultures, and most of the time, the systems are more complex in the real environmental matrices. Millions of microorganisms live in soil,

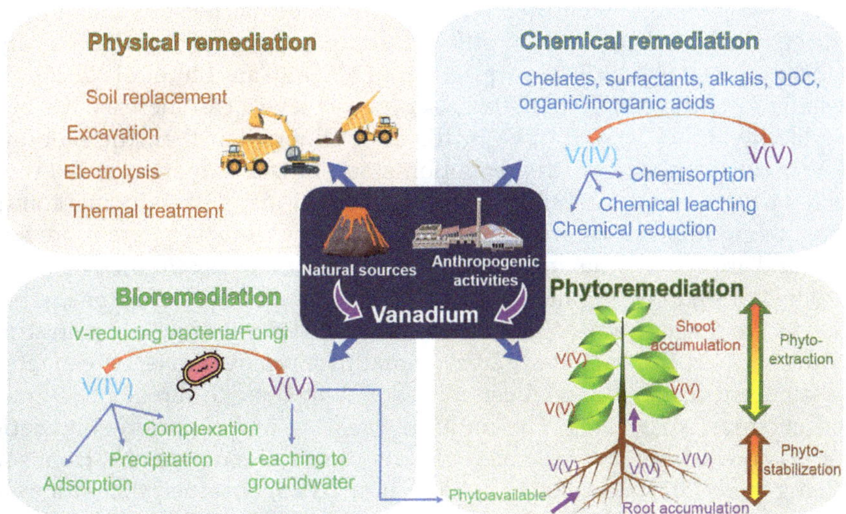

FIGURE 7.1
Graphical presentation of remediation strategies for V remediation in soil.

especially in V mining sites. In Hao et al. (2018), mixtures of two organisms, *Bryobacter* and *Acidobacteriaceae*, proved the capabilities to reduce V(V) by up to 92%. These were compared with the soil media collected from the Uvanite and coal mine areas, where the V content is comparatively different from the V titanomagnetite site (Chen et al., 2015; Hao et al., 2018). *Sphingomonas desiccabilis* synthesizes extracellular substances (EPS) in growth media with thiosulfate as a substrate, further develop the complexation of V(V) within the cells (Zhang, Wang et al., 2019). A few studies have also found that chelation occurs between extracellular substances and V ions to form a stable insoluble globule that improves V bioremediation strategies (Zhang, Cheng et al., 2019a; Zhang, Wang et al., 2019b).

7.3 Factors Affecting Remediation of Vanadium in Soils

Several previous investigations conducted on soil V-remediation have identified that the mobility of V in the soil environment is affected by a number of factors. These factors are mainly associated with the characteristics of the reagent that are used in V-remediation, the oxidation state of V in soil and the properties of the contaminated soil. Soil properties such as the chemical composition, plasticity characteristics, precipitation saturation index and soil pH are most likely to determine the geochemistry of V (Yang et al., 2020; Zou, Xiang et al., 2019). Gan, Liu et al. (2020) observed that soil enriched with inorganic minerals has a higher ability to fix heavy metals including V in the soil medium. Besides this, higher efficiency in V-adsorption on silica indicated chemical reactions as the leading causes for the V-desorption from silica (Gan, Liu et al., 2020). As observed in Shaheen et al. (2019), alkaline pH levels can increase the bioavailability of V(IV) and V(V) in the soil. Other than that, the mobility of V in soil depends on the oxidation state. Lower oxidation states of V such as V(III) are relatively immobile and dominant in reducing conditions, while higher oxidation states such as V(IV) and V(V) are more soluble and mobile in the soil environment (Imtiaz et al., 2015). The other V-remediation affecting factors are associated with the reagent properties such as the reagent type, chemical structure and the reagent concentration (Wang et al., 2015; Yang et al., 2020). Vanadium removal rates investigated for different reagents have been varied due to their different pK_a values and chemical structures. The multiple washing and liquid-to-solid ratio in the soil washing method could also regulate the removal of V from soil (Zhang et al., 2013). For example, Yang et al. (2020) observed that attapulgite and zeolite showed their highest efficiencies in V-remediation at 2% of addition as 65.13% and 41.95%, respectively; hydroxyapatite had its highest efficiency at 10% of addition as 73.69%.

7.4 Limitations and Prospectives

The efficiency of various V-contaminated soil remediation techniques is varied because of their recovery, time consumption, expenditure, and waste production. Considering the soil washing method, acidic and alkaline solutions as reagents have adverse effects on the soil environment by promoting the dissolution of non-contaminant metals. Besides this, the soil washing technique discharges wastewater concentrated with heavy metals that could further pollute soil and water sources. The heavy metal-containing wastewater including V contamination has been investigated by using a number of methods, for example, adsorption, electrolysis, ion exchange, precipitation, biological treatment and solvent extraction (Zhang et al., 2012, 2015). Jiang et al. (2017) have suggested reducing the effluent treating costs by recycling soil washing effluent and recovering metals including V.

The weaknesses of the inorganic and organic amendment application technique for V-remediation can be identified as costly, highly time consuming and requiring a workforce. Furthermore, the phytoremediation technique is an economical approach to reduce the toxicity of V in soil, but it also has a few drawbacks such as low efficiency, a long time duration, and the possibility of further contamination in various trophic levels due to the metal translocation to the aerial parts (Dong et al., 2021; Imtiaz et al., 2015). To avoid this complication, crop species with minimum V-translocation in the aboveground parts need to be screened for using V-remediation from the contaminated soil.

7.5 Conclusions

The V contamination in soil is already considered a global environmental concern that needs further attention. Current V content and situations in the soil and the co-relating factors affecting V's availability in soil were briefly covered in this chapter. It also included all aspects of V-remediation, including physical and chemical remediation as well as bioremediation utilizing potential high V-accumulating plant species and microorganisms. Chemical cleanup solutions are considered to be an effective and immediate response process; however, they are not cost-effective and have a risk for further water pollution. Regarding environmentally friendly perceptions, phytoremediation takes precedence over chemical treatments, even with their slow removal process. Meanwhile, the addition of microbial communities in the soil could improve the bioavailability of V in soil and increase the plant-based phytoremediation process in the contaminated soil. V contamination in soil has been found all over the world, including in both developed and developing countries; therefore, policymakers, scientists and ecologists need to focus urgently on V-remediation technique development and optimization for the contaminated sites.

References

Aihemaiti, A., Gao, Y., Liu, L., Yang, G., Han, S. & Jiang, J. (2020a). Effects of liquid digestate on the valence state of vanadium in plant and soil and microbial community response. *Environ Pollut*, 265, 114916.

Aihemaiti, A., Gao, Y., Meng, Y., Chen, X., Liu, J., Xiang, H., Xu, Y. & Jiang, J. (2020b). Review of plant-vanadium physiological interactions, bioaccumulation, and bioremediation of vanadium-contaminated sites. *Sci Total Environ*, 712, 135637.

Aihemaiti, A., Jiang, J., Li, D., Li, T., Zhang, W. & Ding, X. (2017). Toxic metal tolerance in native plant species grown in a vanadium mining area. *Environ Sci Pollut Res Int*, 24, 26839–26850.

Aihemaiti, A., Jiang, J., Li, D., Liu, N., Yang, M., Meng, Y. & Zou, Q. (2018). The interactions of metal concentrations and soil properties on toxic metal accumulation of native plants in vanadium mining area. *J Environ Manage*, 222, 216–226.

Amuah, E. E. Y., Fei-Baffoe, B. & Kazapoe, R. W. (2021). Emerging potentially toxic elements (strontium and vanadium) in Ghana's pedological studies: Understanding the levels, distributions and potential health implications. A preliminary review. *Environ Challenges*, 5, 100235.

Arwidsson, Z., Elgh-Dalgren, K., Von Kronhelm, T., Sjoberg, R., Allard, B. & Van Hees, P. (2010). Remediation of heavy metal contaminated soil washing residues with amino polycarboxylic acids. *J Hazard Mater*, 173, 697–704.

Atugoda, T., Ashiq, A., Keerthanan, S., Wijekoon, P., Ramanayaka, S. & Vithanage, M. (2021). Biochar amalgamation with clay: Enhanced performance for environmental remediation. *In:* Sarmah, A. K. (ed.) *Advances in Chemical Pollution, Environmental Management and Protection*. Elsevier Inc. 7, 1–37.

Baken, S., Larsson, M. A., Gustafsson, J. P., Cubadda, F. & Smolders, E. (2012). Ageing of vanadium in soils and consequences for bioavailability. *Eur J Soil Sci*, 63, 839–847.

Bandara, T., Herath, I., Kumarathilaka, P., Hseu, Z. Y., Ok, Y. S. & Vithanage, M. (2017). Efficacy of woody biomass and biochar for alleviating heavy metal bioavailability in serpentine soil. *Environ Geochem Health*, 39, 391–401.

Cappuyns, V. & Slabbinck, E. (2012). Occurrence of vanadium in Belgian and European Alluvial soils. *Appl Environ Soil Sci*, 2012, 1–12.

Carpentier, W., De Smet, L., Van Beeumen, J. & Brigé, A. (2005). Respiration and growth of Shewanella oneidensis MR-1 using vanadate as the sole electron acceptor. *J Bacteriology*, 187, 3293–3301.

Carpentier, W., Sandra, K., De Smet, I., Brige, A., De Smet, L. & Van Beeumen, J. (2003). Microbial reduction and precipitation of vanadium by Shewanella oneidensis. *Appl Environ Microbiol*, 69, 3636–3639.

Chen, L., Liu, J. R., Hu, W. F., Gao, J. & Yang, J. Y. (2021). Vanadium in soil-plant system: Source, fate, toxicity, and bioremediation. *J Hazard Mater*, 405, 124200.

Chen, J., Liu, X., Li, L., Zheng, J., Qu, J., Zheng, J., Zhang, X. & Pan, G. (2015). Consistent increase in abundance and diversity but variable change in community composition of bacteria in topsoil of rice paddy under short term biochar treatment across three sites from South China. *Applied Soil Ecology*, 91, 68–79.

Chen, Y., Zhang, S., Xu, X., Yao, P., Li, T., Wang, G., Gong, G., Li, Y. & Deng, O. (2016). Effects of surfactants on low-molecular-weight organic acids to wash soil zinc. *Environ Sci Pollut Res Int*, 23, 4629–4638.

Conte, P., Agretto, A., Spaccini, R. & Piccolo, A. (2005). Soil remediation: Humic acids as natural surfactants in the washings of highly contaminated soils. *Environ Pollut*, 135, 515–522.

Del Carpio, E., Hernandez, L., Ciangherotti, C., Villalobos Coa, V., Jimenez, L., Lubes, V. & Lubes, G. (2018). Vanadium: History, chemistry, interactions with alpha-amino acids and potential therapeutic applications. *Coord Chem Rev*, 372, 117–140.

Dermont, G., Bergeron, M., Mercier, G. & Richer-Lafleche, M. (2008). Soil washing for metal removal: A review of physical/chemical technologies and field applications. *J Hazard Mater*, 152, 1–31.

Di Palma, L. & Mecozzi, R. (2007). Heavy metals mobilization from harbour sediments using EDTA and citric acid as chelating agents. *J Hazard Mater*, 147, 768–775.

Dong, Y. B., Lin, H., Zhao, Y. & Gueret Yadiberet Menzembere, E. R. (2021). Remediation of vanadium-contaminated soils by the combination of natural clay mineral and humic acid. *J Clean Prod*, 279, 123874.

Dong, Y.-B., Liu, Y. & Lin, H. (2018). Leaching behavior of V, Pb, Cd, Cr, and As from stone coal waste rock with different particle sizes. *Int J Miner, Metallurgy, and Materials*, 25, 861–870.

El-Naggar, A., Shaheen, S. M., Chang, S. X., Hou, D., Ok, Y. S. & Rinklebe, J. (2021). Biochar surface functionality plays a vital role in (Im)mobilization and phytoavailability of soil vanadium. *ACS Sustain Chem Eng*, 9, 6864–6874.

Frohne, T., Diaz-Bone, R. A., Du Laing, G. & Rinklebe, J. (2015). Impact of systematic change of redox potential on the leaching of Ba, Cr, Sr, and V from a riverine soil into water. *J Soils Sediments*, 15, 623–633.

Gan, C.-D., Chen, T. & Yang, J.-Y. (2020a). Remediation of vanadium contaminated soil by alfalfa (Medicago sativa L.) combined with vanadium-resistant bacterial strain. *Environ Technol Innovation*, 20, 101090.

Gan, C. D., Chen, T. & Yang, J. Y. (2021). Growth responses and accumulation of vanadium in alfalfa, milkvetch root, and swamp morning glory and their potential in phytoremediation. *Bull Environ Contam Toxicol*, 107, 559–564.

Gan, C. D., Liu, M., Lu, J. & Yang, J. (2020b). Adsorption and desorption characteristics of vanadium (V) on silica. *Water, Air, & Soil Pollution*, 231, 10.

Garbuz, S., Camps-Arbestain, M., Mackay, A., Devantier, B. & Minor, M. (2020). The interactions between biochar and earthworms, and their influence on soil properties and clover growth: A 6-month mesocosm experiment. *Applied Soil Ecology*, 147, 103402.

Ginebra, M., Munoz, C., Calvelo-Pereira, R., Doussoulin, M. & Zagal, E. (2022). Biochar impacts on soil chemical properties, greenhouse gas emissions and forage productivity: A field experiment. *Sci Total Environ*, 806, 150465.

Guagliardi, I., Cicchella, D., De Rosa, R., Ricca, N. & Buttafuoco, G. (2018). Geochemical sources of vanadium in soils: Evidences in a southern Italy area. *Journal of Geochemical Exploration*, 184, 358–364.

Guarino, C., Zuzolo, D., Marziano, M., Baiamonte, G., Morra, L., Benotti, D., Gresia, D., Stacul, E. R., Cicchella, D. & Sciarrillo, R. (2019). Identification of native-metal tolerant plant species in situ: Environmental implications and functional traits. *Sci Total Environ*, 650, 3156–3167.

Gummow, B., Kirsten, W. F., Gummow, R. J. & Heesterbeek, J. A. (2006). A stochastic exposure assessment model to estimate vanadium intake by beef cattle used as sentinels for the South African vanadium mining industry. *Prev Vet Med*, 76, 167–184.

Gunarathne, V., Gunatilake, S. R., Wanasinghe, S. T., Atugoda, T., Wijekoon, P., Biswas, J. K. & Vithanage, M. 2020. Phytoremediation for E-waste contaminated sites. *In:* Prasad, M. N. V., Vithanage, M. & Borthakur, A. (eds.) *Handbook of Electronic Waste Management*. Butterworth-Heinemann.

Hanus-Fajerska, E., Wiszniewska, A. & Kaminska, I. (2021). A dual role of vanadium in environmental systems-beneficial and detrimental effects on terrestrial plants and humans. *Plants (Basel)*, 10.

Hao, L., Zhang, B., Feng, C., Zhang, Z., Lei, Z., Shimizu, K., Cao, X., Liu, H. & Liu, H. (2018). Microbial vanadium (V) reduction in groundwater with different soils from vanadium ore mining areas. *Chemosphere*, 202, 272–279.

He, L., Zhong, H., Liu, G., Dai, Z., Brookes, P. C. & Xu, J. (2019). Remediation of heavy metal contaminated soils by biochar: Mechanisms, potential risks and applications in China. *Environ Pollut*, 252, 846–855.

He, W.-Y., Yang, J.-Y., Li, J., Ai, Y.-W. & Li, J.-X. (2020). Stabilization of vanadium in calcareous purple soil using modified Na-bentonites. *J Clean Prod*, 268, 121978.

Imtiaz, M., Rizwan, M. S., Xiong, S., Li, H., Ashraf, M., Shahzad, S. M., Shahzad, M., Rizwan, M. & Tu, S. (2015). Vanadium, recent advancements and research prospects: A review. *Environ Int*, 80, 79–88.

Jayawardana, D. T., Pitawala, H. M. T. G. A. & Ishiga, H. (2014). Geochemical evidence for the accumulation of vanadium in soils of chronic kidney disease areas in Sri Lanka. *Environ Earth Sci*, 73, 5415–5424.

Jez, E. & Lestan, D. (2016). EDTA retention and emissions from remediated soil. *Chemosphere*, 151, 202–209.

Jiang, B., Xing, Y., Zhang, B., Cai, R., Zhang, D. & Sun, G. (2018). Effective phytoremediation of low-level heavy metals by native macrophytes in a vanadium mining area, China. *Environ Sci Pollut Res Int*, 25, 31272–31282.

Jiang, J., Yang, M., Gao, Y., Wang, J., Li, D. & Li, T. (2017). Removal of toxic metals from vanadium-contaminated soils using a washing method: Reagent selection and parameter optimization. *Chemosphere*, 180, 295–301.

Kasai, M., Yamazaki, J., Kikuchi, M., Iwaya, M. & Sawada, S. (2008). Concentration of vanadium in soil water and its effect on growth and metabolism of rye and wheat plants. *Commun Soil Sci Plant Anal*, 30, 971–982.

Kumpiene, J., Antelo, J., Brännvall, E., Carabante, I., Ek, K., Komárek, M., Söderberg, C. & Wårell, L. (2019). In situ chemical stabilization of trace element-contaminated soil—Field demonstrations and barriers to transition from laboratory to the field—A review. *Appl Geochem*, 100, 335–351.

Larsson, M. A., D'amato, M., Cubadda, F., Raggi, A., Öborn, I., Kleja, D. B. & Gustafsson, J. P. (2015). Long-term fate and transformations of vanadium in a pine forest soil with added converter lime. *Geoderma*, 259–260, 271–278.

Lee, H., Noh, H. J., Yoon, J. K., Lim, J., Lim, G.-H., Kim, H. & Kim, J. (2018). Evaluation of the concentration distribution and the contamination influences for Beryllium, Cobalt, thallium and vanadium in soil around the contaminated sources. *J Soil Groundw. Environ*, 23, 48–59.

Li, Y., Zhang, B., Liu, Z., Wang, S., Yao, J. & Borthwick, A. G. L. (2020). Vanadium contamination and associated health risk of farmland soil near smelters throughout China. *Environ Pollut*, 263, 114540.

Liao, Y. & Yang, J. (2019). Remediation of vanadium contaminated soil by nano-hydroxyapatite. *J Soils Sediments*, 20, 1534–1544.

Lin, H., Liu, J., Dong, Y. & He, Y. (2019). The effect of substrates on the removal of low-level vanadium, chromium and cadmium from polluted river water by ecological floating beds. *Ecotoxicol Environ Saf*, 169, 856–862.

Long, Z., Zhu, H., Bing, H., Tian, X., Wang, Z., Wang, X. & Wu, Y. (2021). Contamination, sources and health risk of heavy metals in soil and dust from different functional areas in an industrial city of Panzhihua City, Southwest China. *J Hazard Mater*, 420, 126638.

Ma, F., Peng, C., Hou, D., Wu, B., Zhang, Q., Li, F. & Gu, Q. (2015). Citric acid facilitated thermal treatment: An innovative method for the remediation of mercury contaminated soil. *J Hazard Mater*, 300, 546–552.

Mamedov, A. I., Huang, C. H., Aliev, F. A. & Levy, G. J. (2017). Aggregate stability and water retention near saturation characteristics as affected by soil texture, aggregate size and polyacrylamide application. *Land Degrad Dev*, 28, 543–552.

Mamedov, A. I., Shainberg, I., Wagner, L. E., Warrington, D. N. & Levy, G. J. (2009). Infiltration and erosion in soils treated with dry PAM, of two molecular weights, and phosphogypsum. *Soil Res*, 47, 788–795.

Marques, A. P. G. C., Rangel, A. O. S. S. & Castro, P. M. L. (2009). Remediation of heavy metal contaminated soils: Phytoremediation as a potentially promising clean-up technology. *Crit Rev Environ Sci Technol*, 39, 622–654.

Oladoye, P. O., Olowe, O. M. & Asemoloye, M. D. (2021). Phytoremediation technology and food security impacts of heavy metal contaminated soils: A review of literature. *Chemosphere*, 132555.

Panichev, N., Mandiwana, K., Moema, D., Molatlhegi, R. & Ngobeni, P. (2006). Distribution of vanadium(V) species between soil and plants in the vicinity of vanadium mine. *J Hazard Mater*, 137, 649–653.

Qian, Y., Gallagher, F. J., Feng, H., Wu, M. & Zhu, Q. (2014). Vanadium uptake and translocation in dominant plant species on an urban coastal brownfield site. *Sci Total Environ*, 476–477, 696–704.

Ray, R., Dutta, B., Mandal, S. K., González, A. G., Pokrovsky, O. S. & Jana, T. K. (2020). Bioaccumulation of vanadium (V), niobium (Nb) and tantalum (Ta) in diverse mangroves of the Indian Sundarbans. *Plant and Soil*, 448, 553–564.

Reijonen, I., Metzler, M. & Hartikainen, H. (2016). Impact of soil pH and organic matter on the chemical bioavailability of vanadium species: The underlying basis for risk assessment. *Environ Pollut*, 210, 371–379.

Rivas-Castillo, A., Orona-Tamayo, D., Gómez-Ramírez, M. & Rojas-Avelizapa, N. G. (2017). Diverse molecular resistance mechanisms of Bacillus megaterium during metal removal present in a spent catalyst. *Biotechnol. Bioprocess Eng*, 22, 296–307.

Shaheen, S. M., Alessi, D. S., Tack, F. M. G., Ok, Y. S., Kim, K. H., Gustafsson, J. P., Sparks, D. L. & Rinklebe, J. (2019). Redox chemistry of vanadium in soils and sediments: Interactions with colloidal materials, mobilization, speciation, and relevant environmental implications- A review. *Adv Colloid Interface Sci*, 265, 1–13.

Shaheen, S. M. & Rinklebe, J. (2018). Vanadium in thirteen different soil profiles originating from Germany and Egypt: Geochemical fractionation and potential mobilization. *Appl Geochem*, 88, 288–301.

Shaheen, S. M., Rinklebe, J., Frohne, T., White, J. R. & Delaune, R. D. (2016). Redox effects on release kinetics of arsenic, cadmium, cobalt, and vanadium in Wax Lake Deltaic freshwater marsh soils. *Chemosphere*, 150, 740–748.

Shi, C., Cui, Y., Lu, J. & Zhang, B. (2020). Sulfur-based autotrophic biosystem for efficient vanadium (V) and chromium (VI) reductions in groundwater. *Chem Eng J*, 395, 124972.

Sikka, R. & Kansal, B. D. (1995). Effect of fly-ash application on yield and nutrient composition of rice, wheat and on pH and available nutrient status of soils. *Bioresour Technol*, 51, 199–203.

Teng, Y., Yang, J., Sun, Z., Wang, J., Zuo, R. & Zheng, J. (2011). Environmental vanadium distribution, mobility and bioaccumulation in different land-use districts in Panzhihua Region, SW China. *Environ Monit Assess*, 176, 605–620.

Tian, L. Y., Yang, J. Y., Alewell, C. & Huang, J. H. (2014). Speciation of vanadium in Chinese cabbage (Brassica rapa L.) and soils in response to different levels of vanadium in soils and cabbage growth. *Chemosphere*, 111, 89–95.

Tian, L. Y., Yang, J. Y. & Huang, J. H. (2015). Uptake and speciation of vanadium in the rhizosphere soils of rape (Brassica juncea L.). *Environ Sci Pollut Res Int*, 22, 9215–9223.

Udovic, M. & Lestan, D. (2012). EDTA and HCl leaching of calcareous and acidic soils polluted with potentially toxic metals: Remediation efficiency and soil impact. *Chemosphere*, 88, 718–724.

Van Marwijk, J., Opperman, D. J., Piater, L. A. & Van Heerden, E. (2009). Reduction of vanadium(V) by Enterobacter cloacae EV-SA01 isolated from a South African deep gold mine. *Biotechnol Lett*, 31, 845–849.

Vithanage, M., Dabrowska, B. B., Mukherjee, A. B., Sandhi, A. & Bhattacharya, P. (2011). Arsenic uptake by plants and possible phytoremediation applications: A brief overview. *Environ. Chem. Lett*, 10, 217–224.

Vithanage, M., Herath, I., Joseph, S., Bundschuh, J., Bolan, N., Ok, Y. S., Kirkham, M. B. & Rinklebe, J. (2017). Interaction of arsenic with biochar in soil and water: A critical review. *Carbon*, 113, 219–230.

Vithanage, M., Mayakaduwa, S. S., Herath, I., Ok, Y. S. & Mohan, D. (2016). Kinetics, thermodynamics and mechanistic studies of carbofuran removal using biochars from tea waste and rice husks. *Chemosphere*, 150, 781–789.

Vithanage, M., Rajapaksha, A. U., Zhang, M., Thiele-Bruhn, S., Lee, S. S. & Ok, Y. S. (2015). Acid-activated biochar increased sulfamethazine retention in soils. *Environ Sci Pollut Res Int*, 22, 2175–2186.

Vocciante, M., Caretta, A., Bua, L., Bagatin, R. & Ferro, S. (2016). Enhancements in ElectroKinetic remediation technology: Environmental assessment in comparison with other configurations and consolidated solutions. *Chem Eng J*, 289, 123–134.

Wang, J. F., Jiang, J., Li, D., Li, T., Li, K. & Tian, S. (2015). Removal of Pb and Zn from contaminated soil by different washing methods: The influence of reagents and ultrasound. *Environ Sci Pollut Res Int*, 22, 20084–20091.

Wang, J. F. & Liu, Z. (1999). Effect of vanadium on the growth of soybean seedlings. *Plant and Soil*, 216, 47–51.

Wang, S., Zhang, B., Li, T., Li, Z. & Fu, J. (2020). Soil vanadium(V)-reducing related bacteria drive community response to vanadium pollution from a smelting plant over multiple gradients. *Environ Int*, 138, 105630.

Wu, C. Y., Asano, M., Hashimoto, Y., Rinklebe, J., Shaheen, S. M., Wang, S. L. & Hseu, Z. Y. (2020). Evaluating vanadium bioavailability to cabbage in rural soils using geochemical and micro-spectroscopic techniques. *Environ Pollut*, 258, 113699.

Wu, Z.-Z., Yang, J.-Y. & Zhang, Y.-X. (2021). Phytoremediation-a sustainable remedial method for soil contaminated by vanadium. *IOP Conf Series: Earth and Env Sci*, 821, 012001.

Wu, Z. Z., Yang, J. Y., Zhang, Y. X., Wang, C. Q., Guo, S. S. & Yu, Y. Q. (2021). Growth responses, accumulation, translocation and distribution of vanadium in tobacco and its potential in phytoremediation. *Ecotoxicol Environ Saf*, 207, 111297.

Yang, J., Gao, X., Li, J., Zuo, R., Wang, J. & Song, L. (2021). The distribution and speciation characteristics of vanadium in typical cultivated soils. *Int J Environ Anal Chem*, 1–14.

Yang, J., Gao, X., Li, J., Zuo, R., Wang, J., Song, L. & Wang, G. (2020). The stabilization process in the remediation of vanadium-contaminated soil by attapulgite, zeolite and hydroxyapatite. *Ecol Eng*, 156, 105975.

Yang, J., Teng, Y., Wang, J. & Li, J. (2011). Vanadium uptake by alfalfa grown in V-Cd-contaminated soil by pot experiment. *Biol Trace Elem Res*, 142, 787–795.

Yang, J., Wang, M., Jia, Y., Gou, M. & Zeyer, J. (2017). Toxicity of vanadium in soil on soybean at different growth stages. *Environ Pollut*, 231, 48–58.

Yang, K., Zhu, L. & Xing, B. (2006). Enhanced soil washing of phenanthrene by mixed solutions of TX100 and SDBS. *Environ Sci Technol*, 40, 4274–4280.

Yu, Y.-Q., Li, J.-X., Liao, Y.-L. & Yang, J.-Y. (2020). Effectiveness, stabilization, and potential feasible analysis of a biochar material on simultaneous remediation and quality improvement of vanadium contaminated soil. *J Clean Prod*, 277, 123506.

Zeng, Q. R., Sauve, S., Allen, H. E. & Hendershot, W. H. (2005). Recycling EDTA solutions used to remediate metal-polluted soils. *Environ Pollut*, 133, 225–231.

Zhang, B., Cheng, Y., Shi, J., Xing, X., Zhu, Y., Xu, N., Xia, J. & Borthwick, A. G. L. (2019). Insights into interactions between vanadium (V) bio-reduction and pentachlorophenol dechlorination in synthetic groundwater. *Chem Eng J*, 375, 121965.

Zhang, B., Feng, C., Ni, J., Zhang, J. & Huang, W. (2012). Simultaneous reduction of vanadium (V) and chromium (VI) with enhanced energy recovery based on microbial fuel cell technology. *J Power Sources*, 204, 34–39.

Zhang, B., Hao, L., Tian, C., Yuan, S., Feng, C., Ni, J. & Borthwick, A. G. (2015). Microbial reduction and precipitation of vanadium (V) in groundwater by immobilized mixed anaerobic culture. *Bioresour Technol*, 192, 410–417.

Zhang, B., Qiu, R., Lu, L., Chen, X., He, C., Lu, J. & Ren, Z. J. (2018). Autotrophic Vanadium(V) Bioreduction in Groundwater by Elemental Sulfur and Zerovalent Iron. *Environ Sci Technol*, 52, 7434–7442.

Zhang, B., Wang, S., Diao, M., Fu, J., Xie, M., Shi, J., Liu, Z., Jiang, Y., Cao, X. & Borthwick, A. G. L. (2019). Microbial Community Responses to Vanadium Distributions in Mining Geological Environments and Bioremediation Assessment. *J Geophys Res Biogeosci*, 124, 601–615.

Zhang, H., Zhang, B., Wang, S., Chen, J., Jiang, B. & Xing, Y. (2020). Spatiotemporal vanadium distribution in soils with microbial community dynamics at vanadium smelting site. *Environ Pollut*, 265, 114782.

Zhang, J., Dong, H., Zhao, L., Mccarrick, R. & Agrawal, A. (2014). Microbial reduction and precipitation of vanadium by mesophilic and thermophilic methanogens. *Chem Geol*, 370, 29–39.

Zhang, T., Liu, J. M., Huang, X. F., Xia, B., Su, C. Y., Luo, G. F., Xu, Y. W., Wu, Y. X., Mao, Z. W. & Qiu, R. L. (2013). Chelant extraction of heavy metals from contaminated soils using new selective EDTA derivatives. *J Hazard Mater*, 262, 464–471.

Zhang, W., Jiang, J., Li, K., Li, T., Li, D. A. & Wang, J. (2018). Amendment of vanadium-contaminated soil with soil conditioners: A study based on pot experiments with canola plants (Brassica campestris L.). *Int J Phytoremediation*, 20, 454–461.

Zou, Q., Li, D., Jiang, J., Aihemaiti, A., Gao, Y., Liu, N. & Liu, J. (2019a). Geochemical simulation of the stabilization process of vanadium-contaminated soil remediated with calcium oxide and ferrous sulfate. *Ecotoxicol Environ Saf*, 174, 498–505.

Zou, Q., Xiang, H., Jiang, J., Li, D., Aihemaiti, A., Yan, F. & Liu, N. (2019b). Vanadium and chromium-contaminated soil remediation using VFAs derived from food waste as soil washing agents: A case study. *J Environ Manage*, 232, 895–901.

8

Potential of Biochar to Immobilize Vanadium in Contaminated Soils

Ali El-Naggar[1,2,3], Ahmed Mosa[4], Avanthi D. Igalavithana[5], Xiao Yang[6], Ahmed H. El-Naggar[7,2], Sabry Shaheen[8,9], Scott X. Chang[3] and Jörg Rinklebe[8]

[1] *State Key Laboratory of Subtropical Silviculture, Zhejiang Agriculture and Forestry University, Hangzhou, China*

[2] *Department of Soil Sciences, Faculty of Agriculture, Ain Shams University, Cairo, Egypt*

[3] *Department of Renewable Resources, University of Alberta, Edmonton, Alberta, Canada*

[4] *Soils Department, Faculty of Agriculture, Mansoura University, Mansoura, Egypt*

[5] *Department of Soil Science, Faculty of Agriculture, University of Peradeniya, Peradeniya, Sri Lanka*

[6] *Key Laboratory of Land Surface Pattern and Simulation, Institute of Geographic Sciences and Natural Resources Research, Chinese Academy of Sciences, Beijing, China*

[7] *Sustainable Natural Resources Management Section, International Centre for Biosaline Agriculture (ICBA), Dubai, United Arab Emirates*

[8] *University of Wuppertal, School of Architecture and Civil Engineering, Institute of Foundation Engineering, Water- and Waste-Management, Laboratory of Soil- and Groundwater-Management, Wuppertal, Germany*

[9] *University of Kafrelsheikh, Faculty of Agriculture, Department of Soil and Water Sciences, Kafr El-Sheikh, Egypt*

CONTENTS

8.1 Introduction 158
8.2 Biochar as an Emergent Candidate in Soil Remediation 159
8.3 Potential of Biochar to (Im)Mobilize V in Soil 160
8.3.1 Biochar Effects on V Immobilization/Mobilization in Soil 160
8.3.2 Factors Governing Biochar Effects on V (Im)Mobilization in Soil 165
8.3.2.1 Factors Related to Soil Properties 165
8.3.2.2 Factors Related to Biochar Properties 168

DOI: 10.1201/9781003173274-8

8.4 Interactions between Biochar and V: Mechanisms for V Remediation....172
8.5 Designer/Modified Biochar for Immobilization of V in Soil....173
8.6 Risk Management of Biochar Application to V Contaminated Soil....175
8.7 Conclusions and Future Research Outlook....177
References....179

8.1 Introduction

Vanadium (V) is the 22nd most abundant element in the earth's crust, with an average concentration of about 110 mg kg^{-1} (Zou et al., 2019). It originates from various pedogenic (e.g., volcanic eruptions and rock weathering) and anthropogenic (e.g., mining, smelting, spent catalysts, combustion of fossil fuel and waste disposal) sources (Aihemaiti, Gao, Liu et al., 2020; Aihemaiti, Gao, Meng e al., 2020; Imtiaz et al., 2015). Moreover, V can accumulate in soil via the excessive use of V-containing fertilizers and sludge generated from several industrial activities and applications (Liu et al., 2019; Xiao et al., 2015). At trace amounts, V is an essential element for human and animal health due to its essential functions in the metabolism of lipids, carbohydrates, and energy production (Rehder, 2016, 2018). However, high concentrations of V in soil could cause contamination, leading to adverse consequences on plants, soil and groundwater and threatening the health of humans and animals (Aihemaiti, Gao, Meng et al., 2020; Li, Zhang et al., 2020), which is becoming a global environmental concern. Therefore, sustainable management strategies for remediation of V contaminated soils are urgently needed to achieve soil security and food safety.

Soil V is present in different fractions such as soluble, exchangeable, bound to carbonate, bound to Mn and Fe (hydr)oxides, complexed with organic matter, bound to sulfides and in non-mobile residues (Shaheen et al., 2019; see Page 73 in this book). The soluble form of V can become immobilized in the soil by its adsorption onto soil particle surfaces. There are several mechanisms underlying the sorption of V(V) onto natural soil colloids, including electrostatic attraction by charges on the outer- and inner-sphere surfaces, complexation with active functional groups, the attraction of valence forces by electron sharing or exchanging, diffusion into inner-sphere surfaces, precipitation and substituting of Al in the octahedral sites of clay minerals (Luo et al., 2017; see Page 50 in this book). The toxicity risk exists when elevated V concentrations are present in the mobile or potentially mobile fractions in soils (Cao et al., 2017), while the risk is low when V is in the non-mobile forms. Therefore, the immobilization of mobile toxic soil elements, including V, has been recommended as a cost-effective strategy for the remediation of contaminated soils. Biochar as a soil amendment has garnered significant attention among different immobilization techniques that can efficiently remediate contaminated soils (Arabi et al., 2021; Chen et al., 2022; El-Naggar, Ahmed et al., 2021; El-Naggar, Chang et al., 2021).

This chapter discusses the potential of biochar as an emergent candidate for the immobilization of V in soils and the responsible mechanisms of interaction between biochar and V in the soil. The promising production types and conditions for producing biochars with a high adsorption efficiency for V in the soil and the management of risk associated with biochar application are also discussed.

8.2 Biochar as an Emergent Candidate in Soil Remediation

A variety of remediation approaches have been developed to combat soil contamination with V. For instance, plants such as bunny cactus (*Opuntia microdasys*), green bristlegrass (Setaria viridis), chickpea (*Cicer arietinum*) and green mustard (*Brassica juncea*) have a high tolerance to and can accumulate high concentrations of V in their tissues, mainly in roots (Aihemaiti et al., 2017; Aihemaiti, Gao, Meng et al., 2020). Hence, high V accumulating plants can be used to remediate V-contaminated soils (Aihemaiti, Gao, Meng et al., 2020). However, phytoremediation may not be economically efficient, especially for large-scale remediation (Palansooriya et al., 2020). Other remediation approaches such as soil washing, soil replacement and electrokinetic extraction were usually expensive and destructive to soil structure and fertility (Karer et al., 2015). Among different mitigation approaches for soil V contamination, biochar application has been suggested as an eco-friendly and cost-effective remediation tool that can be employed to immobilize V in contaminated soils by constricting the mobility and plant uptake of V (Yu et al., 2020).

Biochar is a C-rich soil amendment obtained from thermochemical conversion of biomass under O-limited conditions. Biochar can be produced via different production techniques such as slow pyrolysis, fast pyrolysis, gasification and hydrothermal carbonization. However, the slow pyrolysis method (with a low heating rate and a long residence time) is a widely used technique, owing to the higher biochar yield that can be achieved (~35%). The pyrolysis process also produces other products, that is, bio-oil and syngas (~65%). Production conditions and feedstock properties considerably influence the primary properties of biochars, such as their amorphous C content, porous structure, abundance of surface functional groups, and volatile matter, resident matter and ash contents. Changes in those primary properties lead to significant alterations in biochar's secondary properties such as pH, cation exchange capacity (CEC), surface area, bulk density, stability and surface charge. In general, biochar contains disparate ratios of aliphatic/aromatic organic substances and mainly comprises C, O, H, N and other inorganic elements that are essential for plant growth. Biochar has, typically, a high pH with a large inner surface area, and high porosity, adsorption capacity and CEC. Based on the complex and highly variable nature of biochar, different biochar types have different properties, and hence the reported wide range of biochar functionalities enables its utilization in various applications.

Biochar has received growing interest for its environmental applications, including mitigation of climate change, production of clean energy, sustainable management of agricultural and animal wastes and remediation of contaminated soils and waters (El-Naggar, Rajapaksha et al., 2018; Elkhlifi et al., 2021; Man et al., 2021; Zhou et al., 2021). When applied and aged in soil, biochar can enhance aeration, water holding capacity, organic C content, microbial communities and activities and nutrient availability in soil, resulting in enhanced soil quality, especially in low fertility soils (El-Naggar, Lee et al., 2019).

Biochar has also been proven as an efficient tool for remediation of soils contaminated with several toxic elements. The potential for biochar to adsorb toxic elements in soil depends on the physicochemical characteristics and the surface functionality of biochar (El-Naggar, Lee et al., 2019; Shaheen et al., 2018). Immobilization of toxic elements in biochar-treated soils also depends on the behavior of the toxic element, other co-existing toxic elements, and soil properties (Shaheen et al., 2018).

8.3 Potential of Biochar to (Im)Mobilize V in Soil

The use of biochar for the remediation of V contamination has been reported (Fan et al., 2020; Ghanim, Murnane et al., 2020; Meng et al., 2018); however, those studies focused on V contamination in wastewater and groundwater, not in soil systems. It is worth noting that, compared to other toxic elements, only a few studies have documented the potential for biochar to immobilize/mobilize V in contaminated soils (Figure 8.1). However, the available preliminary results provide encouraging insights for researchers to further study biochar as a remediation tool for V contamination in the soil, as we discuss in the following sections.

8.3.1 Biochar Effects on V Immobilization/Mobilization in Soil

The effectiveness of biochar on remediation of V contaminated soil was investigated by Yu et al. (2020), using biochars produced from three feedstocks (i.e., corn (*Zea mays* L.) cob, rice (*Oryza sativa* L.) straw and wheat (*Triticum*

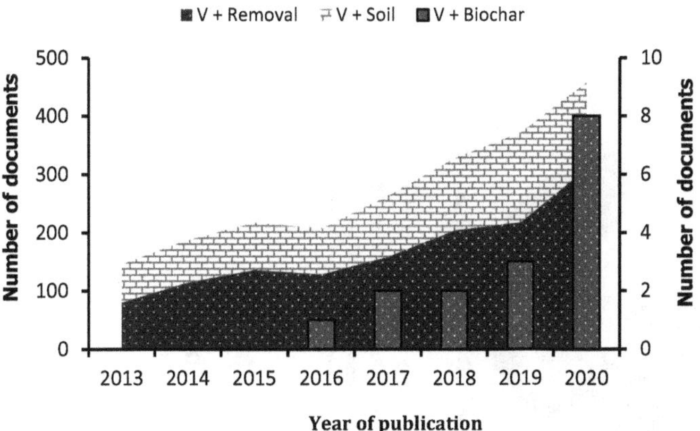

FIGURE 8.1
Number of documents extracted from SCOPUS for publications between 2013–2020 according to keywords: (i) Vanadium + removal, (ii) vanadium + soil and (iii) vanadium + biochar.

aestivum L.) straw) at two pyrolysis temperatures (i.e., 350 and 650°C) for 2 hours. The produced biochar was sieved using different sieve sizes, that is, 60-mesh (<0.25 mm), 120-mesh (<0.125 mm) and 200-mesh (<0.075 mm) (Figure 8.2). The rice straw biochar produced at 650 °C and sieved with a 200-mesh screen had pH of 10.37, 32% organic C content, 62.7 $m^2\ g^{-1}$ specific surface area, a porous structure, abundant functional groups and low V content (2.7 mg kg^{-1}). Thereafter, they tested the efficiency of biochar in mitigating V contamination in soils spiked with V at different concentrations (i.e., 0, 150, 300, 600, and 1,000 mg V kg^{-1} soil) and aged for six weeks prior to the application of biochar. The application of the previously mentioned rice straw biochar led to significant decreases in bioavailable V (soil V extracted by ethylenediaminetetraacetic acid) in different soils over the incubation experiment (ten weeks), and this impact was more pronounced, with the increase in addition rate of biochar from 1 to 3% and with the higher initial content of V in soil from 150 to 1,000 mg V kg^{-1} soil (Figure 8.3). The maximum amount of immobilized V was up to 225.6 mg V kg^{-1} soil in the highly contaminated soil (1,000 mg V kg^{-1} soil) treated with a 3% level of the rice straw biochar. This capacity to immobilize V was attributed to the role of specific surface area, microporosity and organic C content of biochar in the adsorption of V onto biochar surfaces. However, in the same study, the water-soluble V concentrations in biochar-treated soils had reversed trends with the results of bioavailable V concentrations (Figure 8.4). The increased water-soluble V was attributed to increased soil pH with the biochar addition, leading to higher V mobility into the soil solution. However, this increase in water-soluble V after biochar addition gradually decreased over time to reach its minimum level after ten weeks of incubation. The reverse effects of biochar

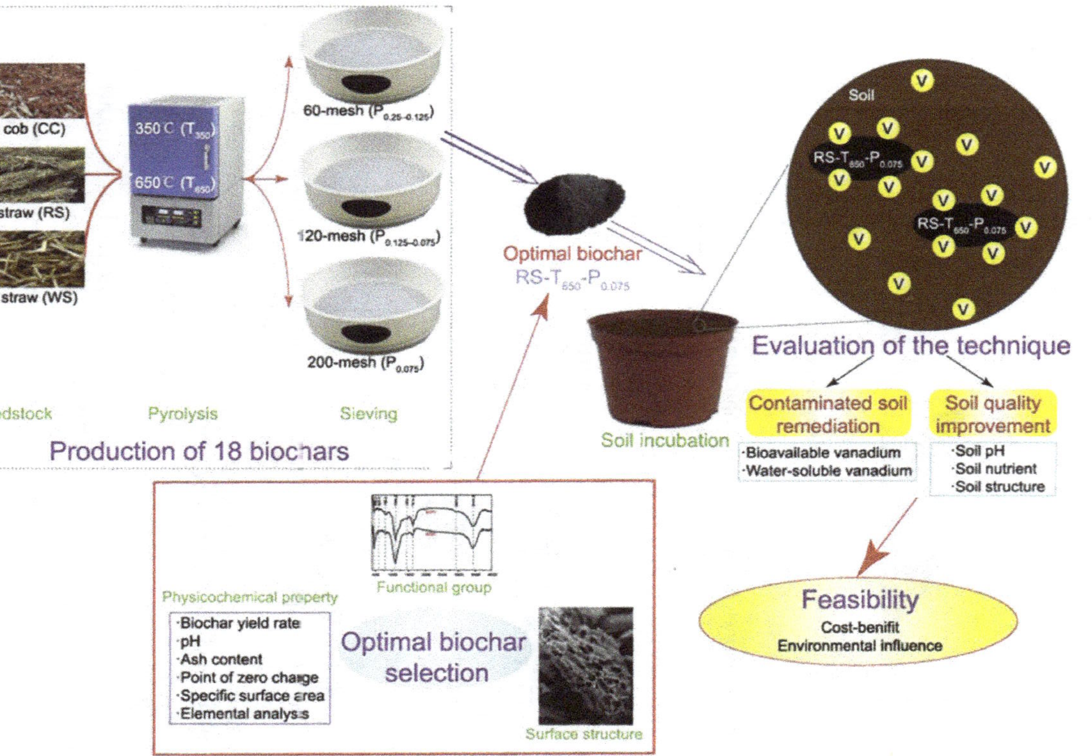

FIGURE 8.2
Production of biochar under different conditions: feedstocks (i.e., corn cob, rice straw and wheat straw), pyrolysis temperatures (i.e., 350 and 650°C), and sieve sizes (i.e., 60-mesh (<0.25 mm), 120-mesh (<0.125 mm) and 200-mesh (<0.075 mm)); and its application for remediation of V contaminated soil.

Source: Reproduced from Yu et al. (2020), with permission from the publisher.

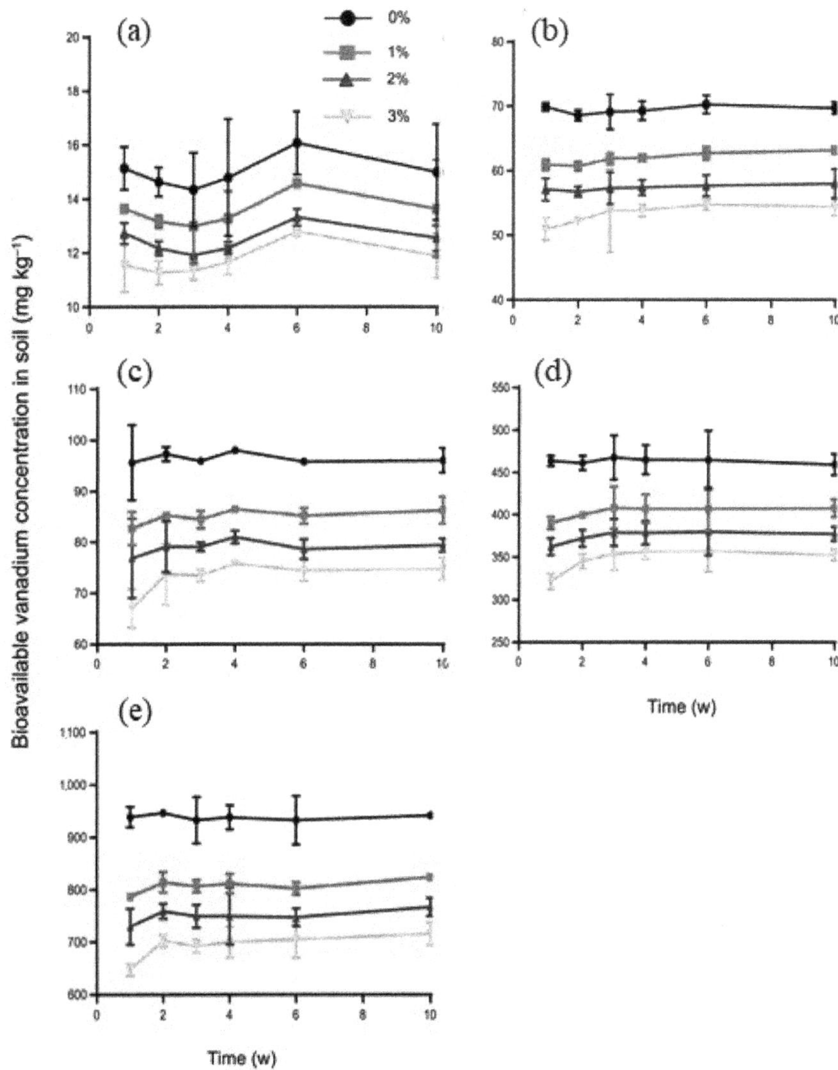

FIGURE 8.3
Bioavailable V concentrations in soils treated with rice straw biochar at 0, 1, 2 and 3% and incubated for ten weeks.
Source: Reproduced from Yu et al. (2020), with permission from the publisher.

on reducing bioavailable V and increasing water-soluble V were explained by the interactions between biochar-induced soluble organic matter and the V in soil, highlighting biochar's vital role in influencing V mobility and bioavailability. More technically, soluble V can be affected differently by dissolved aliphatic and aromatic C components in biochar-treated soils.

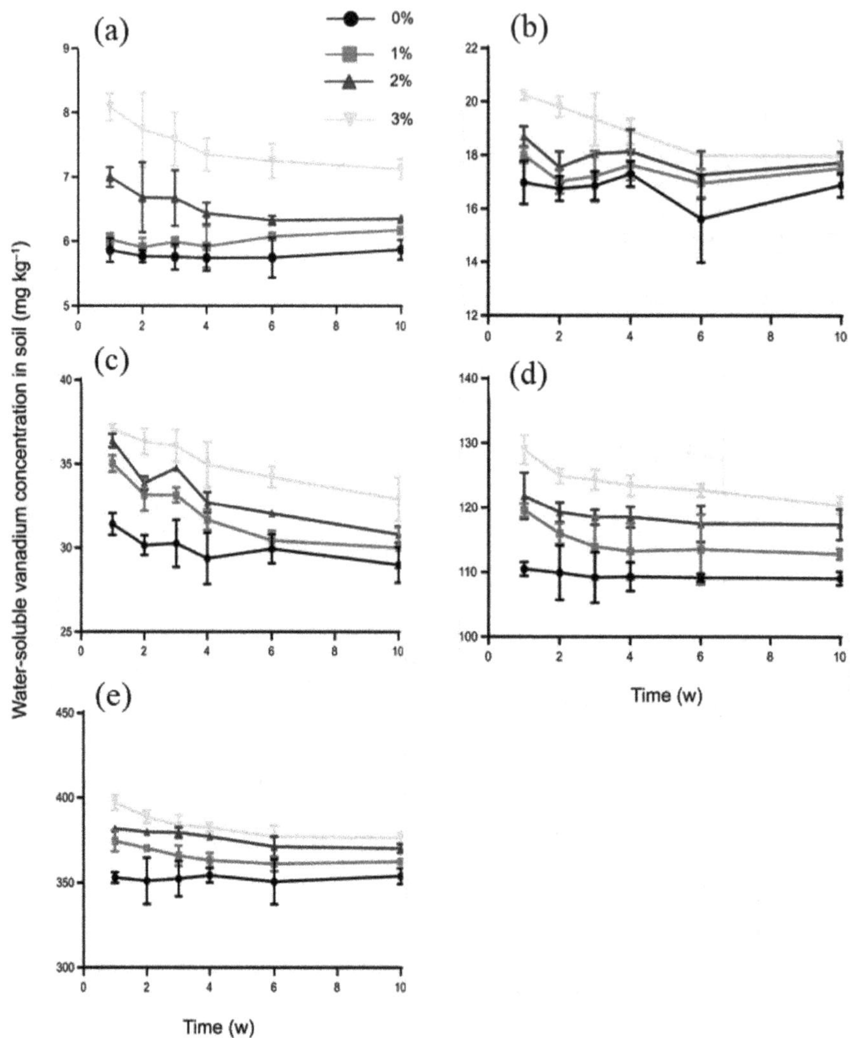

FIGURE 8.4
Water-soluble V concentrations in soil treated with rice straw biochar at 0, 1, 2 and 3% and incubated for ten weeks.
Source: Reproduced from Yu et al. (2020), with permission from the publisher.

The release of dissolved aliphatic organic C in soil stimulates the release of associated V, while the increased aromatic organic C components in soil facilitate the formation of complexes with V, reducing its bioavailability in soil (El-Naggar, Shaheen et al., 2021).

The biochar surface functional groups and organic/inorganic components affect the solubility and phytoavailability of V in soil (El-Naggar, Shaheen et al., 2021). The study of El-Naggar, Shaheen et al. (2021) aimed to test the impact of biochar derived from rice straw and wood residues with different surface functionality and porosity properties applied at three doses (0, 2.5, and 5%) on V immobilization, solubility and phytoavailability in a contaminated acidic soil. They found that the addition of wood biochar reduced V solubility by 46% when applied at a rate of 2.5% to the soil (Figure 8.5) and simultaneously decreased the soluble + exchangeable V fraction by 31.9% and decreased V concentrations in both roots and shoots of sorghum (*Sorghum bicolor* L.) and corn, as compared to the control. However, the other biochar types exhibited adverse effects on V remediation as they enhanced V solubility and phytoavailability in soil. The study reported that the superior performance of the highly functionalized wood biochar over the other biochar types was attributed to a higher abundance of O-containing functional groups, higher surface reactivity, acidity and hydrophilicity, which all facilitated the biochar's role in V immobilization.

8.3.2 Factors Governing Biochar Effects on V (Im)Mobilization in Soil

8.3.2.1 Factors Related to Soil Properties

The (im)mobilization and plant uptake of V are governed by several factors such as soil pH, cation exchange capacity, organic matter, Fe and Mn oxides and microbial activity (Aihemaiti, Gao, Meng et al., 2020; Panichev et al., 2006; Shaheen et al., 2019; Shaheen and Rinklebe, 2018). Vanadium exists in the soil in different fractions comprising bioavailable (water-soluble and exchangeable forms), non-residual (organic matter-, sulfide-, carbonate- and Fe/Mn oxides-bound forms) and residual (Shaheen et al., 2019). Vanadium shows multiple oxidations states according to soil physicochemical characteristics: (i) V(V) under most environmental conditions with a tendency to bind with Al/Fe/Ti (hydr)oxides, (ii) V(IV) under low pH values and slightly reducing conditions with a high affinity to form complexes with soil organic matter, and (iii) V(III) as the major octahedral cation in soils located under highly anoxic and sulfidic conditions (Gustafsson, 2019). Vanadium speciation in the soil is highly sensitive to alterations in soil E_H since V(V) is reduced to V(IV) by electron donations from soil constitutions (e.g., organic matter), and V(IV) might be further reduced to V(III) by strong reductants (e.g., sulfides) (Chen et al., 2021). The high pH value causes V(V) species to be dominant and increases the phytotoxicity potential of V in soil (Reijonen et al., 2016). Soil pH and V solubility was positively related (El-Naggar, Shaheen et al., 2021; Shaheen et al., 2019; Yu et al., 2020). For instance, increasing soil pH from 5.1 to 6.5 increased V solubility in soil from 157.8 to 975.5 mg kg^{-1} (Figure 8.6). A long-term (10 years) column leaching experiment was carried out to assess the effect of simulated acid rain on the leaching risk of V from contaminated paddy soil collected from a smelting area (Xiao et al., 2017). Vanadium concentrations in leachates were 49.0, 71.3, and 61.1 μg L^{-1} at pH values of 4.5, 5.6 and 7.0, respectively.

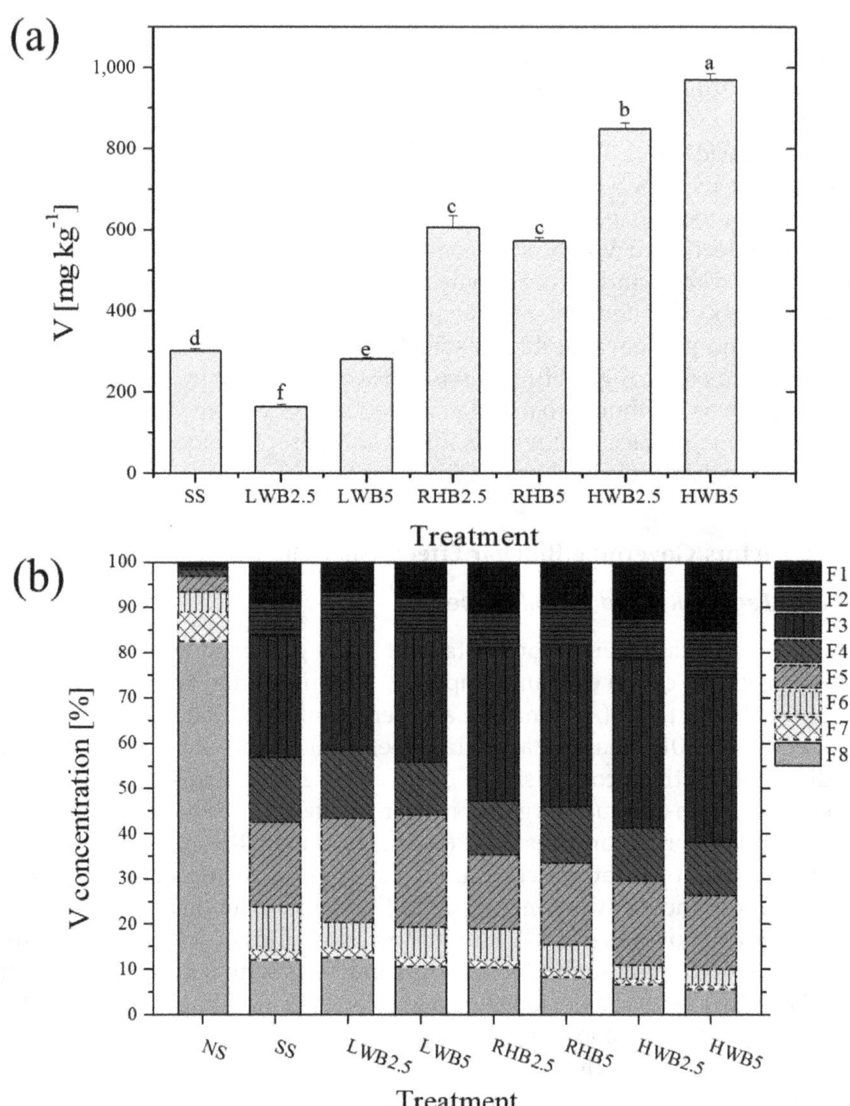

FIGURE 8.5
Water-soluble V (a), and geochemical fractions of V (b) in soils treated with different biochar types (rice hull biochar (RHB), wood biochar (250 °C) and wood biochar (500 °C)), and rates (i.e., 0, 2.5 and 5%): original soil (NS); V-spiked soil (SS); V-spiked soil + LWB (2.5%) (LWB2.5); V-spiked soil + LWB (5%) (LWB5); V-spiked soil + RHB (2.5%) (RHB2.5); V-spiked soil + RHB (5%) (RHB5); V-spiked soil + HWB (2.5%) (HWB2.5); and V-spiked soil + HWB (5%) (HWB5).

F1: soluble + exchangeable, F2: carbonate, F3: Mn oxide, F4: organic matter, F5: sulfide, F6: amorphous Fe oxide, F7: crystalline Fe oxide, and F8: residue.

Source: Reproduced from El-Naggar, Shaheen et al. (2021), with permission from the publisher.

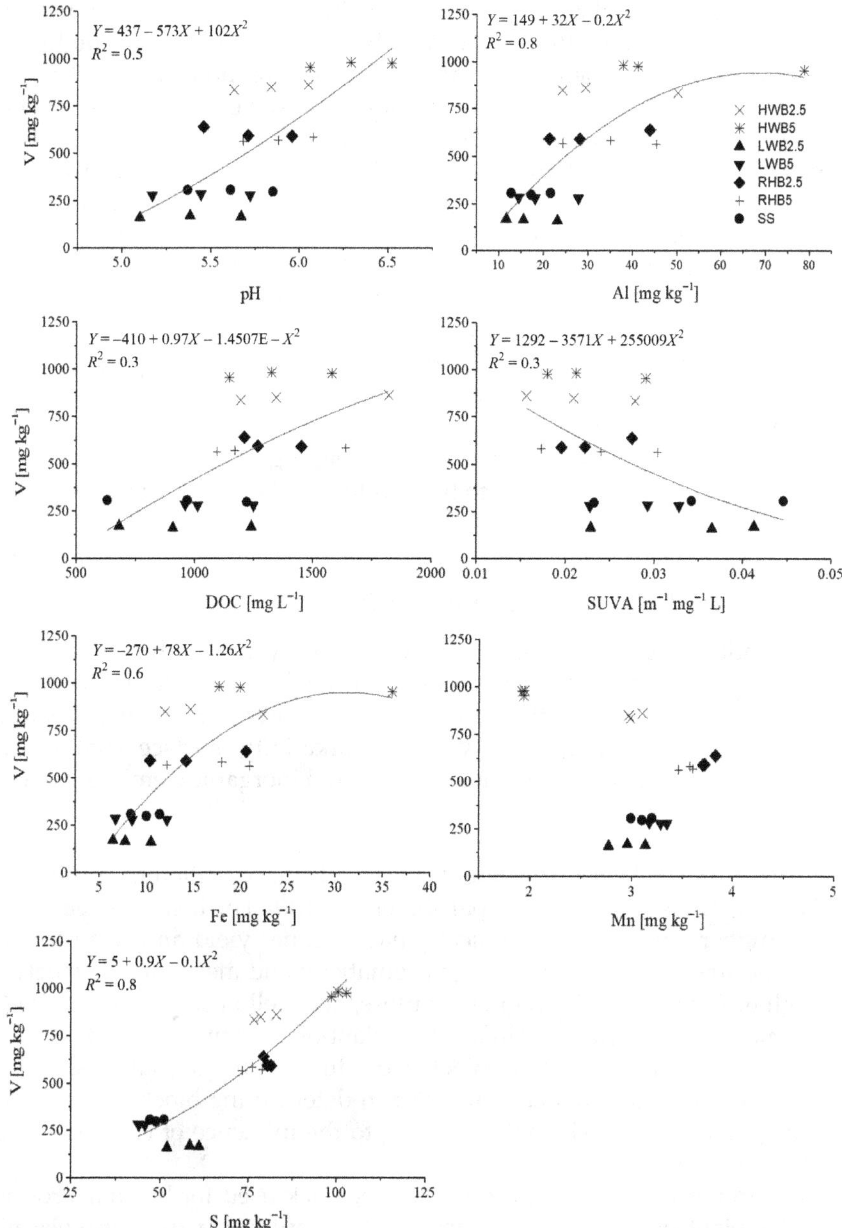

FIGURE 8.6
Relationships between water-soluble V and some soil parameters such as pH, DOC (dissolved organic carbon, SUVA (specific UV absorbance), used as an indicator of aromaticity of C compounds in the soil solution), Fe, Mn, Al and S.

Source: Reproduced from El-Naggar, Shaheen et al. (2021), with permission from the publisher. Copyright 2021 American Chemical Society.

Soil organic matter also plays a major role in controlling the phytotoxicity of V because organic matter acts as a sorbent and reductant for the highly toxic V(V) form (Reijonen et al., 2016). The two components of dissolved organic C (i.e., aliphatic and aromatic) show opposite effects on V solubility in soil (El-Naggar, Shaheen et al., 2021; Yu et al., 2020), as increased dissolved aliphatic components enhanced V solubility, whereas increased dissolved aromatic components decreased V solubility (Figure 8.6). In addition, the V(V) fraction bound to Fe/Mn oxides comprises a significant portion (>30%) of the total V in soils having different constitutions (minerals, organics and organo-minerals) (Agnieszka and Barbara, 2012). El-Naggar, Shaheen et al. (2021) reported positive relationships between the solubility of V and the contents of Fe, Al and S, but not with Mn, in the soil solution (Figure 8.6). Soil biota play a substantial role in controlling the bioavailability of V in soil. In this context, some V-resistant bacterial genera (e.g., *Arthrobacter, Clostridium, Geobacter and Pseudomonas*) might pose a scavenging effect against the phytotoxicity of V by reducing V(V) into the less toxic V(IV) form (Gan et al., 2020; Wang et al., 2020).

8.3.2.2 Factors Related to Biochar Properties

The potential for biochar to immobilize V in soil varies with biochar characteristics. The biochar characteristics are formed mainly via the feedstock type and production condition, affecting the biochar aromaticity/aliphaticity, hydrophilicity/hydrophobicity, acidity/alkalinity, surface charge and area, CEC, surface porosity and morphology and inorganic elements content.

8.3.2.2.1 Feedstock

Biochars produced from different feedstock have different properties (El-Naggar, Lee et al., 2018; Rajapaksha et al., 2019). For instance, feedstock with a higher lignin content typically has a higher yield and fixed C content, large specific surface area, high aromaticity and mesoporous structure in biochar. Feedstock with higher cellulose/hemicellulose content typically poses microporous structures in biochar. Plant-derived biochar tends to have a higher total surface area than biochar produced from animal wastes. The type of feedstock is also a critical factor in determining biochar properties such as pH, CEC and ash content, owing to the influence of co-existed elements in the feedstock.

The V in biochar originates from the feedstock used for biochar production. For plant biomass, V is commonly accumulated in their tissues via atmospheric deposition and/or wastewater irrigation, while food intake for animals is the primary source of V existence in their residues (Qiu et al., 2015). The V concentration in biochar is usually low. We extracted and analyzed data of 52 individual observations collected from different published papers to figure out the effect of feedstock type and pyrolysis temperature on V concentration in biochar (Figure 8.7). The V concentration was higher in biochars derived from manure and sludge compared to woody and herbaceous feedstocks (30.5 versus 3.7 mg kg^{-1}). Algal-based biochar produced via pyrolysis at 450 °C contained < 1 mg kg^{-1} of V. Vanadium concentration in biochar was 6.85 and 6.10 mg kg^{-1} in rabbit manure waste biochar produced via slow pyrolysis at 450 and 600°C, respectively, and 5.26 and 6.06 mg kg^{-1} in hydrochar produced from the same feedstock but via the hydrothermal carbonization method at 190 and 240°C, respectively (Cárdenas-Aguiar et al., 2020). In another study, the V concentration was < 5 mg kg^{-1} in various biochars produced from cotton (*Gossypium* L.) straw, soybean (*Glycine max* L.) straw, potato (*Solanum tuberosum* L.) straw, leaf (*Folium* L.), rice straw, wheat straw, maize straw, nut and wood dust at 300, 450 and 600 °C, while V concentration ranged within 15–35 mg kg^{-1} for swine manure and cow manure biochars and 55–65 mg kg^{-1} for chicken manure biochar (Qiu et al., 2015). In general, the low V concentrations in biochars give biochar another advantage for its use in remediation of V contamination.

8.3.2.2.2 Production Conditions

Biochar production conditions, including pyrolysis temperature, heating rate and residence time, determine the properties of biochar and its potential to immobilize V in soil. Biochar is usually produced via slow pyrolysis at temperatures ranging between 250 and 800°C. Low pyrolysis temperature (e.g., 250–450 °C) is conducive to the formation and preservation of surface functional groups (e.g., -COOH, -C=O, -CO- and -OH) and the macrostructure of the feedstock (Sizmur et al., 2017). At high pyrolysis temperatures (e.g., >450 °C), a significant quantity of O-containing functional groups are lost, C content and aromatic groups are increased and H and O contents are decreased, while the biochar develops porous structure and high specific surface area (Figure 8.8). However, at pyrolysis temperature >700 °C, surface area and functionality can be reduced due to the high carbonization, and the porous structure can be collapsed (Yang et al., 2015). In addition, raising pyrolysis temperature increases V concentration in biochar. Among 52 individual observations at different pyrolysis temperatures, V concentrations in biochar were 9.0, 12.9 and 26.0 mg kg^{-1} at pyrolysis temperatures of <450, 450–600 and >600°C, respectively (Figure 8.7).

The high heating rate also tends to form lower pore volume and less specific surface area (Rodriguez et al., 2020). The electrostatic attraction of V(V) onto biochar surfaces is the initial mechanism toward its remediation, and the surface charge of biochar plays a vital role in this regard. The prevalence of active functional groups on biochars produced at relatively low pyrolysis temperatures (≤500 °C) exhibited a higher electronegativity with a high CEC and low anionic exchangeable capacity (AEC); however, the high pyrolysis temperatures (≥700 °C) produced biochars with low CEC and high AEC values due to the dominance of positively charged surfaces arising from non-hydrolyzable bridging of oxonium groups (Banik et al., 2018). Preservation of O-containing functional groups in biochar produced at low pyrolysis tem-

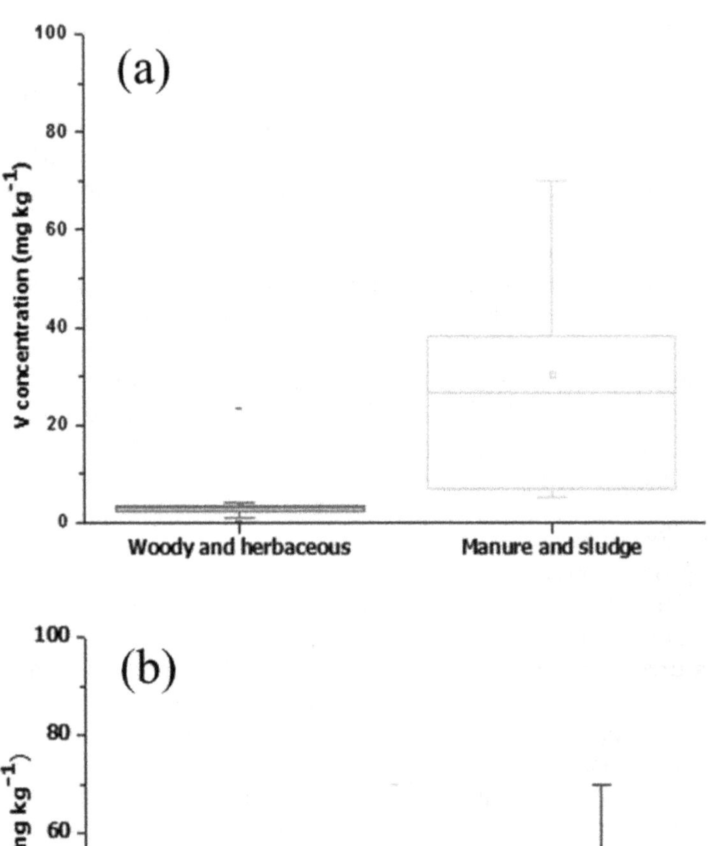

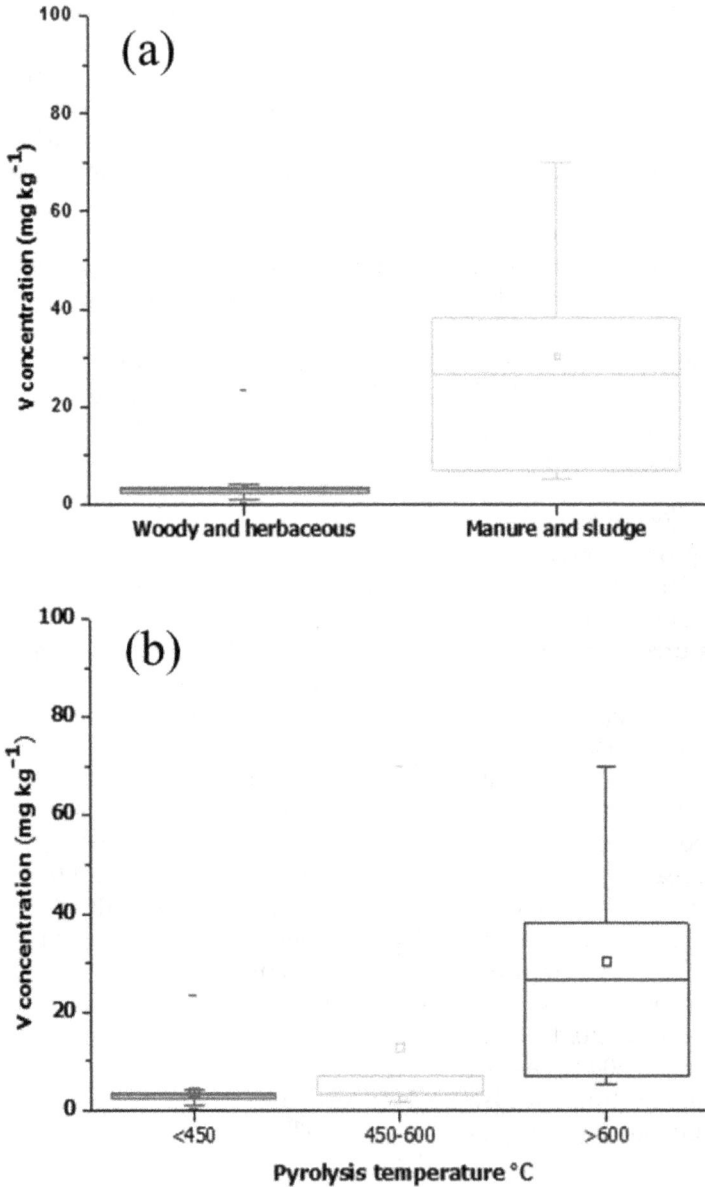

FIGURE 8.7
Effects of feedstock type (a) and pyrolysis temperature (b) on V concentration (mg kg^{-1}) in biochar. Data based on 52 individual observations extracted from five studies (Cárdenas-Aguiar et al., 2020; Meng et al., 2018; Naeem et al., 2007; Qiu et al., 2015; von Gunten et al., 2017). The box chart is represented by the median (centerline), mean (dot), lower and upper quartiles (the lower and upper borders of the box, respectively) and whiskers-error bars (the minimum and maximum observations).

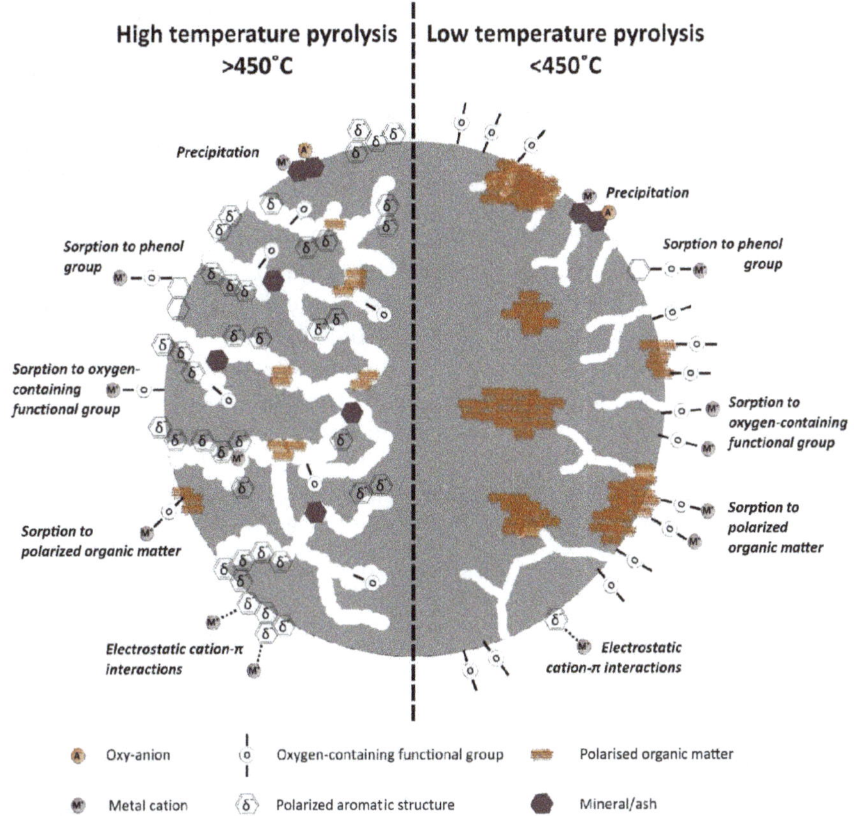

FIGURE 8.8
Effects of pyrolysis temperature on the potential of biochar to adsorb cationic toxic elements on its surfaces.
Source: Reproduced from Sizmur et al. (2017), with permission from the publisher.

peratures is favored for immobilization of V in soil. Surface complexation onto O-containing functional groups is an essential mechanism for V immobilization. El-Naggar, Shaheen et al. (2021) demonstrated that the application of wood biochar produced at 250°C having high active functional groups decreased the phytoavailability of V and its uptake by corn and sorghum. Conversely, the application of biochar with low abundance of active functional groups increased V extractability and its uptake potentials by plants.

8.3.2.2.3 Redox Activity

The potentiality of biochar to serve as a redox material for electron donation has been widely investigated for metals/metalloids remediation (El-Naggar, Shaheen et al., 2018; Liu et al., 2020; Rinklebe et al., 2020; Xu et al., 2019). In this context, biochar's electron-donating/ accepting capacities would determine the transformations between the highly toxic V(V) and the less toxic V(IV) forms. The electron-donating capacity of biochar is feedstock-dependent since the biomasses with higher hemicellulose and/or cellulose contents exhibit higher donating capacity relative to other biomasses with higher lignin content (Zhang et al., 2018). The thermochemical conversion of hemicellulose and cellulose tend to generate effective electron-donating moieties (e.g., alcohols, aldehyde, ketone and acids); however, thermochemical conversion of lignin generates electron acceptors (e.g., carbonyls, lactones and quinonyl) (Brebu et al., 2011; McBeath et al., 2014; Shen and Gu, 2009). The content of dissolved organic matter is also an important component controlling the electron-donating capacity of biochar (Prévoteau et al., 2016). Pyrolysis temperature also plays a cardinal role in affecting the electron-donating capacity of biochar. According to Zhang et al. (2018), the electron-donating capacity of biochar showed higher values at low and intermediate pyrolysis temperatures (200 – 400°C) due to the presence of active functional groups. However, this donating capacity showed higher values at relatively higher pyrolysis temperatures (>650 °C) given the conjugated π-electron system of the aromatic structure.

8.4 Interactions between Biochar and V: Mechanisms for V Remediation

Biochar has shown high potential for immobilizing V in soil, and the governing mechanisms are dependent on the V species in the soil (Figure 8.9). Solution pH is one of the most vital parameters influencing the V adsorption process on biochar because it modifies both the speciation of V and the surface charge of biochar. For instance, Ghanim, Murnane et al. (2020) observed higher V adsorption at pH 4 onto biochar surfaces which had pH_{pzc} of 6.1 and 6.3, which is attributed to that V species remaining as anionic (i.e., $H_2VO_4^-$, $H_2V_2O_7^{2-}$, and HVO_4^{2-}) when pH ≥ 4, and biochar surface is positively charged below the pH_{pzc} while enhancing the electrostatic attraction. However, V adsorption was significantly reduced when solution pH increased (to pH > 6), as biochar surface became negatively charged above their pH_{pzc}.

Other than the electrostatic attraction, pore filling, ion exchange, surface precipitation and complexation are potential adsorption mechanisms that biochar uses to immobilize V (Amin et al., 2019; Meng et al., 2018). It has been observed that increased pore volume enhances V adsorption on biochar surfaces. Complexation with elements on biochar surface is another mechanism of V adsorption onto biochar, and zero-valent iron and zinc are two elements that have a high affinity to complexation with V (Bello et al., 2019; Meng et al., 2018). The surface functional groups of biochar also play an essential role in the adsorption of V. The functional groups of C-H, O-H, N-H, and C=O

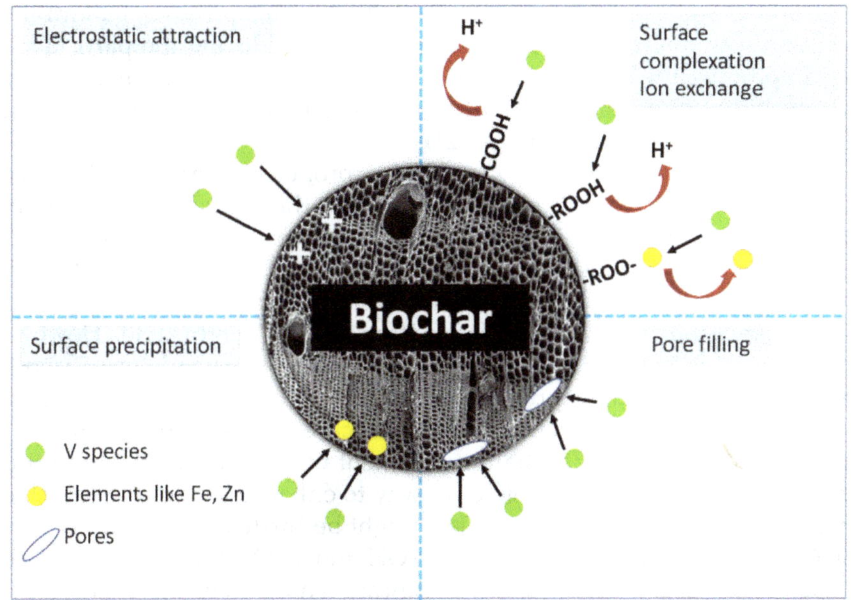

FIGURE 8.9
Mechanisms for the adsorption of vanadium by biochars.

on biochar surface strongly interact with V via electrostatic attraction, ion-exchange, and surface complexation (Meng et al., 2018). Surface precipitation of V on biochar has been reported to associate with some elements like zinc and iron (Fan et al., 2020; Meng et al., 2018).

8.5 Designer/Modified Biochar for Immobilization of V in Soil

Although pristine biochar is capable of adsorbing V onto its surfaces, the amount of V adsorbed remains relatively low in most cases, and further enhancement is required to increase the biochar adsorption capacity. Biochar properties can be modified with simple methods to improve their performance in V adsorption. Functionalizing designer biochars with higher electron-donating capacity increases the effectiveness of relevant redox reactions of V(V) reduction. The modified Hummer's method has a high efficacy for maximizing the electron-donating capacity of engineered relative to the pristine biochar (0.590 Vs. 0.244 mmol $e^- g^{-1}$ biochar) (Chacón et al., 2020). The size of biochar particles affects the formation of aggregates between soil and biochar and the potentiality of biochar to serve as a battery in biogeochemical electron transfer cycles (Li, Shao et al., 2020). For this purpose, ball milling has been introduced as a classical technique for reducing the particle size and improving the functionality of engineered biochars (e.g., maximizing the specific surface area and originating new active edge sites (Amusat et al., 2021; Kumar et al., 2020; Xu et al., 2021).

Surface area and pore volume are biochar properties that affect V adsorption, as discussed earlier. More importantly, these two biochar properties can be enhanced by using physical or chemical modification methods (Azargohar and Dalai, 2008; Ghanim, O'Dwyer et al., 2020; Shim et al., 2015). For instance, activating biochar surface via steam or acid/alkali treatments may enhance the capacity for surface adsorption of V (Figure 8.10). However, the potential of steam activation is likely weaker than the latter methods. Despite steam activation promoting the surface area and porosity, it can reduce the abundance of O-containing functional groups on biochar surfaces (Sizmur et al., 2017). The improvement of surface functional groups has been identified as a very effective way to enhance V adsorption by biochar. The redox properties of biochar might be further enhanced by grafting O-containing functional groups (C-OH and C=O) onto biochar surfaces (Chacón et al., 2020). The chemical modification method using KOH can be a very effective method to increase the abundance of the O-containing surface functional groups, providing additional benefits for V adsorption (Ghanim, O'Dwyer et al., 2020). Acid activation is also a promising method that can enhance the biochar capacity to adsorb V, owing to the high abundance of carboxyl groups in the modified biochar surfaces (Sizmur et al., 2017); however, no one has studied the performance of acid-modified biochar in V immobilization in soil.

Impregnating certain elements such as Zn, Zr and Cs can be effective in enhancing V adsorption. For instance, Meng et al. (2018) observed higher V removal from water by a biochar modified with Zn due to surface complexation and surface precipitation of V with Zn as well as electrostatic attraction, along with higher surface area than other unmodified and modified biochars tested. Moreover, modified biochars have a very high adsorption capacity for Zn, Zr and Cs in a wide range of solution pH (i.e., 4–8) compared to the unmodified biochar (Meng et al., 2018). In addition, the impregnation of nZVI showed a very high removal rate of V from aqueous solution, and V adsorption on biochar surfaces occurred, likely due to the interaction between nZVI/BC and V via electrostatic attraction, redox and precipitation processes (Fan et al., 2020). The incorporation of heteroatom dopants (e.g., O, N and S) into the biochar matrix was investigated for V(V) decontamination (Wu et al., 2021). Organo-element dopants enhanced the electro-positivity of biochar toward V(V) attraction by hard Lewis bases, and O-doped biochar was the most efficient form, with a sorption capacity of 70 mg g^{-1} (Wu et al., 2021). However, further research is needed to develop modification methods that enhance the potential for biochars to immobilize V in different soil types.

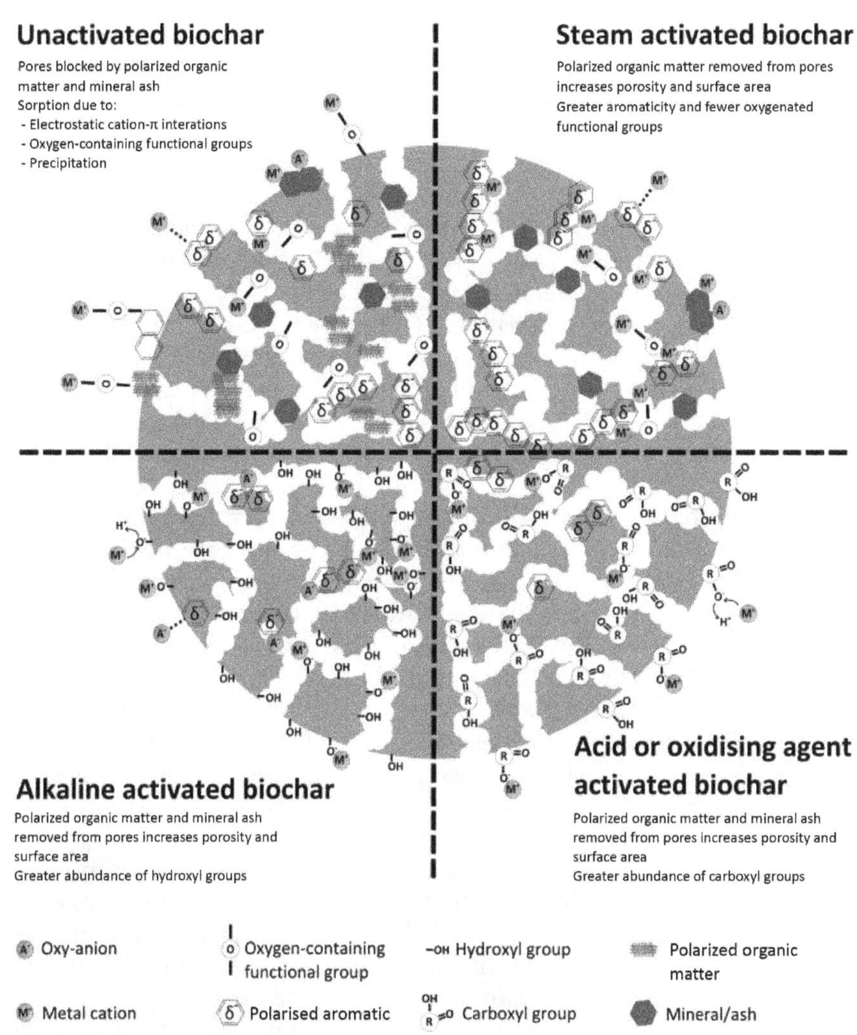
Unactivated biochar
Pores blocked by polarized organic matter and mineral ash
Sorption due to:
- Electrostatic cation-π interations
- Oxygen-containing functional groups
- Precipitation
Steam activated biochar
Polarized organic matter removed from pores increases porosity and surface area
Greater aromaticity and fewer oxygenated functional groups
Alkaline activated biochar
Polarized organic matter and mineral ash removed from pores increases porosity and surface area
Greater abundance of hydroxyl groups
Acid or oxidising agent activated biochar
Polarized organic matter and mineral ash removed from pores increases porosity and surface area
Greater abundance of carboxyl groups
Oxy-anion
Metal cation
Oxygen-containing functional group
Polarised aromatic
Hydroxyl group
Carboxyl group
Polarized organic matter
Mineral/ash

FIGURE 8.10
Physical and chemical modification methods for biochar.
Source: Reproduced from Sizmur et al. (2017), with permission from the publisher.

8.6 Risk Management of Biochar Application to V Contaminated Soil

Recent reports showed that biochar application might pose potential risks to the ecosystem. Although biochar is commonly used to improve soil fertility and to remediate contaminated soils, it may have contrasting effects in some cases. Biochar ageing in the soil might cause an inhibitory effect on soil biota (e.g., fungi and earthworms) and reduce the underground root biomass (Anyanwu et al., 2018). Biochar even could increase the mobilization of potentially toxic elements under certain conditions (Rinklebe et al., 2020; Shaheen et al., 2018). For example, biochar increased the mobility and phytoavailability of toxic elements and their distribution in the dissolved rather than the colloidal fraction under dynamic redox conditions (El-Naggar, Shaheen et al., 2018, 2019). In addition, attention should be paid when biochar is applied to a multi-element contaminated soil, where biochar showed less efficient remediation performance (Ahmad et al., 2017). Moreover, some of the modification methods are not eco-friendly and could even result in biochar being a source of contamination, such as biochar activation using inorganic acids (Lonappan et al., 2019). Therefore, we need to ensure that methods for biochar modification are safe and do not cause secondary contaminations.

Application of biochar to soil might stimulate the nitrification process, producing a high concentration of nitrate, which might be exposed to leaching to the groundwater (Pan et al., 2016). Biochar application might also lead to leaching of phosphate and sulfate to groundwater or increased methane efflux to the atmosphere (He et al., 2019). In addition, biochar application might reduce nutrient supply potentials of soil through reaction with phytoavailable plant nutrients to form stable organo-mineral complexes (El-Naggar, Awad et al., 2018; Yang et al., 2019; Zhou et al., 2021). Another potential risk to human health can be caused by biochar dust during the application of biochar to soils. Low-density biochar particles may become suspended in the air and inhaled by humans, causing respiratory illness. However, risks of health damage by biochar dust can be minimized by following proper application methods such as deep application, application of relatively large particle-sized biochar (~2 mm) and maintaining a medium level of soil moisture during the biochar application (El-Naggar, Lee et al., 2019). The biochar can also be moistened before transportation to the field to reduce the dust level during the application. In some cases, biochar could also be a source of contaminants such as polycyclic aromatic hydrocarbons (PAHs), polychlorinated dibenzodioxins and dibenzofurans (PCDD/DF) (He et al., 2019), which could be generated during the pyrolysis process.

Recent research has also highlighted the inhibitory effect of biochar on seed germination, growth of seedlings and crop productivity due to environmentally persistent free radicals (EPFRs) in biochar (Luo et al., 2021; Odinga et al., 2020; Ruan et al., 2019). The EPFRs remain stable for at least five years, generating reactive O species (ROS) in plant cells (Sigmund et al., 2021). Major factors that influence PAH concentrations in biochar during its production are pyrolysis temperature and residence time (El-Naggar, El-Naggar et al., 2019). Thus, environmental risk assessment is required before considering large-scale applications of biochar. In order to ensure that the added biochar does not pose any risk to the ecosystem, different parameters, including potentially toxic elements, PAHs and PCDD/DF contents, should be tested in the soils treated with different types of modified biochars at the end of the field remediation experiments.

8.7 Conclusions and Future Research Outlook

Biochar is a candidate material that can be used for the immobilization of V in soils. Although studies on the topic are scarce, the available preliminary results provide encouraging insights for researchers to further study the use of biochar as a tool for the remediation of V contamination in the soil. The potential of biochar to reduce V availability on soil is influenced by several factors related to (1) biochar properties, such as those that are affected by feedstock type and production condition, in which biochars with a higher ash content (e.g., manure biochar), abundance of O-containing functional groups and electron-donating capacity are preferred; and (2) soil properties, such as soil pH, E_H, contents of Fe, Mn, Al, S and dissolved aliphatic and aromatic organic C have direct/indirect effects on the mobilization and phytoavailability of V in soil.

The potential of biochar to immobilize V in the soil can be further enhanced via various modification methods such as acid/alkali activation and particle size reduction by ball milling. However, more research on those methods is required to ensure their role in enhancing biochar performance and assure that they do not pose any toxicological effects for organisms or transfer any pollutants to the soil (Figure 8.11). Considering that the research on the potential of biochar to immobilize V is still in its infancy, some future outlooks for studies are suggested as shown below:

i. An array of agricultural crop residues should be evaluated to serve as potential candidates for efficient biochar engineering towards V decontamination.
ii. The optimum engineering conditions for biochar production should be identified to ensure the proper functionality of biochar in V-contaminated soils.
iii. The optimum soil application rate should be identified to ensure its applicability under large-scale applications.
iv. Environmental risk assessment is further needed to warrant that the large-scale application of biochar will not cause significant hazards to the ecosphere.
v. Fabrication of designated biochar with a high electropositivity for improving the sorption of negatively charged vanadate (V(V)).

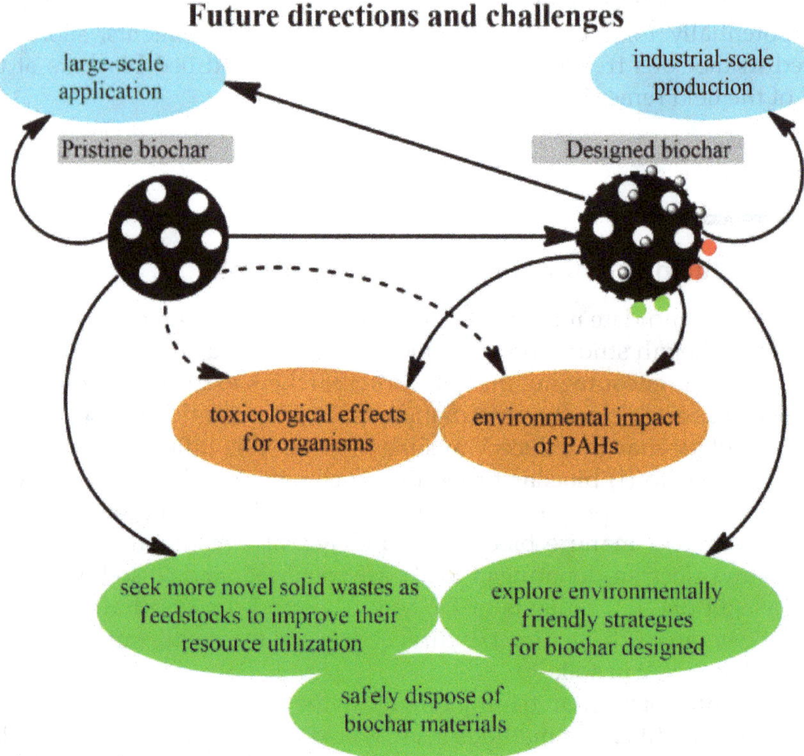

FIGURE 8.11
Future directions and challenges of biochar research.
Source: Reproduced from Deng et al. (2020), with permission from the publisher.

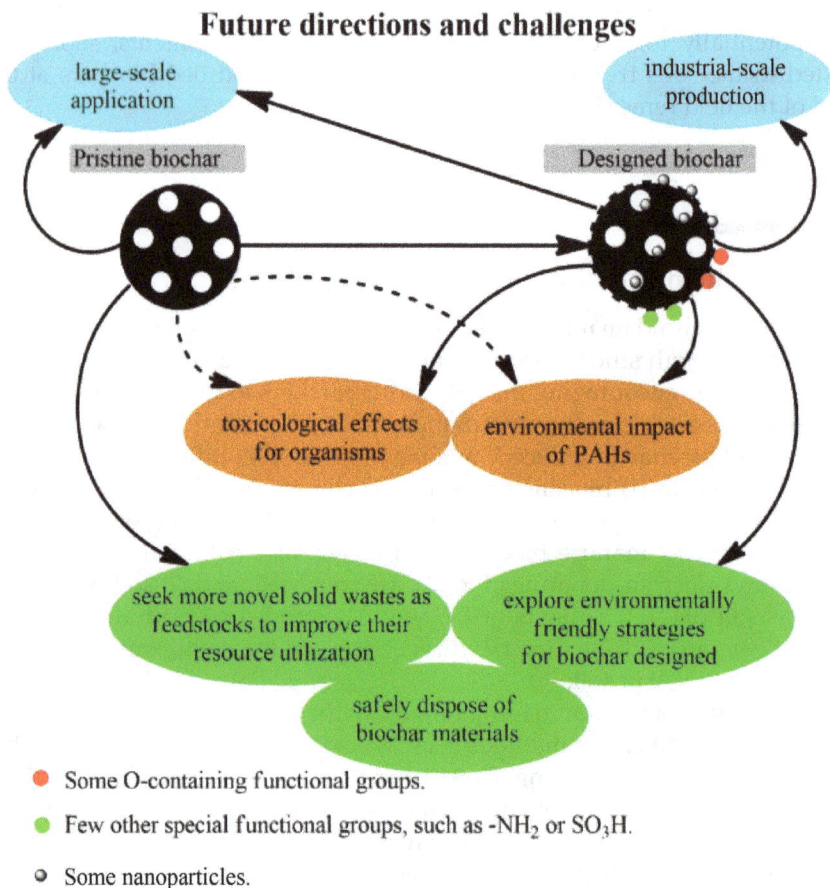

vi. Designating aged acidized biochars with low alkaline nature to increase the reduction rate of V(V) into V(IV).
vii. Preparing biochars with enriched V-reducing bacterial genera to increase V (V) reduction to V(IV).
viii. Engineering biochars with higher electron-donating capacity, with emphasis on feedstocks with higher hemicellulose and/or cellulose contents.
ix. The vertical mobility of V-loaded biochars should be studied in laboratory column experiments to understand the potential kinetics of V in the soil matrix.

x. Studies on the redox-mediated transformation of V in soil are still very limited, and further investigations are strongly needed, using several application rates and feedstock types.

xi. The fingerprint of dissolved organic C content of different biochars should be investigated, given the crucial effect of dissolved organic C on V phytoavailability.

References

Agnieszka, J., Barbara, G., 2012. Chromium, nickel and vanadium mobility in soils derived from fluvioglacial sands. *J. Hazard. Mater.* 237–238, 315–322. https://doi.org/10.1016/j.jhazmat.2012.08.048.

Ahmad, M., Lee, S.S., Lee, S.E., Al-Wabel, M.I., Tsang, D.C.W., Ok, Y.S., 2017. Biochar-induced changes in soil properties affected immobilization/mobilization of metals/metalloids in contaminated soils. *J. Soils Sediments.* 17, 717–730. https://doi.org/10.1007/s11368-015-1339-4.

Aihemaiti, A., Gao, Y., Liu, L., Yang, G., Han, S., Jiang, J., 2020. Effects of liquid digestate on the valence state of vanadium in plant and soil and microbial community response. *Environ. Pollut.* 265, 114916. https://doi.org/10.1016/j.envpol.2020.114916.

Aihemaiti, A., Gao, Y., Meng, Y., Chen, X., Liu, J., Xiang, H., Xu, Y., Jiang, J., 2020. Review of plant-vanadium physiological interactions, bioaccumulation, and bioremediation of vanadium-contaminated sites. *Sci. Total Environ.* 712, 135637. https://doi.org/10.1016/j.scitotenv.2019.135637.

Aihemaiti, A., Jiang, J., Li, D., Li, T., Zhang, W., Ding, X., 2017. Toxic metal tolerance in native plant species grown in a vanadium mining area. *Environ. Sci. Pollut. Res.* 24, 26839–26850. https://doi.org/10.1007/s11356-017-0250-5.

Amin, M.T., Alazba, A.A., Shafiq, M., 2019. Application of biochar derived from date palm biomass for removal of lead and copper ions in a batch reactor: Kinetics and isotherm scrutiny. *Chem. Phys. Lett.* 722, 64–73. https://doi.org/10.1016/j.cplett.2019.02.018.

Amusat, S.O., Kebede, T.G., Dube, S., Nindi, M.M., 2021. Ball-milling synthesis of biochar and biochar-based nanocomposites and prospects for removal of emerging contaminants: A review. *J. Water Process Eng.* https://doi.org/10.1016/j.jwpe.2021.101993.

Anyanwu, I.N., Alo, M.N., Onyekwere, A.M., Crosse, J.D., Nworie, O., Chamba, E.B., 2018. Influence of biochar aged in acidic soil on ecosystem engineers and two tropical agricultural plants. *Ecotoxicol. Environ. Saf.* 153, 116–126. https://doi.org/10.1016/j.ecoenv.2018.02.005.

Arabi, Z., Rinklebe, J., El-Naggar, A., Hou, D., Sarmah, A.K., Moreno-Jiménez, E., 2021. (Im)mobilization of arsenic, chromium, and nickel in soils via biochar: A meta-analysis. *Environ. Pollut.* 286, 117199. https://doi.org/10.1016/j.envpol.2021.117199.

Azargohar, R., Dalai, A.K., 2008. Steam and KOH activation of biochar: Experimental and modeling studies. *Microporous Mesoporous Mater.* 110, 413–421. https://doi.org/10.1016/j.micromeso.2007.06.047.

Banik, C., Lawrinenko, M., Bakshi, S., Laird, D.A., 2018. Impact of pyrolysis temperature and feedstock on surface charge and functional group chemistry of biochars. *J. Environ. Qual.* 47, 452–461. https://doi.org/10.2134/jeq2017.11.0432.

Bello, A., Leiviskä, T., Zhang, R., Tanskanen, J., Maziarz, P., Matusik, J., Bhatnagar, A., 2019. Synthesis of zerovalent iron from water treatment residue as a conjugate with kaolin and its application for vanadium removal. *J. Hazard. Mater.* 374, 372–381. https://doi.org/10.1016/j.jhazmat.2019.04.056.

Brebu, M., Cazacu, G., Chirila, O., 2011. Pyrolysis of lignin—A potential method for obtaining chemicals and/or fuels. *Cellul. Chem. Technol.* 45, 43–50.

Cao, X., Diao, M., Zhang, B., Liu, H., Wang, S., Yang, M., 2017. Spatial distribution of vanadium and microbial community responses in surface soil of Panzhihua mining and smelting area, China. *Chemosphere.* 183, 9–17. https://doi.org/10.1016/j.chemosphere.2017.05.092.

Cárdenas-Aguiar, E., Suárez, G., Paz-Ferreiro, J., Askeland, M.P.J., Méndez, A., Gascó, G., 2020. Remediation of mining soils by combining Brassica napus growth and amendment with chars from manure waste. *Chemosphere.* 261, 127798. https://doi.org/10.1016/j.chemosphere.2020.127798.

Chacón, F.J., Sánchez-Monedero, M.A., Lezama, L., Cayuela, M.L., 2020. Enhancing biochar redox properties through feedstock selection, metal preloading and post-pyrolysis treatments. *Chem. Eng. J.* 395, 125100. https://doi.org/10.1016/j.cej.2020.125100.

Chen, H., Gao, Y., El-Naggar, A., Niazi, N.K., Sun, C., Shaheen, S.M., Hou, D., Yang, X., Tang, Z., Liu, Z., Hou, H., Chen, W., Rinklebe, J., Pohořelý, M., Wang, H., 2022. Enhanced sorption of trivalent antimony by chitosan-loaded biochar in aqueous solutions: Characterization, performance and mechanisms. *J. Hazard. Mater.* 425, 127971. https://doi.org/10.1016/J.JHAZMAT.2021.127971.

Chen, L., Liu, J. rong, Hu, W. fang, Gao, J., Yang, J. yan, 2021. Vanadium in soil-plant system: Source, fate, toxicity, and bioremediation. *J. Hazard. Mater.* https://doi.org/10.1016/j.jhazmat.2020.124200.

Deng, R., Huang, D., Wan, J., Xue, W., Wen, X., Liu, X., Chen, S., Lei, L., Zhang, Q., 2020. Recent advances of biochar materials for typical potentially toxic elements management in aquatic environments: A review. *J. Clean. Prod.* https://doi.org/10.1016/j.jclepro.2019.119523.

El-Naggar, A., Ahmed, N., Mosa, A., Niazi, N.K., Yousaf, B., Sharma, A., Sarkar, B., Cai, Y., Chang, S.X., 2021. Nickel in soil and water: Sources, biogeochemistry, and remediation using biochar. *J. Hazard. Mater.* 419, 126421. https://doi.org/10.1016/j.jhazmat.2021.126421.

El-Naggar, A., Awad, Y.M., Tang, X.-Y., Liu, C., Niazi, N.K., Jien, S.-H., Tsang, D.C.W., Song, H., Ok, Y.S., Lee, S.S., 2018. Biochar influences soil carbon pools and facilitates interactions with soil: A field investigation. *L. Degrad. Dev.* 29, 2162–2171. https://doi.org/10.1002/ldr.2896.

El-Naggar, A., Chang, S.X., Cai, Y., Lee, Y.H., Wang, J., Wang, S.L., Ryu, C., Rinklebe, J., Sik Ok, Y., 2021. Mechanistic insights into the (im)mobilization of arsenic, cadmium, lead, and zinc in a multi-contaminated soil treated with different biochars. *Environ. Int.* 156, 106638. https://doi.org/10.1016/j.envint.2021.106638.

El-Naggar, A., El-Naggar, A.H., Shaheen, S.M., Sarkar, B., Chang, S.X., Tsang, D.C.W., Rinklebe, J., Ok, Y.S., 2019. Biochar composition-dependent impacts on soil nutrient release, carbon mineralization, and potential environmental risk: A review. *J. Environ. Manage.* 241, 1–10. https://doi.org/10.1016/j.jenvman.2019.02.044.

El-Naggar, A., Lee, S.S., Awad, Y.M., Yang, X., Ryu, C., Rizwan, M., Rinklebe, J., Tsang, D.C.W., Ok, Y.S., 2018. Influence of soil properties and feedstocks on biochar potential for carbon mineralization and improvement of infertile soils. *Geoderma*. 332, 100–108. https://doi.org/10.1016/j.geoderma.2018.06.017.

El-Naggar, A., Lee, S.S., Rinklebe, J., Farooq, M., Song, H., Sarmah, A.K., Zimmerman, A.R., Ahmad, M., Shaheen, S.M., Ok, Y.S., 2019. Biochar application to low fertility soils: A review of current status, and future prospects. *Geoderma*. 337, 536–554. https://doi.org/10.1016/j.geoderma.2018.09.034.

El-Naggar, A., Rajapaksha, A.U., Shaheen, S.M., Rinklebe, J., Ok, Y.S., 2018. Potential of biochar to immobilize nickel in contaminated soils. In: *Nickel in Soils and Plants*. CRC Press, pp. 293–318. https://doi.org/10.1201/9781315154664-13.

El-Naggar, A., Shaheen, S.M., Chang, S.X., Hou, D., Ok, Y.S., Rinklebe, J., 2021. Biochar surface functionality plays a vital role in (im)mobilization and phytoavailability of soil vanadium. ACS Sustain. *Chem. Eng*. 6864–6874. https://doi.org/10.1021/acssuschemeng.1c01656.

El-Naggar, A., Shaheen, S.M., Ok, Y.S., Rinklebe, J., 2018. Biochar affects the dissolved and colloidal concentrations of Cd, Cu, Ni, and Zn and their phytoavailability and potential mobility in a mining soil under dynamic redox-conditions. *Sci. Total Environ*. 624, 1059–1071. https://doi.org/10.1016/j.scitotenv.2017.12.190.

El-Naggar, A., Shaheen, S.M., Hseu, Z.-Y.Z.-Y., Wang, S.-L.S.-L., Ok, Y.S.Y.S., Rinklebe, J., 2019. Release dynamics of As, Co, and Mo in a biochar treated soil under pre-definite redox conditions. *Sci. Total Environ*. 657, 686–695. https://doi.org/10.1016/J.SCITOTENV.2018.12.026.

Elkhlifi, Z., Kamran, M., Maqbool, A., El-Naggar, A., Ifthikar, J., Parveen, A., Bashir, S., Rizwan, M., Mustafa, A., Irshad, S., Ali, S., Chen, Z., 2021. Phosphate-lanthanum coated sewage sludge biochar improved the soil properties and growth of ryegrass in an alkaline soil. *Ecotoxicol. Environ. Saf.* 216, 112173. https://doi.org/10.1016/j.ecoenv.2021.112173.

Fan, C., Chen, N., Qin, J., Yang, Y., Feng, C., Li, M., Gao, Y., 2020. Biochar stabilized nano zero-valent iron and its removal performance and mechanism of pentavalent vanadium(V(V)). *Colloids Surfaces A Physicochem. Eng. Asp*. 599, 124882. https://doi.org/10.1016/j.colsurfa.2020.124882.

Gan, C. dan, Chen, T., Yang, J. yan, 2020. Remediation of vanadium contaminated soil by alfalfa (Medicago sativa L.) combined with vanadium-resistant bacterial strain. *Environ. Technol. Innov*. 20, 101090. https://doi.org/10.1016/j.eti.2020.101090.

Ghanim, B., Murnane, J.G., O'Donoghue, L., Courtney, R., Pembroke, J.T., O'Dwyer, T.F., 2020. Removal of vanadium from aqueous solution using a red mud modified saw dust biochar. *J. Water Process Eng*. 33, 101076. https://doi.org/10.1016/j.jwpe.2019.101076.

Ghanim, B., O'Dwyer, T.F., Leahy, J.J., Willquist, K., Courtney, R., Pembroke, J.T., Murnane, J.G., 2020. Application of KOH modified seaweed hydrochar as a biosorbent of vanadium from aqueous solution: Characterisations, mechanisms and regeneration capacity. *J. Environ. Chem. Eng*. 8, 104176. https://doi.org/10.1016/j.jece.2020.104176.

Gustafsson, J.P., 2019. Vanadium geochemistry in the biogeosphere—speciation, solid-solution interactions, and ecotoxicity. *Appl. Geochem*. https://doi.org/10.1016/j.apgeochem.2018.12.027.

He, L., Zhong, H., Liu, G., Dai, Z., Brookes, P.C., Xu, J., 2019. Remediation of heavy metal contaminated soils by biochar: Mechanisms, potential risks and applications in China. *Environ. Pollut.* 252, 846–855. https://doi.org/10.1016/j.envpol.2019.05.151.

Imtiaz, M., Rizwan, M.S., Xiong, S., Li, H., Ashraf, M., Shahzad, S.M., Shahzad, M., Rizwan, M., Tu, S., 2015. Vanadium, recent advancements and research prospects: A review. *Environ. Int.* 80, 79–88. https://doi.org/10.1016/j.envint.2015.03.018.

Karer, J., Wawra, A., Zehetner, F., Dunst, G., Wagner, M., Pavel, P.B., Puschenreiter, M., Friesl-Hanl, W., Soja, G., 2015. Effects of biochars and compost mixtures and inorganic additives on immobilisation of heavy metals in contaminated soils. *Water. Air. Soil Pollut.* 226, 1–12. https://doi.org/10.1007/s11270-015-2584-2.

Kumar, M., Xiong, X., Wan, Z., Sun, Y., Tsang, D.C.W., Gupta, J., Gao, B., Cao, X., Tang, J., Ok, Y.S., 2020. Ball milling as a mechanochemical technology for fabrication of novel biochar nanomaterials. *Bioresour. Technol.* https://doi.org/10.1016/j.biortech.2020.123613.

Li, S., Shao, L., Zhang, H., He, P., Lü, F., 2020. Quantifying the contributions of surface area and redox-active moieties to electron exchange capacities of biochar. *J. Hazard. Mater.* 394, 122541. https://doi.org/10.1016/j.jhazmat.2020.122541.

Li, Y., Zhang, B., Liu, Z., Wang, S., Yao, J., Borthwick, A.G.L., 2020. Vanadium contamination and associated health risk of farmland soil near smelters throughout China. *Environ. Pollut.* 263, 114540. https://doi.org/10.1016/j.envpol.2020.114540.

Liu, L., Wang, L., Su, S., Yang, T., Dai, Z., Qing, M., Xu, K., Hu, S., Wang, Y., Xiang, J., 2019. Leaching behavior of vanadium from spent SCR catalyst and its immobilization in cement-based solidification/stabilization with sulfurizing agent. *Fuel.* 243, 406–412. https://doi.org/10.1016/j.fuel.2019.01.160.

Liu, P., Ptacek, C.J., Blowes, D.W., Finfrock, Y.Z., Liu, Y.Y., 2020. Characterization of chromium species and distribution during Cr(VI) removal by biochar using confocal micro-X-ray fluorescence redox mapping and X-ray absorption spectroscopy. *Environ. Int.* 134, 105216. https://doi.org/10.1016/j.envint.2019.105216.

Lonappan, L., Liu, Y., Rouissi, T., Brar, S.K., Surampalli, R.Y., 2019. Development of biochar-based green functional materials using organic acids for environmental applications. *J. Clean. Prod.* 118841. https://doi.org/10.1016/j.jclepro.2019.118841.

Luo, K., Pang, Y., Wang, D., Li, X., Wang, L., Lei, M., Huang, Q., Yang, Q., 2021. A critical review on the application of biochar in environmental pollution remediation: Role of persistent free radicals (PFRs). *J. Environ. Sci. (China).* https://doi.org/10.1016/j.jes.2021.02.021.

Luo, X., Yu, L., Wang, C., Yin, X., Mosa, A., Lv, J., Sun, H., 2017. Sorption of vanadium (V) onto natural soil colloids under various solution pH and ionic strength conditions. *Chemosphere.* 169, 609–617. https://doi.org/10.1016/j.chemosphere.2016.11.105.

Man, Y., Wang, B., Wang, J., Slaný, M., Yan, H., Li, P., El-Naggar, A., Shaheen, S.M., Rinklebe, J., Feng, X., 2021. Use of biochar to reduce mercury accumulation in Oryza sativa L: A trial for sustainable management of historically polluted farmlands. *Environ. Int.* 153, 106527. https://doi.org/10.1016/j.envint.2021.106527.

McBeath, A. V., Smernik, R.J., Krull, E.S., Lehmann, J., 2014. The influence of feedstock and production temperature on biochar carbon chemistry: A solid-state 13C NMR study. *Biomass and Bioenergy.* 60, 121–129. https://doi.org/10.1016/j.biombioe.2013.11.002.

Meng, R., Chen, T., Zhang, Y., Lu, W., Liu, Yanting, Lu, T., Liu, Yanjun, Wang, H., 2018. Development, modification, and application of low-cost and available biochar derived from corn straw for the removal of vanadium(v) from aqueous solution and real contaminated groundwater. *RSC Adv.* 8, 21480–21494. https://doi.org/10.1039/c8ra02172d.

Naeem, A., Westerhoff, P., Mustafa, S., 2007. Vanadium removal by metal (hydr)oxide adsorbents. *Water Res.* 41, 1596–1602. https://doi.org/10.1016/j.watres.2007.01.002.

Odinga, E.S., Waigi, M.G., Gudda, F.O., Wang, J., Yang, B., Hu, X., Li, S., Gao, Y., 2020. Occurrence, formation, environmental fate and risks of environmentally persistent free radicals in biochars. *Environ. Int.* https://doi.org/10.1016/j.envint.2019.105172.

Palansooriya, K.N., Shaheen, S.M., Chen, S.S., Tsang, D.C.W., Hashimoto, Y., Hou, D., Bolan, N.S., Rinklebe, J., Ok, Y.S., 2020. Soil amendments for immobilization of potentially toxic elements in contaminated soils: A critical review. *Environ. Int.* 134, 105046. https://doi.org/10.1016/j.envint.2019.105046.

Pan, F., Hu, J., Suo, L., Wang, X., Ji, Y., Meng, L., 2016. Effect of corn stalk and its biochar on N2O emissions from latosol soil. *J. Agro-Environment Sci.* 35, 396–402.

Panichev, N., Mandiwana, K., Moema, D., Molatlhegi, R., Ngobeni, P., 2006. Distribution of vanadium(V) species between soil and plants in the vicinity of vanadium mine. *J. Hazard. Mater.* 137, 649–653. https://doi.org/10.1016/j.jhazmat.2006.03.006.

Prévoteau, A., Ronsse, F., Cid, I., Boeckx, P., Rabaey, K., 2016. The electron donating capacity of biochar is dramatically underestimated. *Sci. Rep.* 6, 1–11. https://doi.org/10.1038/srep32870.

Qiu, M., Sun, K., Jin, J., Han, L., Sun, H., Zhao, Y., Xia, X., Wu, F., Xing, B., 2015. Metal/metalloid elements and polycyclic aromatic hydrocarbon in various biochars: The effect of feedstock, temperature, minerals, and properties. *Environ. Pollut.* 206, 298–305. https://doi.org/10.1016/j.envpol.2015.07.026.

Rajapaksha, A.U., Ok, Y.S., El-Naggar, A., Kim, H., Song, F., Kang, S., Tsang, Y.F., 2019. Dissolved organic matter characterization of biochars produced from different feedstock materials. *J. Environ. Manage.* 233. https://doi.org/10.1016/j.jenvman.2018.12.069.

Rehder, D., 2016. Perspectives for vanadium in health issues. *Future Med. Chem.* https://doi.org/10.4155/fmc.15.187.

Rehder, D., 2018. Vanadium in health issues. *ChemTexts.* 4, 20. https://doi.org/10.1007/s40828-018-0074-z.

Reijonen, I., Metzler, M., Hartikainen, H., 2016. Impact of soil pH and organic matter on the chemical bioavailability of vanadium species: The underlying basis for risk assessment. *Environ. Pollut.* 210, 371–379. https://doi.org/10.1016/j.envpol.2015.12.046.

Rinklebe, J., Shaheen, S.M., El-Naggar, A., Wang, H., Du Laing, G., Alessi, D.S., Sik Ok, Y., 2020. Redox-induced mobilization of Ag, Sb, Sn, and Tl in the dissolved, colloidal and solid phase of a biochar-treated and un-treated mining soil. *Environ. Int.* 140, 105754. https://doi.org/10.1016/j.envint.2020.105754.

Rodriguez, J.A., Lustosa Filho, J.F., Melo, L.C.A., de Assis, I.R., de Oliveira, T.S., 2020. Influence of pyrolysis temperature and feedstock on the properties of biochars produced from agricultural and industrial wastes. *J. Anal. Appl. Pyrolysis.* 149, 104839. https://doi.org/10.1016/j.jaap.2020.104839.

Ruan, X., Sun, Y., Du, W., Tang, Y., Liu, Q., Zhang, Z., Doherty, W., Frost, R.L., Qian, G., Tsang, D.C.W., 2019. Formation, characteristics, and applications of environmentally persistent free radicals in biochars: A review. *Bioresour. Technol.* https://doi.org/10.1016/j.biortech.2019.02.105.

Shaheen, S.M., Alessi, D.S., Tack, F.M.G., Ok, Y.S., Kim, K.H., Gustafsson, J.P., Sparks, D.L., Rinklebe, J., 2019. Redox chemistry of vanadium in soils and sediments: Interactions with colloidal materials, mobilization, speciation, and relevant environmental implications—A review. *Adv. Colloid Interface Sci.* 265, 1–13. https://doi.org/10.1016/j.cis.2019.01.002.

Shaheen, S.M., El-Naggar, A., Wang, J., Hassan, N.E.E., Niazi, N.K., Wang, H., Tsang, D.C.W., Ok, Y.S., Bolan, N., Rinklebe, J., 2018. Biochar as an (om)mobilizing agent for the potentially toxic elements in contaminated soils. In: *Biochar from Biomass and Waste*. Elsevier, pp. 255–274. https://doi.org/10.1016/b978-0-12-8117 29-3.00014-5.

Shaheen, S.M., Rinklebe, J., 2018. Vanadium in thirteen different soil profiles originating from Germany and Egypt: Geochemical fractionation and potential mobilization. *Appl. Geochem.* 88, 288–301. https://doi.org/10.1016/j.apgeochem.2017.02.010.

Shen, D.K., Gu, S., 2009. The mechanism for thermal decomposition of cellulose and its main products. *Bioresour. Technol.* 100, 6496–6504. https://doi.org/10.1016/j.biortech.2009.06.095.

Shim, T., Yoo, J., Ryu, C., Park, Y.K., Jung, J., 2015. Effect of steam activation of biochar produced from a giant Miscanthus on copper sorption and toxicity. *Bioresour. Technol.* 197, 85–90. https://doi.org/10.1016/j.biortech.2015.08.055.

Sigmund, G., Santín, C., Pignitter, M., Tepe, N., Doerr, S.H., Hofmann, T., 2021. Environmentally persistent free radicals are ubiquitous in wildfire charcoals and remain stable for years. *Commun. Earth Environ.* 2, 1–6. https://doi.org/10.1038/s43247-021-00138-2.

Sizmur, T., Fresno, T., Akgül, G., Frost, H., Moreno-Jiménez, E., 2017. Biochar modification to enhance sorption of inorganics from water. *Bioresour. Technol.* https://doi.org/10.1016/j.biortech.2017.07.082.

von Gunten, K., Alam, M.S., Hubmann, M., Ok, Y.S., Konhauser, K.O., Alessi, D.S., 2017. Modified sequential extraction for biochar and petroleum coke: Metal release potential and its environmental implications. *Bioresour. Technol.* 236, 106–110. https://doi.org/10.1016/j.biortech.2017.03.162.

Wang, S., Zhang, B., Li, T., Li, Z., Fu, J., 2020. Soil vanadium(V)-reducing related bacteria drive community response to vanadium pollution from a smelting plant over multiple gradients. *Environ. Int.* 138, 105630. https://doi.org/10.1016/j.envint.2020.105630.

Wu, B., Ifthikar, J., Oyekunle, D.T., Jawad, A., Chen, Zhuqi, Chen, Zhulei, Sellaoui, L., Bouzid, M., 2021. Interpret the elimination behaviors of lead and vanadium from the water by employing functionalized biochars in diverse environmental conditions. *Sci. Total Environ.* 148031. https://doi.org/10.1016/j.scitotenv.2021.148031.

Xiao, X.Y., Jiang, Z., Guo, Z., Wang, M., Zhu, H., Han, X., 2017. Effect of simulated acid rain on leaching and transformation of vanadium in paddy soils from stone coal smelting area. *Process Saf. Environ. Prot.* 109, 697–703. https://doi.org/10.1016/j.psep.2017.05.006.

Xiao, X.Y., Yang, M., Guo, Z.H., Jiang, Z.C., Liu, Y.N., Cao, X., 2015. Soil vanadium pollution and microbial response characteristics from stone coal smelting district. *Trans. Nonferrous Met. Soc. China (English Ed)*. 25, 1271–1278. https://doi.org/10.1016/S1003-6326(15)63727-X.

Xu, X., Huang, H., Zhang, Y., Xu, Z., Cao, X., 2019. Biochar as both electron donor and electron shuttle for the reduction transformation of Cr(VI) during its sorption. *Environ. Pollut*. 244, 423–430. https://doi.org/10.1016/j.envpol.2018.10.068.

Xu, X., Xu, Z., Huang, J., Gao, B., Zhao, L., Qiu, H., Cao, X., 2021. Sorption of reactive red by biochars ball milled in different atmospheres: Co-effect of surface morphology and functional groups. *Chem. Eng. J*. 413, 127468. https://doi.org/10.1016/j.cej.2020.127468.

Yang, G., Wang, Z., Xian, Q., Shen, F., Sun, C., Zhang, Y., Wu, J., 2015. Effects of pyrolysis temperature on the physicochemical properties of biochar derived from vermicompost and its potential use as an environmental amendment. *RSC Adv*. 5, 40117–40125. https://doi.org/10.1039/c5ra02836a.

Yang, X., Tsibart, A., Nam, H., Hur, J., El-Naggar, A., Tack, F.M.G., Wang, C.-H., Lee, Y.H., Tsang, D.C.W., Ok, Y.S., 2019. Effect of gasification biochar application on soil quality: Trace metal behavior, microbial community, and soil dissolved organic matter. *J. Hazard. Mater*. 365. https://doi.org/10.1016/j.jhazmat.2018.11.042.

Yu, Y., Li, J., Liao, Y., Yang, J., 2020. Effectiveness, stabilization, and potential feasible analysis of a biochar material on simultaneous remediation and quality improvement of vanadium contaminated soil. *J. Clean. Prod*. 277, 123506. https://doi.org/10.1016/j.jclepro.2020.123506.

Zhang, Y., Xu, X., Cao, L., Ok, Y.S., Cao, X., 2018. Characterization and quantification of electron donating capacity and its structure dependence in biochar derived from three waste biomasses. *Chemosphere*. 211, 1073–1081. https://doi.org/10.1016/j.chemosphere.2018.08.033.

Zhou, R., El-Naggar, A., Li, Y., Cai, Y., Chang, S.X., 2021. Converting rice husk to biochar reduces bamboo soil N_2O emissions under different forms and rates of nitrogen additions. *Environ. Sci. Pollut. Res*. https://doi.org/10.1007/s11356-021-12744-w.

Zou, Q., Li, D., Jiang, J., Aihemaiti, A., Gao, Y., Liu, N., Liu, J., 2019. Geochemical simulation of the stabilization process of vanadium-contaminated soil remediated with calcium oxide and ferrous sulfate. *Ecotoxicol. Environ. Saf.* 174, 498–505. https://doi.org/10.1016/j.ecoenv.2019.02.082.

9

Plant Uptake and Ecotoxicity of Vanadium: The Role of Soil Chemistry

Jon Petter Gustafsson

Department of Soil and Environment, Swedish University of Agricultural Sciences, Uppsala, Sweden

CONTENTS

9.1 Introduction: Vanadium in the Soil-Plant System 187
9.2 Effect of Chemical Variables on Vanadium Uptake and Ecotoxicity. Review of Current Evidence 190
9.3 Predicting Vanadium Ecotoxicity in Soil: Towards a Conceptual Model 197
9.4 Conclusions and Outlook 200
References 200

9.1 Introduction: Vanadium in the Soil-Plant System

Vanadium is an element known for its complicated redox chemistry, with at least three, possibly four, oxidation states that are stable under natural conditions, the most oxidized of which is vanadium(V), which predominates under aerobic conditions prevailing in the unsaturated zone of many soils (see Chapter 5 in this book). Under low-temperature environmental conditions, vanadium(V) forms tetrahedral dihydrogen vanadate ions, $H_2VO_4^-$, over a wide pH range (between 3.6 and 8.7) (Gustafsson, 2019). Vanadate has similar, although not identical, chemical properties as *o*-phosphate, which is dominated by dihydrogen phosphate, $H_2PO_4^-$, under mildly acidic conditions up to pH 7.2, and by monohydrogen phosphate, HPO_4^{2-}, above this pH value. Under slightly reducing conditions and especially at low pH, for example in organic soils, vanadium instead occurs primarily as vanadium(IV), which in its free form occurs as the oxocation vanadyl, VO^{2+} (Shaheen et al., 2019). The vanadyl cation is very reactive and readily forms strong complexes with many organic ligands. Hence, plants in the soil environment usually encounter

DOI: 10.1201/9781003173274-9

vanadium as either vanadate or vanadyl. The more reduced vanadium(III) is stable under anoxic conditions such as in many freshwater and sea sediments, but is not likely to be of importance in the unsaturated zone of soils (Gustafsson, 2019).

Vanadium has a few known biological functions. Perhaps most significantly, it is present in vanadium nitrogenase, an "alternative" nitrogenase, which is used by cyanobacteria present in, for example, lichens for N_2 fixation (Darnajoux et al., 2014). Moreover, it is present also in vanadium haloperoxidases, which are important for oxidizing halides in algae (Rehder, 2015). Possibly, at low concentrations vanadium may also be beneficial for the growth of some plants including pepper plants, although the mechanisms are not fully understood (García-Jiménez et al., 2018). However, although there are uncertainties, the general picture is that vanadium is not an essential element for most organisms including plants. Instead, vanadium can be toxic at high concentrations.

Studies in the 1990s showed that for most plants, ecotoxic effects of vanadium appear when the soil solution concentration of V is in the low mg L^{-1} range (Kaplan, Adriano et al., 1990; Kaplan, Sajwan et al. 1990; Gil et al. 1995). Consistent with this, Larsson et al. (2013) obtained EC_{50} (median effective concentration) values ranging from 2.7 to 15.4 mg V L^{-1} for barley root length, from 1.5 to 4.0 mg V L^{-1} for barley shoot growth, and from 0.8 to 1.3 mg V L^{-1} for tomato shoot growth, with EC_{10} (10% effective concentration) values about half of these. The structural similarity between the vanadate and *o*-phosphate oxyanions explains the main mode of ecotoxicity of vanadium, which is to interfere with the phosphorus metabolism in various ways. A well-established fact is that vanadate oxyanions are able to prevent enzymes such as phosphatase, ribonuclease, and ATPase from hydrolyzing the ester bonds of different types of organic P compounds through forming a complex with the active sites of the enzymes (Seargeant and Stinson, 1979; Crans et al., 2004). The toxic action of vanadium is, however, complicated further by the redox chemistry of vanadium. First, also vanadium(IV) is toxic and is also able to deactivate phosphatase enzymes. In this case the mechanism is not fully elucidated, but may involve the action of negatively charged $VO(OH)_3^-$ ions, which may be stabilized at circumneutral pH through its strong ability to complex the phosphatase enzymes in a similar manner as $H_2VO_4^-$ (Crans et al., 2004; Costa Pessoa et al., 2015). Second, the reduction of vanadium to vanadium(IV) represents a possible detoxification mechanism, as vanadium(IV) forms strong complexes with a range of organic ligands, which enables plants to store vanadium in for example, their roots in vacuoles or on cell walls before it is translocated further into aboveground biomass (Hou et al., 2014; Gurau et al., 2015). Consistent with this, many studies in recent years have shown that the vanadium concentration is highest in plant roots and considerably lower in other parts of the plant, such as in shoots and in leaves (Saco et al., 2013; Hou et al., 2014; Yang and Tang, 2015; Imtiaz et al., 2018; Hou et al., 2020; Wu et al., 2021). The degree of translocation of vanadium from roots to other biomass parts varies not only with plant species but also with the vanadium concentration. The latter is supported by the results from a V-contaminated brownfield site in New Jersey, which showed that the ratio of vanadium in shoots to vanadium in roots increased considerably as the vanadium concentration was increased (Qian et al., 2014).

Soil chemical properties are important determinants of vanadium ecotoxicity, not least because both vanadium(IV) and vanadium(V) bind strongly to soil particle surfaces, rendering a large fraction of soil vanadium unavailable for uptake. Vanadate is strongly bound to high-surface area iron and aluminum (hydr)oxide-type mineral phases, such as goethite, ferrihydrite, amorphous $Al(OH)_3$ and allophane/imogolite, particularly at low pH values (Peacock and Sherman, 2004; Larsson, Persson et al., 2017; Gustafsson, 2019). Moreover, organic matter serves as a possible sink of vanadium(IV) in organic matter-containing soils (Templeton and Chasteen, 1980; Larsson, D'Amato et al., 2015). Hence, when the EC_{50} value is expressed on a dry soil basis, a strong dependence on the V sorption properties of the soil is expected. This is illustrated by the EC_{50} values obtained by Larsson et al. (2013) for five different soils (Figure 9.1), according to which the EC_{50} value for three different plant assays was related to the so-called Freundlich sorption strength.

Relationships such as the one shown in Figure 9.1 may, however, be obscured by various factors. One example is the so-called ageing process of

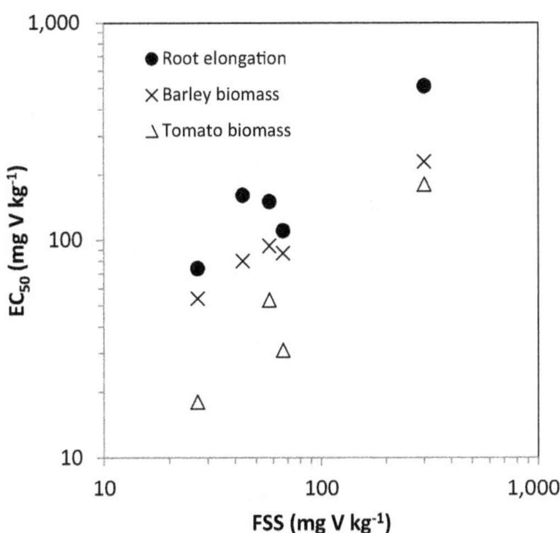

FIGURE 9.1
EC_{50} values for barley root growth ("Root elongation"), growth of biomass for barley ("Barley biomass") and biomass growth for tomato ("Tomato biomass") as a function of FSS expressed as mg V kg^{-1}. FSS is defined as the concentration of V bound to the soil when dissolved V is equal to 2.5 mg V L^{-1}.

Source: The data used to draw the figure were taken from Larsson et al. (2013).

the applied vanadium, which is expected to occur mainly because of slow diffusion of the adsorbed anion into the interior of the soil particles, leading to a lower bioavailability with time (see, e.g., Barrow, 1983). As Baken et al. (2012) showed, $CaCl_2$-extractable V decreased twofold between 14 and 100 days after spiking soils with vanadate salts, and the EC_{50} value, when expressed on a mg kg^{-1} basis, increased. Moreover, the uptake and ecotoxicity may in themselves be dependent on soil chemical factors such as pH and dissolved P (Ji et al., 2020). As the latter authors argued, if the soil chemical factors governing vanadium uptake and ecotoxicity were known, it would, in principle, be possible to derive a biotic ligand-type model able to predict ecotoxicity for terrestrial plants. Biotic ligand models (BLMs) were first developed for copper (di Toro et al., 2001) and have since then been applied also to other cationic metals such as zinc, nickel and lead (van Sprang et al., 2016). To date, most BLMs were derived for aquatic organisms, although efforts to develop BLMs for terrestrial organisms have been tested (e.g., Thakali et al., 2006). In the remainder of this chapter, we will first investigate how different soil chemical parameters may affect vanadium uptake and ecotoxicity to plants. This is done by compiling available literature that has somehow addressed this aspect. Later on, we will demonstrate how a vanadium BLM might be formulated and apply a vanadium BLM for *Tritium aestivum* to four different soils to find out how variations in soil chemistry are expected to affect vanadium uptake and ecotoxicity.

9.2 Effect of Chemical Variables on Vanadium Uptake and Ecotoxicity. Review of Current Evidence

Factors affecting vanadium uptake and ecotoxicity to plants have been studied both in hydroponic systems (i.e., without any soil present) and in various kinds of pot or field experiments where soil has been present (Table 9.1). Most commonly, the uptake of vanadium has been related to the vanadium concentration of soil extracts or in the dissolved phase. Table 9.1 and Table 9.2 do not aim to provide a complete list of all studies where this has been investigated; rather, they should be seen as lists of representative examples. Of course, as could be expected, the general picture is that the uptake and ecotoxicity of vanadium is, in some sense, related to the concentration of vanadium, although there are exceptions: Yang et al. (2017), in their uptake experiments with soybean, noted that the V concentration of the beans actually decreased at higher V additions, while that of roots increased, as expected. This was attributed to the low degree of translocation of V from the root to other parts of the plant, which appears to be increasingly prevented at higher V concentrations, a finding that also is substantiated by other authors (Qian et al., 2014; Wu et al., 2021).

TABLE 9.1

Studies Investigating the Effect of Chemical Properties on Vanadium Uptake or Ecotoxicity of Plants. I. Experimental Conditions

Reference	Study Type	No. Soil Samples	Soil pH (±s.d.)	Plant(s) Studied	Properties Studied
Aihemaiti et al., 2018	soil	11	8.51 ± 0.43	*Setaria viridis, Kochia scoparia, Chenopodium album*	Soil V, pH, organic matter, Olsen-P
El Naggar et al., 2021	soil	1	4.75	*Brassica napus, Pisum sativum, Hordeum vulgare, Zea mays*	Biochar additions, pH
Hou et al., 2019	hydroponic	-	-	*Zea mays*	Mercury
Imtiaz et al., 2017	hydroponic	-	-	*Cicer arietinum*	Phosphate
Ji et al., 2020	hydroponic	-	-	*Triticum aestivum*	Other anions
Larsson et al., 2013	soil	5	6.56 ± 0.86	*Hordeum vulgare, Solanum lycopersium*	V sorption, Fe+Al, pH, clay, CEC
Larsson, Baken et al., 2015	soil	2	6.1	*Hordeum vulgare*	Soil V, dissolved V
Olness et al., 2005	hydroponic	-	-	*Cuphea viscosissima* × *Cuphea lanceolata*	Magnesium
Qian et al., 2014	soil	22	5.93 ± 0.62	Six species	pH, TOC, soil V
Schwartz et al., 1991	hydroponic	-	-	leaves from *Commelina communis*	Chloride
Smith et al., 2013	soil	1	4.7	*Lactuta sativa, Elymus virginicus, Panicum virgatum, Lycopus americanus, Prunella vulgaris*	Nutrient status
Tian et al., 2014	soil	1	n.d.	*Brassica rapa*	Soil V
Tian et al., 2015	soil	2	8.2	*Brassica juncea*	Water-soluble V, total V
Welch, 1973	hydroponic	-	-	*Hordeum vulgare*	pH, other anions
Wu et al., 2020	soil	94	6.63 ± 1.47	*Brassica chinensis*	Extractable V
Wu et al., 2021	hydroponic	-	-	*Nicotiana tabacum*	Added V
Yang et al., 2017	soil	1	6.54	*Glycine max*	Added V

n.d. = not determined

TABLE 9.2

Studies Investigating the Effect of Chemical Properties on Vanadium Uptake or Ecotoxicity of Plants. II. Main Findings

Reference	Main Findings
Aihemaiti et al., 2018	Vanadium uptake was significantly related to soil V for *Setaria viridis*, no other significant relationships were observed.
El Naggar et al., 2021	In general, addition of three biochars to V contaminated soil caused increased V uptake, which was attributed to increases in pH and dissolved V, causing higher V uptake and toxicity. The exception was one neutral wood biochar, which caused reduced V uptake when applied at low (but not high) concentrations.
Hou et al., 2019	The results suggest a complex interaction between V and Hg concerning uptake and accumulation in maize seedlings, where a certain level of V can induce Hg uptake by thiols.
Imtiaz et al., 2017	There was a highly significant effect on V uptake from P in chickpea (i.e., higher P decreased V uptake), and regression equations were developed that described the interaction.
Ji et al., 2020	Vanadium ecotoxicity (EC_{50} values) was significantly related to pH (higher pH led to lower V uptake in hydroponic systems), P was able to inhibit V uptake at higher concentrations. Further, V ecotoxicity was significantly inversely related to HCO_3^-, but this could result from the pH effect. No significant interactions with SO_4 were observed.
Larsson et al., 2013	Vanadium ecotoxicity (EC_{50} values, mg/kg) was strongly correlated to the Freundlich sorption strength, not significantly correlated to other soil properties.
Larsson et al., 2015	Vanadium uptake (barley leaf shoots) was significantly correlated with dissolved V, as evidenced by $CaCl_2$ extraction, but less with total V in soil.
Olness et al., 2005	Increasing Mg in solution increased the sensitivity of cuphea to V toxicity, for unknown reasons.
Qian et al., 2014	Vanadium uptake was significantly correlated to HNO_3-extractable V but not to pH or to TOC. The V translocation efficiency from root to shoot decreased with increasing V content in the soil.
Schwartz et al., 1991	At high Cl- concentrations, vanadate was found not to inhibit the stomatal function of plants, probably because of competition during uptake to guard cells.

Reference	Main Findings
Smith et al., 2013	Addition of 20:20:20 N:P:K fertilizer decreased the ecotoxic effects of V. The EC_{50} values increased for all five plant species investigated.
Tian et al., 2014	Nonsignificant positive relationships between extractable V fractions and root V. Low NH_4VO_3 additions stimulated plant growth, probably because of an improved N status.
Tian et al., 2015	Vanadium uptake was related to soil V, above all water-soluble V. Roots accumulated much more V than shoots and seed.
Welch, 1973	Vanadium uptake to barley roots was strongly affected by pH, with the highest uptake at low pH (4) and very low uptake above 10. Further, there was some inhibition of uptake from other anions. In the presence of 50 µM P, V uptake decreased by 27%.
Wu et al., 2020	Vanadium uptake to Chinese cabbage was significantly correlated with 0.5 M HCO_3-extractable V ($r = 0.71^{**}$) and with 0.05 M EDTA ($r = 0.57^{*}$), but not with 0.01 M $CaCl_2$ ($r = 0.54$) or 0.1 M HCl ($r = 0.03$).
Wu et al., 2021	Vanadium uptake was closely dependent on the vanadium addition to hydroponic systems with tobacco plants, and the degree of translocation from roots to other plant parts was much reduced at higher vanadium additions.
Yang et al., 2017	Increased soil V concentrations caused a larger accumulation of V in the root of soybean, particularly on the cell wall. However, V translocation from the root at higher V additions was inefficient, and the V concentration of the beans actually decreased at higher V additions.

As shown by many authors, vanadium uptake to roots and ecotoxicity is most strongly related to relatively soluble fractions of vanadium (Larsson et al., 2013; Larsson, Baken et al., 2015; Qian et al., 2014; Tian et al., 2015; Wu et al., 2020). This is also in line with the ageing study of Baken et al. (2012), who showed that the ecotoxic response was smaller (i.e., EC_{50} values higher) for plants grown in soil where V had been added 5 to 11 months before the experiment, leading to lower dissolved V. Hence, V uptake from the soil solution to the root appears to be the driving factor in V bioavailability to plants.

There are relatively few studies that have addressed the importance of other chemical factors. Ji et al. (2020) studied the pH dependence of vanadate ecotoxicity to wheat (*Tritum aestivum*) in hydroponic systems. As Figure 9.2

shows, there was a close correspondence between the log EC_{50} of $\{H_2VO_4^-\}$ (i.e., the aqueous activity of dihydrogen vanadate) and pH, implying a higher ecotoxic effect at low pH. This observation confirms the work of Welch (1973), who studied vanadate uptake to barley roots (*Hordeum vulgare*) as a function of pH. The highest uptake was observed at low pH (Figure 9.2). When expressing the uptake on a logarithmic scale and when calculating the regression line for the relationship between pH and log uptake (excluding the pH points at pH > 8.8, where $H_2VO_4^-$ is no longer the dominating V species), the slope (−0.12) is equivalent to the one shown for the pH dependent EC_{50} values

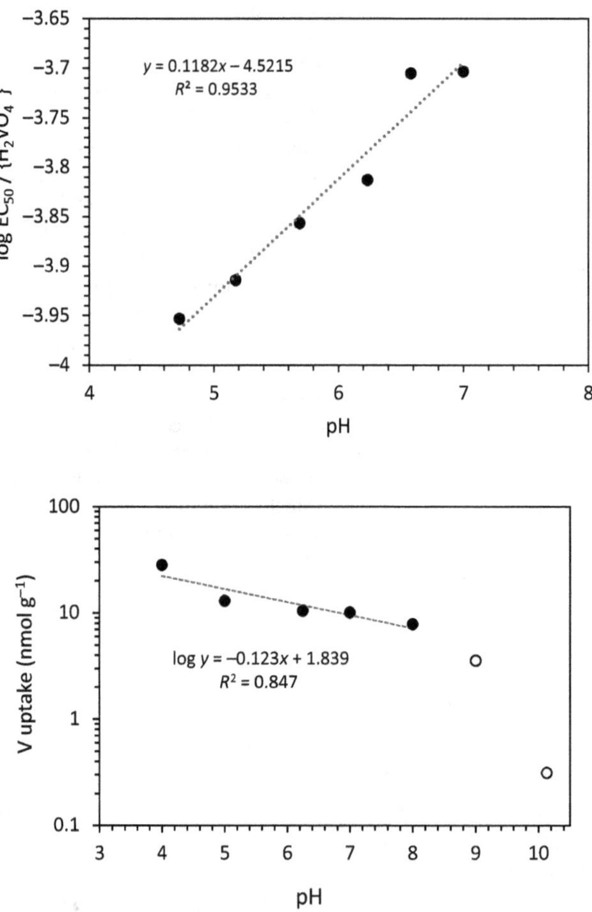

FIGURE 9.2
Top: pH dependence of log EC_{50} for vanadate toxicity to wheat (wheat root length; data from Ji et al., 2020), where the EC_{50} is expressed on the basis of $\{H_2VO_4^-\}$. Bottom: pH dependence of vanadium uptake to barley roots (data from Welch, 1973). The two highest pH points were not included in the regression as they are outside of the predominance field of $H_2VO_4^-$.

for wheat (0.12). Although the close correspondence between these figures may be purely coincidental, the result nevertheless suggests that vanadate uptake and ecotoxicity is moderately pH-dependent, with maximum uptake and effect (as reflected by low EC_{50}) at low pH.

Because an important part of the ecotoxic effect of vanadium is considered a result of its interaction with *o*-phosphate, it is of interest to compare the pH-dependence of vanadate uptake to that of *o*-phosphate. It seems clear that *o*-phosphate is predominantly taken up as $H_2PO_4^-$ (Sentenac and Grignon, 1985; Rausch and Bucher, 2002). Again, most studies report a weak pH dependence of $H_2PO_4^-$ uptake similar to that observed for vanadate, that is, that the uptake of $H_2PO_4^-$ is more efficient at low pH (e.g., Sentenac and Grignon, 1985). This has been interpreted as evidence for H^+-assisted $H_2PO_4^-$ transport into the cells of the root symplasm (Rausch and Bucher, 2002).

Due to this similarity in uptake behavior, it could be tempting to expect a strong interaction between vanadate and *o*-phosphate such that increasing *o*-phosphate in solution would lead to less vanadium uptake (and ecotoxicity). However, Ji et al. (2020) showed that vanadate uptake to wheat roots appeared to be insensitive to *o*-phosphate, when $\{H_2PO_4^-\}$ was lower than $\sim 0.1 \times 10^{-3}$ (equivalent to a concentration of slightly more than 0.1 mM). This was in contrast to the effect of *o*-phosphate on As(V) ecotoxicity, for which EC_{50} increased nearly threefold over the range from $\{H_2PO_4^-\} = 0.01 \times 10^{-3}$ to 0.08×10^{-3}. Only at higher $\{H_2PO_4^-\}$ was EC_{50} of vanadate affected in the expected way (Figure 9.3).

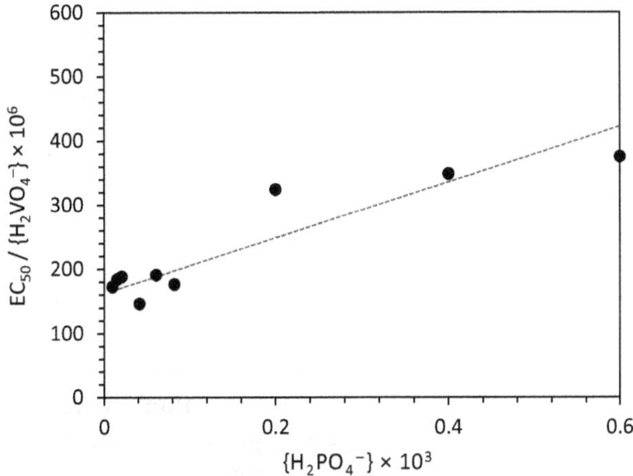

FIGURE 9.3
Relationship between EC_{50} of $\{H_2VO_4^-\}$ and $\{H_2PO_4^-\}$ for V uptake to wheat roots (data from Ji et al., 2020). The dotted line is the fit using the biotic ligand model with log $K_{BL/P}$ set to 23 (see text).

In another "hydroponic" study focusing on the interaction between vanadate and *o*-phosphate concerning uptake to chickpea roots and shoots, Imtiaz et al. (2017) observed reduced V uptake over the whole range of *o*-phosphate concentrations used. However, in this case, the lowest applied P concentration was 0.146 mM, and thus the results of Imtiaz et al. (2017) need not be contradictory to the results of Ji et al. (2020). In addition, the V-P interaction may not need to be the same concerning V uptake as for V ecotoxicity.

The reason behind the relatively modest influence of P on V ecotoxicity is probably related both to differences in uptake mechanisms and to V redox chemistry. Vanadium appears to be complexed to different constituents, for example, polysaccharides such as pectin, in the cell walls of roots. This probably represents the most important uptake mechanism for vanadium (Hou et al., 2014; Gurau et al., 2015; Hou et al., 2020). Although both vanadate(V) and vanadyl(IV) may be complexed to these constituents, vanadyl(IV) is expected to be complexed more strongly to such ligands. This, together with the presence of redox-active groups on these ligands, facilitates the conversion of some vanadium(V) to the vanadium(IV) oxidation state (Gurau et al., 2015). Further, as was mentioned earlier, the vanadium may also be transported into vacuoles, where thiol-containing reductants such as glutathione and catechol may convert a large part of the remaining vanadate to vanadium(IV), which is stored as organic complexes (e.g., Crans et al., 2010). These reactions are important detoxification mechanisms, and further, they provide an explanation both to the low degree of translocation of vanadium from roots and probably also to the relatively modest interaction with *o*-phosphate. Similar types of reactions also occur in other parts of the plant, after translocation from the roots. For example, Arsic et al. (2020) showed that vanadate, once it entered the leaf tissue, showed a very similar behavior as *o*-phosphate and could even be used as an *o*-phosphate tracer. However, after 24 hours most of the vanadate had been reduced to vanadium(IV). As was mentioned earlier, the interpretation of these reactions as evidence for detoxification are complicated by the fact that also certain vanadium(IV) species are able to interact with phosphatase and other enzymes. To conclude, many details remain unclear of the exact mechanisms on how the vanadium detoxification mechanisms in plants operate and the extent to which they govern the final ecotoxicity of vanadium, and also of to what extent such mechanisms are species-dependent.

Apart from the previously described interactions with H^+ and *o*-phosphate, vanadium uptake could be influenced also by other chemical variables. For example, magnesium was found to aggravate vanadium toxicity (Olness et al., 2001, 2005), and in another study, mercury and vanadium were found to interact in complex ways, which affect the uptake of both (Hou et al., 2019). However, the nature and extent of these interactions remain poorly known, and additional research needs to be carried out before general statements can be made on their importance.

9.3 Predicting Vanadium Ecotoxicity in Soil: Towards a Conceptual Model

As noted by Ji et al. (2020), so far there have been few, if any, efforts to develop BLMs for oxyanionic constituents. Therefore, they designed a study to provide data for toxic effects on wheat roots, which can be used to derive a BLM. Their data on the roles of H^+ and *o*-phosphate on vanadate ecotoxicity have been shown in the previous section (Figures 9.2 and 9.3). These data can be used to derive a preliminary BLM. As a starting point for the model, we apply the approach of Deleebeeck et al. (2007), who developed models for Ni^{2+} ecotoxicity to fish. Using this model approach, we define an ecotoxicity parameter Q, which is constant under all water chemistry conditions. In the case of vanadium, Q is defined as follows:

$$Q = -\log \frac{\{HVO_4^{\ 2-}\}}{1 + K_{P\text{-}BL}\{PO_4^{\ 3-}\}\{H^+\}^2} + S_{pH} \cdot pH \tag{9.1}$$

Here, $\{HVO_4^{2-}\}$ and $\{PO_4^{3-}\}$ are used rather than $\{H_2VO_4^-\}$ and $\{H_2PO_4^-\}$, which were used in Figures 9.2 and 9.3. The reason for this was to facilitate the implementation of these reactions into Visual MINTEQ (Gustafsson, 2020), which was used as the calculation engine in the current section. In Visual MINTEQ, HVO_4^{2-} and PO_4^{3-} are the so-called components (or master species) for vanadate and *o*-phosphate, respectively. In Equation 9.1, the *o*-phosphate BLM equilibrium constant, $K_{P\text{-}BL}$, is defined from:

$$K_{P\text{-}BL} = \frac{[BL\text{-}H_2PO_4]}{[BL^+]\{PO_4^{\ 3-}\}\{H^+\}^2} \tag{9.2}$$

where BL^+ represents a biotic ligand site. The conceptualization of the biotic ligand sites and complexes and their mathematical treatment are described elsewhere (di Toro et al., 2001). Further, the term S_{pH} in Equation 9.1 represents the slope of the logarithm of a toxicity endpoint, for example, EC_{50}, versus pH. To optimize the BLM, the first step is to determine S_{pH} from the data shown in Figure 9.2. In the second step, the data of Figure 9.3 are used to optimize $K_{P\text{-}BL}$ by use of Equation 9.2. Finally, these values are inserted into Equation 9.1 to determine the value of Q. Alternatively, Q can be determined directly from a plot of $-\log\{HVO_4^{2-}\}$ at EC_{50} vs. pH, where Q is the intercept. For the EC_{50} of wheat roots, this results in the following values:

$$S_{pH} = 1.1182; \log K_{P\text{-}BL} = 23; Q = 13.27$$

The value of S_{pH} is the same as the slope shown in Figure 9.2, although the value is exactly 1 unit higher, since S_{pH} is determined on the basis of HVO_4^{2-}

rather than $H_2VO_4^-$. The fits of the optimized BLM to the data of Figures 9.2 and 9.3 are shown as dotted lines in the figures. The overall good fits to the data suggest that the described BLM approach may be a promising tool. However, to be used as a risk assessment tool for plant ecotoxicity, additional data sets on for example, pH and P-dependent toxicity need to be assembled for different taxonomic groups of plants including effects in different assays, that is, both on for example, root length and shoot growth.

In any case, by use of Visual MINTEQ it is possible to use the optimized BLM for wheat root toxicity to illustrate how pH-dependent sorption is likely to influence the ecotoxicity in a soil where vanadium(V) (as vanadate) is the most stable oxidation state of vanadium in the soil solution. For this assessment, we use four soils from the previous work of Larsson, Hadialhejazi et al. (2017), in which pH-dependent sorption data were used to develop extended Freundlich equations describing vanadate adsorption. The general form of the extended Freundlich equation for vanadium is (Larsson, Hadialhejazi et al., 2017):

$$Q_{ads} = K_F\,([V]\,\{H^+\}^{0.36})^m \qquad \text{(Equation 9.3)}$$

where Q_{ads} is the adsorbed amount of vanadium (mol kg^{-1} dry soil), [V] is the total dissolved vanadium (mol L^{-1}), while K_F and m are coefficients that are optimized. Equation 9.3 can be used to predict the adsorbed amount of vanadium at EC_{50} and as a function of pH between pH 4.7 and 8, which was the range of pHs used by Ji et al. (2020). For all four soils in this example, $\{HVO_4^{2-}\}$ at EC_{50} was calculated from Equation 9.1, then Visual MINTEQ was used to calculate the total dissolved vanadate concentration ([V] in Equation 9.3) at an ionic strength of 0.003 mol L^{-1}, and finally Equation 9.3 was used to calculate the amount of adsorbed vanadium at EC_{50}. The resulting predictions are seen in Figure 9.4. When interpreting this result, it is important to keep in mind that the adsorbed amounts represent V that is geochemically active on a time scale of one week (which was the equilibration time in the adsorption experiments), that is, "ageing" reactions are not considered (Baken et al., 2012), and V bound to primary minerals, which may be a significant part of soil V (Larsson, Hadialhejazi et al., 2017), is not included either.

As seen in Figure 9.4, there are very large differences between the soils in terms of their sensitivity to vanadium ecotoxicity. The content of oxalate-extractable Fe and Al, which indicate the content of reactive Fe and Al (hydr)ous oxide surfaces, is one important governing parameter that explains these differences. While Kloten Bs had a very high value of oxalate-extractable Fe+Al (994 mmol kg^{-1}), Säby had an intermediate content (125 mmol kg^{-1}), while Pustnäs and Zwijnaarde both had low contents (54 and 59 mmol kg^{-1}, respectively). Moreover, the results in Figure 9.4 suggest a slight pH-dependence such that the soils may contain a slightly higher amount of V at 50% toxicity when the pH is low. This pH-dependence is opposite to the one observed for vanadate toxicity in the solution phase (Figure 9.2). The

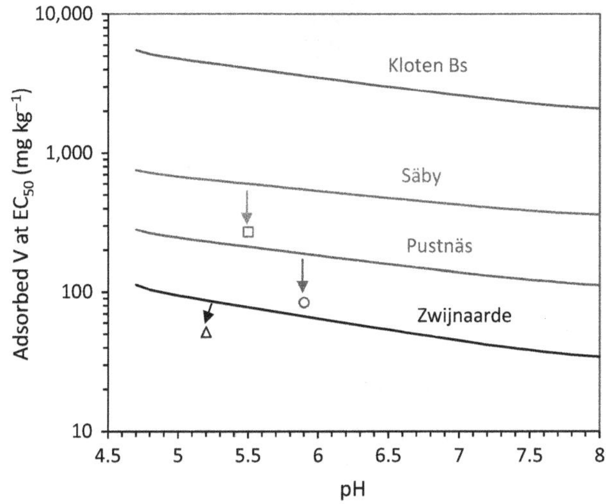

FIGURE 9.4
Lines: predicted concentration of adsorbed V at EC_{50} for wheat root length for four soils and as a function of pH, as obtained by combining the vanadate BLM with the extended Freundlich equation for V adsorption. The arrows point to observations of barley shoot growth (symbols), as determined by Larsson et al. (2013).

difference is explained by pH-dependent sorption. Equation 9.3 shows that the adsorbed amount is proportional to the term $\{H^+\}^{-0.36}$. Since the models were optimized for the pH range when $H_2VO_4^-$ is the dominating V species, this means that the sorption increases 0.36 log units for every unit decrease in pH, while log EC_{50} instead decreases 0.12 log units. The larger pH dependence of sorption will therefore be the determining factor governing the pH-dependence trend of adsorbed V at EC_{50} as a function of pH.

For comparison, the adsorbed amounts at EC_{50} are shown also for barley shoot growth, which was analyzed for three of the soils by Larsson et al. (2013). As Figure 9.4 shows, the barley shoot growth EC_{50}s are consistently lower than those of wheat root lengths, showing a higher tolerance to V in the case of the wheat root length assay.

In the calculations of this example, total dissolved P was kept at a low value (10 µmol L^{-1}), which affected the result only to a very low extent. Currently, P is not included in the empirical Freundlich model used, and for this reason we cannot fully consider the effect of P in the model used in this example. However, consideration of P is not likely to have substantial effects on the predicted adsorbed V at EC_{50}. As was previously discussed, *o*-phosphate had only small effects on V ecotoxicity at values below 100 µmol L^{-1}. Similarly, *o*-phosphate does not have strong effects on vanadate sorption (Larsson, Hadialhejazi et al., 2017). In addition, these two effects are likely to counteract each other, as increasing *o*-phosphate will both prevent uptake of vanadate to plants and at the same time lead to an increased solubility of vanadate in the soil solution because of desorption of vanadate from soil particle surfaces. For these reasons, the P term in the suggested vanadate-BLM is likely to be of any importance only in, for example, overfertilized soils where dissolved P is in the mg L^{-1} range. It is important to keep in mind, however, that so far this conclusion is based only on one high-quality data set (the one of Ji et al., 2020), and additional data would be needed to confirm these trends.

9.4 Conclusions and Outlook

Risk assessments of vanadium-contaminated soils, focusing on the risk of plant ecotoxicity, would benefit from use of a vanadium BLM or a similar type of model, in which vanadium ecotoxicity can be predicted from soil chemical parameters. Such a BLM requires knowledge on toxic effects for plants in different taxonomic groups and also of chemical interactions that affect vanadium uptake and ecotoxicity. As for the chemical interactions that are in focus in this chapter, the available evidence for wheat indicates that for the soil solution, the EC_{50} values of vanadium increase with increasing pH. When expressing the EC_{50} values on a dry soil basis, the inverse pH dependence (decreasing EC_{50} with increasing pH) is suggested, at least for wheat, which is explained by the effect of vanadate sorption onto soil particles, which has an opposite pH dependence that is stronger. The effect of *o*-phosphate on vanadium ecotoxicity is predicted to be small or insignificant under most soil chemical conditions. The vanadate BLM used also suggests, perhaps not surprisingly, that different soils have very different sensitivity to vanadium ecotoxicity.

However, these conclusions are based on very few data and need independent confirmation from other studies where results for a larger range of plants and/or assays have been analyzed. Moreover, the significance of the interactions with other chemical constituents (i.e., other than pH and *o*-phosphate) requires additional study. This is particularly true for magnesium (Olness et al., 2001, 2005). Thus, the development of a more generic model to predict vanadium uptake and ecotoxicity from soil chemistry is still at a very early stage.

References

Aihemaiti, A., Jiang, J., Li, D., Liu, N., Yang, M., Meng, Y., Zou, Q. 2018. The interactions of metal concentrations and soil properties on toxic metal accumulation of native plants in vanadium mining area. *Journal of Environmental Management* 222, 216–226, https://doi.org/10.1016/j.jenvman.2018.05.081.

Arsic, M., Le Tougaard, S., Persson, D.P., Martens, H.J., Doolette, C.L., Lombi, E., Schjoerring, J.K., Husted, S. 2020. Bioimaging techniques reveal foliar phosphate uptake pathways and leaf phosphorus status. *Plant Physiology* 183, 1472–1483, https://doi.org/10.1104/pp.20.00484.

Baken, S., Larsson, M.A., Gustafsson, J.P., Cubadda, F., Smolders, E. 2012. Ageing of vanadium in soils and consequences for bioavailability. *European Journal of Soil Science* 63, 839–847, https://doi.org/10.1111/j.1365-2389.2012.01491.x.

Barrow, N.J. 1983. A mechanistic model for describing the sorption and desorption of phosphate by soil. *Journal of Soil Science* 34, 733–750, https://doi.org/10.1111/j.1365-2389.1983.tb01068.x.

Costa Pessoa, J.C., Garribba, E., Santos, M.F.A., Santos-Silva, T. 2015. Vanadium and proteins: Uptake, transport, structure, activity and function. *Coordination Chemistry Reviews* 301–302, 49–86, https://doi.org/10.1016/j.ccr.2015.03.016.

Crans, D.C., Smee, J.J., Gaidamauskas, E., Yang, L. 2004. The chemistry and biochemistry of vanadium and the biological activities exerted by vanadium compounds. *Chemical Reviews* 104, 849–902, https://doi.org/10.1021/cr020607t.

Crans, D.C., Zhang, B., Gaidamauskas, E., Keramidas, A.D., Willsky, G.R., Roberts, C.R. 2010. Is vanadate reduced to thiols under biological conditions? Changing the redox potential of V(V)/V(IV) by complexation in aqueous solution. *Inorganic Chemistry* 49, 4245–4256, https://doi.org/10.1021/ic100080k.

Darnajoux, R., Constantin, J., Miadlikowska, J., Lutzoni, F., Bellenger, J.-P. 2014. Is vanadium a biometal for boreal cyanolichens? *New Phytologist* 202, 765–771, https://doi.org/10.1111/nph.12777.

Deleebeeck, N.M.E., de Schamphelaere, K.A.C., Janssen, C.R. 2007. A bioavailability model predicting the toxicity of nickel to rainbow trout (*Oncorhynchus mykiss*) and fathead minnow (*Pimephales promelas*) in synthetic and natural waters. *Ecotoxicology and Environmental Safety* 67, 1–13, https://doi.org/10.1016/j.ecoenv.2006.10.001.

Di Toro, D.M., Allen, H.E., Bergman, H.L., Meyer, J.S., Paquin, P.R., Santore, R.C. 2001. Biotic ligand model of the acute toxicity of metals. 1. Technical basis. *Environmental Toxicology and Chemistry* 20, 2383–2396, https://doi.org/10.1002/etc.5620201034.

El-Naggar, A., Shaheen, S.M., Chang, S.X., Hou, D., Ok, Y.S., Rinklebe, J. 2021. Biochar surface functionality plays a vital role in (im)mobilization and phytoavailability of soil vanadium. *ACS Sustainable Chemistry and Engineering in Press*, https://doi.org/10.1021/acssuschemeng.1c01656.

García-Jiménez, A., Trejo-Téllez, L.I., Guillén-Sánchez, D., Gómez-Merino, F.C. 2018. Vanadium stimulates pepper plant growth and flowering, increases concentrations of amino acids, sugars and chlorophylls, and modifies nutrient concentrations. *PloS One* 13(8), e0201908, https://doi.org/10.1371/journal.pone.0201908.

Gil, J., Alvarez, C.E., Martinez, M.C., Perez, N. 1995. Effect of vanadium on lettuce growth, cationic nutrition, and yield. *Journal of Environmental Science and Health* A30, 73–87, https://doi.org/10.1080/10934529509376186.

Gurau, G., Palma, A., Lauro, G.P., Mele, E., Senette, C., Manunza, B., Delana, S. 2015. Detoxification processes from vanadate at the root apoplasm activated by caffeic and polygalacturonic acids. *PloS One* 10(10), e0141041, https://doi.org/10.1371/journal.pone.0141041.

Gustafsson, J.P. 2019. Vanadium geochemistry in the biogeosphere—speciation, solid-solution interactions, and ecotoxicity. *Applied Geochemistry* 102, 1–25, https://doi.org/10.1016/j.apgeochem.2018.12.027.

Gustafsson, J.P. 2020. Visual MINTEQ version 3.1. Web. http://vminteq.lwr.kth.se.

Hou, M., Huo, Y., Yang, X., He, Z. 2020. Chemical form and subcellular distribution of vanadium in corn seedlings. *Microchemical Journal* 153, 104468, https://doi.org/10.1016/j.microc.2019.104468.

Hou, M., Li, M., Yang, X., Pan, R. 2019. Responses of nonprotein thiols to stress of vanadium and mercury in maize (*Zea Mays* L.) seedlings. *Bulletin of Environmental Contamination and Toxicology* 102, 425–431, https://doi.org/10.1007/s00128-019-02553-w.

Hou, M., Lu, C., wie, K. 2014. Accumulation and speciation of vanadium in *Lycium* seedling. *Biological Trace Element Research* 159, 373–37 8, https://doi.org/10.1007/s12011-014-0014-8.

Imtiaz, M., Ashraf, M., Rizwan, M.S., Nawaz, M.A., Rizwan, M., Mehmood, S., Yousaf, B., Yuan, Y., Ditta, A., Mumtaz, M.A., Ali, M., Mahmood, S., Tu, S. 2018. Vanadium toxicity in chickpea (*Cicer arietinum* L.) grown in red soil: Effects on cell death, ROS and antioxidative systems. *Ecotoxicology and Environmental Safety* 158, 139–144, https://doi.org/10.1016/j.ecoenv.2018.04.022.

Imtiaz, M., Rizwan, M.S., Mushtaq, M.A., Yousaf, B., Ashraf, M., Ali, M., Yousuf, A., Rizwan, M., Din, M., Dai, Z., Xiong, S., Mehmood, S., Tu, S. 2017. Interactive effects of vanadium and phosphorus on their uptake, growth and heat shock proteins in chickpea genotypes under hydroponic conditions. *Environmental and Experimental Botany* 134, 72–81, http://dx.doi.org/10.1016/j.envexpbot.2016.11.003.

Ji, J., He, E., Qiu, H., Peijnenburg, W.J.G.M., van Gestel, C.A.M., Cao, X. 2020. Effective modelling framework for quantifying the potential impact of coexisting anions on the toxicity of arsenate, selenite, and vanadate. *Environmental Science and Technology* 54, 2379–2388, https://doi.org/10.1021/acs.est.9b06837.

Kaplan, D.I., Adriano, D.C., Carlson, C.L., Sajwan, K.S. 1990. Vanadium-toxicity and accumulation by beans. *Water Air Soil Pollution* 49, 81–91, https://doi.org/10.1007/BF00279512.

Kaplan, D.I., Sajwan, K.S., Adriano, D.C., Gettier, S. 1990. Phytoavailability and toxicity of beryllium and vanadium. *Water Air Soil Pollution* 53, 203–212, https://doi.org/10.1007/BF00170737.

Larsson, M.A., Baken, S., Gustafsson, J.P., Hadialhejazi, G., Smolders, E. 2013. Vanadium bioavailability and toxicity to soil microorganisms and plants. *Environmental Toxicology and Chemistry* 32, 2266–2273, https://doi.org/10.1002/etc.2322.

Larsson, M.A., Baken, S., Smolders, E., Cubadda, F., Gustafsson, J.P. 2015. Vanadium bioavailability in soils amended with blast furnace slag. *Journal of Hazardous Materials* 296, 158–165, https://doi.org/10.1016/j.jhazmat.2015.04.034.

Larsson, M.A., D'Amato, M., Cubadda, F., Raggi, A., Öborn, I., Kleja, D.B., Gustafsson, J.P. 2015. Long-term fate and transformations of vanadium in a pine forest soil with added converter lime. *Geoderma* 259–260, 271–278, https://doi.org/10.1016/j.geoderma.2015.06.012.

Larsson, M.A., Hadialhejazi, G., Gustafsson, J.P. 2017. Vanadium sorption by mineral soils: Development of a predictive model. *Chemosphere* 168, 925–932, https://doi.org/10.1016/j.chemosphere.2016.10.117.

Larsson, M.A., Persson, I., Sjöstedt, C., Gustafsson, J.P. 2017. Vanadate complexation to ferrihydrite: X-ray absorption spectroscopy and CD-MUSIC modelling. *Environmental Chemistry* 14, 141–150, https://doi.org/10.1071/EN16174.

Olness, A., Gesch, R., Forcella, F., Archer, D., Rinke, J. 2005. Importance of vanadium and nutrient ionic ratios on the development of hydroponically grown cuphea. *Industrial Crops and Products* 21, 165–171, https://doi.org/10.1016/j.indcrop.2004.02.005.

Olness, A., Palmquist, D., Rinke, J. 2001. Ionic ratios and crop performance: II. Effects of interactions amongst vanadium, phosphorus, magnesium, and calcium on soybean yield. *Journal of Agronomy and Crop Science* 187, 47–52, https://doi.org/10.1046/j.1439-037X.2001.00499.x.

Peacock, C.L., Sherman, D.M., 2004. Vanadium(V) adsorption onto goethite (α-FeOOH) at pH 1.5 to 12: A surface complexation model based on ab initio molecular geometries and EXAFS spectroscopy. *Geochimica et Cosmochimica Acta* 68, 1723–1733, https://doi.org/10.1016/j.gca.2003.10.018.

Qian, Y., Gallagher, F.J., Feng, H., Wu, M., Zhu, Q. 2014. Vanadium uptake and translocation in dominant plant species on an urban coastal brownfield site. *Science of the Total Environment* 476–477, 696–704, https://doi.org/10.1016/j.scitotenv.2014.01.049.

Rausch, C., Bucher, M. 2002. Molecular mechanisms of phosphate transport in plants. *Planta* 216, 23–37, https://doi.org/10.1007/s00425-002-0921-3.

Rehder, D., 2015. The role of vanadium in biology. *Metallomics* 7, 730–742, https://doi.org/10.1039/c4mt00304g.

Saco, D., Martín, S., San José, P. 2013. Vanadium distribution in roots and leaves of *Phaseolus vulgaris*: Morphological and ultrastructural effects. *Biologia Plantarum* 57, 128–132, https://doi.org/10.1007/s10535-012-0133-z.

Schwartz, A., Illan, N., Assmann, S.M. 1991. Vanadate inhibition of stomatal opening in epidermal peels of *Commelina communis*. *Planta* 183, 590–596, https://doi.org/10.1007/BF00194281.

Seargeant, L.E., Stinson, R.A. 1979. Inhibition of human alkaline phosphatases by vanadate. *Biochemical Journal* 181, 247–250, https://doi.org/10.1042/bj1810247.

Sentenac, H., Grignon, C. 1985. Effect of pH on orthophosphate uptake by corn roots. *Plant Physiology* 77, 136–141, https://doi.org/10.1104/pp.77.1.136.

Shaheen, S.M., Alessi, D.S., Tack, F.M.G., Ok, Y.S., Kim, K.-H., Gustafsson, J.P., Sparks, D.L., Rinklebe, J. 2019. Redox chemistry of vanadium in soils and sediments: Interactions with colloidal materials, mobilization, speciation, and relevant environmental implications—a review. *Advances in Colloid and Interface Science* 265, 1–13, https://doi.org/10.1016/j.cis.2019.01.002.

Smith, P.G., Boutin, C., Knopper, L. 2013. Vanadium pentoxide phytotoxicity: Effects of species selection and nutrient concentration Arch. *Environmental Contamination and Toxicology* 64, 87–96, https://doi.org/10.1007/s00244-012-9806-z.

Templeton, G.D., Chasteen, N.D. 1980. Vanadium-fulvic acid chemistry: Conformational and binding studies by electron spin probe techniques. *Geochimica et Cosmochimica Acta* 44, 741–752, https://doi.org/10.1016/0016-7037(80)90163-5.

Thakali, S., Allen, H.E., di Toro, D.M., Ponizovsky, A.A., Rooney, C.P., Zhao, F.J., McGrath, S.P. 2006. A terrestrial biotic ligand model. 1. Development and application to Cu and Ni toxicities to barley root elongation in soils. *Environmental Science and Technology* 40, 7085–7093, https://doi.org/10.1021/es061171s.

Tian, L., Yang, J., Alewell, C., Huang, J.H. 2014. Speciation of vanadium in Chinese cabbage (*Brassica rapa* L.) and soils in response to different levels of vanadium in soils and cabbage growth. *Chemosphere* 111, 89–95, https://doi.org/10.1016/j.chemosphere.2014.03.051.

Tian, L., Yang, J., Huang, J.H. 2015. Uptake and speciation of vanadium in the rhizosphere soils of rape (*Brassica juncea* L.). *Environmental Science and Pollution Research* 22, 9215–9223, https://doi.org/10.1007/s11356-014-4031-0.

van Sprang, P.A., Nys, C., Blust, R.J.P., Chowdhury, J., Gustafsson, J.P., Janssen, C.J., de Schamphelaere, K.A.C. 2016. The derivation of effects threshold concentrations of lead for European freshwater ecosystems. *Environmental Toxicology and Chemistry* 35, 1310–1320, https://doi.org/10.1002/etc.3262.

Welch, R.M. 1973. Vanadium uptake by plants. Absorption kinetics and the effect of pH, metabolic inhibitors, and other anions and cations. *Plant Physiology* 51, 828–832, https://doi.org/10.1104/pp.51.5.828.

Wu, C.Y., Asano, M., Hashimoto, Y., Rinklebe, J., Shaheen, S.M., Wang, S.L., Hseu, Z.Y. 2020. Evaluating vanadium bioavailability to cabbage in rural soils using geochemical and micro-spectroscopic techniques. *Environmental Pollution* 258, 113699, https://doi.org/10.1016/j.envpol.2019.113699.

Wu, Z.Z., Yang, J.Y., Zhang, Y.X., Wang, C.Q., Guo, S.S., Yu, Y.Q. 2021. Growth responses, accumulation, translocation and distribution of vanadium in tobacco and its potential in phytoremediation. *Ecotoxicology and Environmental Safety* 207, 11129 7, https://doi.org/10.1016/j.ecoenv.2020.111297.

Yang, J.Y., Tang, Y. 2015. Accumulation and biotransformation of vanadium in *Opuntia microdasys*. *Bulletin of Environmental Contamination and Toxicology* 94, 448–452.

Yang, J.Y., Wang, M., Jia, Y.B., Gou, M., Zeyer, J. 2017. Toxicity of vanadium in soil on soybean at different growth stages. *Environmental Pollution* 231, 48–58, https://doi.org/10.1016/j.envpol.2017.07.075.

10

Vanadium in Plants: Present Scenario and Future Prospects

Sheikh Mansoor[1¥], Tawseef Rehman Baba[2¥], Syed Inam ul Haq[3¥], Iqra F. Khan[2¥], Sofora Jan[4], Sadaf Rafiq[5], Jörg Rinklebe[6] and Parvaiz Ahmad[7,8]

[1] *Division of Biochemistry, Sher-e-Kashmir University of Agricultural Sciences and Technology of Jammu, Jammu and Kashmir, India*

[2] *Division of Fruit Sciences, Faculty of Horticulture, Sher-e-Kashmir University of Agricultural Sciences and Technology of Kashmir, Srinagar, Jammu and Kashmir, India*

[3] *Division of Plant Biotechnology, Sher-e-Kashmir University of Agricultural Sciences and Technology of Kashmir, Srinagar, Jammu and Kashmir, India*

[4] *Division of Genetics and Plant Breeding, Faculty of Agriculture, Sher-e-Kashmir University of Agricultural Sciences and Technology of Kashmir, Srinagar, Jammu and Kashmir, India*

[5] *Division of Floriculture and Landscape Architecture, Sher-e-Kashmir University of Agricultural Sciences and Technology of Kashmir, Srinagar, Jammu and Kashmir, India*

[6] *University of Wuppertal, School of Architecture and Civil Engineering, Institute of Foundation Engineering, Water and Waste Management, Laboratory of Soil and Groundwater Management, Wuppertal, Germany*

[7] *Botany and Microbiology Department, College of Science, King Saud University, Riyadh, Saudi Arabia*

[8] *Department of Botany, Government Degree College Pulwama, Pulwama, Jammu and Kashmir, India*

¥ *authors with equal contribution*

CONTENTS

10.1 Introduction 206
10.2 Sources of Vanadium 207
10.3 Bioavailability of Vanadium 207
10.4 Translocation of Vanadium in Plants 207
10.5 Vanadium Bioaccumulation and Interactions in Plants 209
    10.5.1 Vanadium Bioaccumulation 209
    10.5.2 Plant Vanadium Interactions 210

DOI: 10.1201/9781003173274-10

10.5.3 Effect of Vanadium on Plant Growth and Development....212
10.6 Phytoremediation....214
10.7 Bioremediation....216
10.8 Conclusion and Future Prospectus....217
References....217

10.1 Introduction

Vanadium (Z=23) is a transition metal having ubiquitous distribution in the environment. Andrés Manuel Del Río discovered vanadium in 1801 while analyzing a lead mineral found in Mexico and named it erythronium. However, it was declared to be impure chromium and not a new element by a French chemist, Collet-Descostils, in 1805. It was later rediscovered and identified by Sefstrom in 1830 from a mineral extracted in Sweden, and he named it vanadium. The metal has a steel gray color and has been named after the goddess of beauty and fertility (Dutta and Lodhari 2018). Vanadium is found in almost all organisms in trace amounts. Vanadium ranks 22th among the elements and fifth among transition metals in terms of abundance in the crust.

China is the largest producer of vanadium in the world. Vanadium is found in different oxidation states in nature as –1, 0, +2, +3, +4 and +5, the most abundantly present being V(+4) and V(+5) (Amorim, Welz et al. 2007). Average concentration of V ranges from 10 to 220 mg/kg in different soils (Imtiaz, Rizwan et al. 2015). In the oceans, vanadium ranks second in abundance among the transition metals (molybdate being the most abundant). The approximate concentration of V in oceans is 30–35 nM existing as Mn+ H_2VO_4 (where Mn+ = dissolved cations). In fresh water, the concentration of vanadium is about 10 nM (Del Carpio, Hernández et al. 2018). V is found in all plants in low concentrations. The initial studies identified V as a highly toxic element to plants. However, Bertrand and Lee (1950) reported that at low concentrations (10 ng/g V) in soil, V showed positive influence on plant growth. Low levels of V in soils (lt; 2 mg/kg) increases chlorophyll synthesis, aids potassium utilization by the plants and improves nitrogen fixation (Tripathi, Mani et al. 2018). At higher concentrations, V causes chlorosis, impairs plant growth, alters physiological and biochemical activities, inhibits chlorophyll and protein production and increases production of ROS (Hou, Li et al. 2019; Chen, Liu et al. 2020). Vanadate (V+5) can inhibit the plasma membrane hydrogen (H+)-translocation ATPase which plays an important role in nutrient uptake by plant cells, thus interfering with the nutrient uptake of plants (Vachirapatama, Jirakiattiku et al. 2011).

10.2 Sources of Vanadium

Natural as well as anthropogenic sources contribute to the V content in soil and environment. The mechanical and chemical weathering of parental rocks and the presence of V containing minerals in subsoil constitute the natural source of vanadium in soils. Higher concentrations of vanadium are found in basic rocks compared to acidic or neutral rocks (Barceloux and Barceloux 1999). The primary V containing minerals present in soils include titanomagnetite, vanadinite, roscoelite, bravoite and patronite (Fang, Wu et al. 2017). Fossil fuels like coal and petroleum contain vanadium; hence the increased use of coal and petroleum has resulted in a higher influx of vanadium into the environment through combustion and refinery processes. In aquatic environments, the natural sources that introduce V are rock weathering and sediment leaching. Fertilizers, sewage sludge and domestic wastewater discharge into the aquatic environments also lead to vanadium enrichment of water bodies (Shi, Mangal et al. 2016).

10.3 Bioavailability of Vanadium

Soil factors like redox-status, pH, soil solution composition, organic matter content and presence of iron, aluminum and manganese (hydr)oxides influence the bioavailability of V in soils (Reijonen 2017). Of the various oxidation states of V, the higher oxidation states, that is, V(+3), V(+4) and V(+5) are prevalent in the biosphere. V(+5) and V(+4) are the stable forms found in oxidizing conditions, whereas V(+3) is an unstable form dominant in highly reducing conditions. Under oxidizing conditions, V(+3) is readily oxidized to V(+4) or V(+5) (Shaheen, Alessi et al. 2019). Chemical availability of V(+5) is reduced by its adsorption onto the surface of (hydr)oxides. This can be reversed by anions such as –OH, phosphate, fluoride and arsenate that replace V(+5) from sorption sites by releasing H_2VO_{4-} back to soil solution, making it available for plant uptake (Bradl 2004; Reijonen 2017; Shaheen, Alessi et al. 2019).

10.4 Translocation of Vanadium in Plants

Vanadium (V) behavior in the soil-plant environment can be comprehended by absorption and translocation of Vanadium by plants. In recent years, owing to the dramatic rise in vanadium mining processes and their harmful effects at elevated levels on plant and human health, many studies have concentrated on vanadium uptake by plants (Del Carpio, Hernández et al. 2018). Vanadium pentoxide (V_2O_5) is the most commonly existing and usable form of vanadium, along with ammonium metavanadate (NH_4VO_3), sodium metavanadate ($NaVO_3$) and sodium orthovanadate (Na_3VO_4). Numerous reports indicate that vanadium transport or translocation and bioaccumulation depend upon its two oxidation states, viz., tetravalent and pentavalent forms, of which the latter is more mobile and more active in the vanadium biogeochemical cycle and exhibits higher toxicity to both plants and animals (Imtiaz, Rizwan et al. 2015; Roychoudhury 2020).

Vanadium translocation is a complex process involving the release of vanadium ions from the soil away from the rhizosphere into soil pore water (Teng, Yang et al. 2011; Yang, Teng et al. 2011) and absorption of these vanadium ions by the plant roots (Welch 1973; Khan, Kazi et al. 2013), followed by accumulation in the root tissues, which are then translocated to aerial parts of plant (Figure 10.1) (Weis and Weis 2004; Pilon-Smits 2005). In addition, numerous biogeochemical factors such as vanadium soil fractions and soil properties (pH and total organic carbon) play a key role in the accumulation and translocation of vanadium in plants (Weis and Weis 2004; Pilon-Smits 2005; Teng, Yang et al. 2011; Khan, Kazi et al. 2013; Chen, Liu et al. 2020). According to Qian, Gallagher et al. (2014), "soil vanadium concentration is the key factor affecting accumulation and root to shoot translocation of Vanadium in plants." The plant roots absorb vanadium primarily through

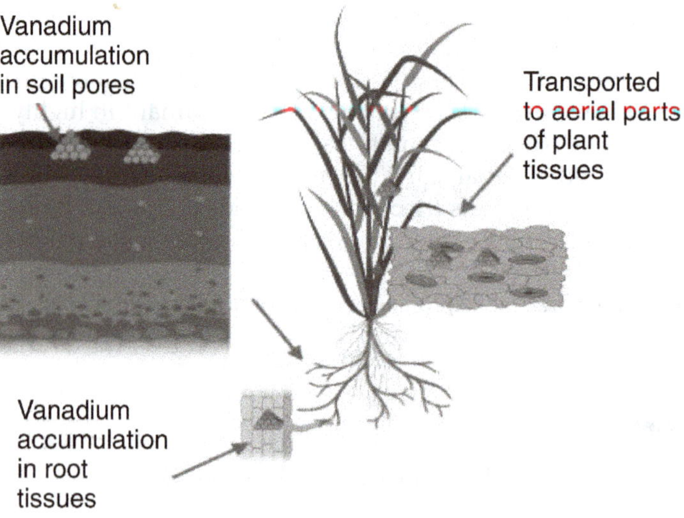

FIGURE 10.1
The vanadium pathway is a critical process that involves the release of vanadium ions from the soil into soil pore water, accumulation by root system and deposition in root tissues before being translocated to aerial areas of the plant.

some carriers that are responsible for assimilation of essential/vital plant nutrients, and this accumulation is reliant on plant species, with certain species being better accumulators than others (Chen, Liu et al. 2020). Moreover, vanadium uptake is an active transport mechanism, thus involving plant energy expenditure (Figure 10.1) (Kubes, Skalicky et al. 2019; Sun, Li et al. 2019). According to various studies, the tolerance mechanisms of plants to vanadium stress also influence the vanadium accumulation and translocation in plants (Tian, Yang et al. 2014; Tian, Yang et al. 2015). Generally, higher vanadium concentration is found in root tissues of plants, followed by stems and leaves (Qian, Gallagher et al. 2014; Tian, Yang et al. 2015). Moreover, vanadium translocation from roots to aerial or aboveground parts is limited and mainly depends on vanadium speciation inside the root tissue (Chen, Liu et al. 2020). After penetration of heavy metals including vanadium into the roots, they either accrete in the root tissues or are translocated to the aerial parts of plant (shoot and leaves) (Chen, Liu et al. 2020), and this translocation greatly depends on various factors such as plant species, transpiration rate and the availability of metals in the soil-plant environment. Moreover, vanadium mobility within the plants and competition with other elements might also affect its translocation (Chen, Liu et al. 2020). Potential vanadium accumulator plants pass elevated volumes of vanadium from roots to aboveground parts through symplasmic passage in the xylem (Thakur, Singh et al. 2016; Drava, Cornara et al. 2019). The infusion of vanadium from the soil pore water into the root followed by incorporation into the xylem takes place primarily in two phases. The first phase is the transport of vanadium from the soil water across the plasma membrane of the root cell walls (Thakur, Singh et al. 2016), followed by incorporation of these vanadium ions into the stele via symplasmic-transport, where they are afterwards freed into the xylem (Ghosh and Singh 2005). The transport of vanadium in the xylem is regulated by membrane carrier proteins, and the flow of vanadium through the casparian strip is an active process of transport (Drava, Cornara et al. 2019; Kumar, AlMomin et al. 2019). Vanadium may pass through various forms of cation channels in the cell membrane designed for other mineral element ions, such as those of phosphorus, while within a cell (Chen, Liu et al. 2020).

10.5 Vanadium Bioaccumulation and Interactions in Plants

10.5.1 Vanadium Bioaccumulation

Plants have developed diverse aggregation and exclusion processes in response to higher levels of toxic metal concentrations. In certain cells, tissues or organelles, toxic metals are dispersed during the aggregation process, thus reducing their spread in essential and vulnerable organs. These toxic metals may be transferred from roots to shoots during exclusion, thus limiting their absorption by roots (Hall 2002; Wang, Liu et al. 2008). According to researchers, aggregation or accumulation is the key solution technique to vanadium toxicity in a majority of plants (primarily in roots) when grown in media containing vanadium. According to Yang, Wang et al. (2017), there is a linear relationship between vanadium concentration in roots and its level in the growth media when used below threshold point for apparent toxicity in plants. Various plants, viz., *C. arietinum, G. max, O. microdasys, S. lycopersicum, S. tuberosum, M. sativa, Z. mays, Triticum, B. oleracea* etc., demonstrated high resistance and stored significant quantities of vanadium both in roots as well as aboveground tissues. Growing of such crops with high potential to absorb vanadium from soils enriched or contaminated with vanadium may be the reason for serious health issues both in animals as well as humans. On the other hand, *B. juncea, B. rapa* subsp. pekinensis and *L. ferocissimum* suck up meager quantities of vanadium when grown in media containing high levels of vanadium concentrations, thus making these plants suitable for vanadium enriched and polluted soils for reclamation purposes without any health concerns (Aihemaiti, Gao et al. 2020). In plants, vanadium accumulates mainly in cell walls at the subcellular level, which might be due to presence of higher concentration of proteins and polysaccharides along with many ligands in its composition (-OH, $-NH_2$ and -COOH). These ligands play a major role in the accumulation process through formation of chelates with vanadium ions and other heavy metals, thereby reducing their translocation and toxicity (Allan and Jarrell 1989; Rauser 1999). Chelation of vanadium ions in the root cell wall with polar compounds, development of stable Ca compounds in the root tissues and compartmentalization in root and stem vacuoles decrease the translocation of free vanadium ions, which might be the reason for low levels of aggregation of vanadium in aboveground plant parts (Aihemaiti, Gao et al. 2020).

10.5.2 Plant Vanadium Interactions

Low vanadium levels in plants have favorable impacts on plant development and growth. Excessive V, however, has been shown to induce various detrimental effects on plant development, growth and physiological parameters (Chen, Liu et al. 2020). The effects of vanadium toxicity and tolerance mechanisms are depicted in (Figure 10.2). According to the report by García-Jiménez, Trejo-Téllez et al. (2018), vanadium application @ 0.59 mg L^{-1} increased physiological and biochemical parameters such as plant height, stem width, number of leaves, amino acids and total sugars, etc. in pepper plants (*Capsicum annuum* L.). Moreover, the use of vanadium below 40 mg L^{-1} improved the sweet basil's (*Ocimum basilicum* L.) shoot and root dry mass (Akoumianaki-Ioannidou, Barouchas et al. 2016). Less than 2 mg kg^{-1} of vanadium greatly increased chlorophyll synthesis, nitrogen fixation and even potassium use, according to Imtiaz, Rizwan et al. (2015). Higher

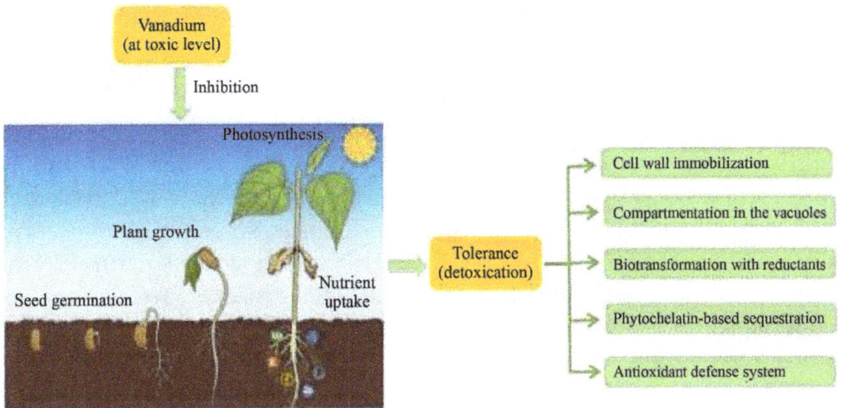

FIGURE 10.2
Vanadium toxicity and tolerance mechanisms (Chen, Liu et al. 2020).

concentrations, however, affect plants due to chlorosis and stunted growth. It was reported that plants grown in contaminated vanadium soils were lower in height than plants grown out of vanadium mines. Application of V caused toxic effects in onion roots and increased genotoxic effects as V uptake increased (Marcano, Carruyo et al. 2006). Further, it was found that the use of vanadium @ 2.0 mg L^{-1} impaired tobacco plant development. However, through self-adaptive regulation, tobacco plants showed comparatively decent vanadium tolerance and have the capacity to positively act as a phytostabilizer in cleaning the vanadium-contaminated environment. Yang, Wang et al. (2017) reported that soybean germination and biomass production were inhibited by high levels of V concentration. However, plants may produce self-defense systems to tolerate V toxicity in response to increasing concentrations of vanadium in soil. According to several studies, the efficiency of water usage and uptake capacity was reported to suppress in vanadium treated plants, which might result from a decrease in root surface area and root length, as roots are responsible for water absorption. Moreover, it was associated with attenuation of photosynthetic assimilation, stomatal conductance, intercellular carbon dioxide (CO_2) and transpiration, which provides the force to lift water from the root to leaf and generates sap flow to transport minerals and nutrients to aerial parts of the plant (Aihemaiti, Gao et al. 2020). Vanadium was also found to retard the growth of Chinese green mustard and tomato plants, especially for nutrient solutions containing more than 40 mg/l ammonium metavanadate (NH_4VO_3) (Vachirapatama, Jirakiattiku et al. 2011). In general, low levels of V positively regulates plant growth and development, but excessive V accumulation in plant tissue induces toxicity by retarding different physiological and biochemical parameters.

10.5.3 Effect of Vanadium on Plant Growth and Development

Vanadium is a common element in the global environment. It is one of the top five transitional elements (Baken, Larsson et al. 2012). V deposits are abundant in China, Russia, South Africa and the United States. China is the world's largest producer and consumer of V, accounting for 57% of global V production (GCVIR 2014–2017). The primary source of V is parent rocks such as shale, phosphate deposits, asphalt deposits, some uranium ores and titanium ferrous magnetites (Adriano 2001). In addition to the natural source, the burning of fossils and innumerable industrial processes are also accountable for additional V in the environment (Baken, Larsson et al. 2012). Plants quickly absorb the V available in the soil, but its impact on plants relies on its concentration. V facilitates plant growth at a lower concentration (10 ng V g^{-1} soil) and strengthens nitrogen fixation, chlorophyll synthesis and potassium utilization (Shaheen, Alessi et al. 2019). Application of doses below 0.05 mg L^{-1} V enhanced the quality and productivity of maize (*Zea mays*) (Singh 1971). Likewise, in tomato plants (*Solanum lycopersicum*), application of 250 ng mL^{-1}V increased height, number of leaves and flowers and concentration of chlorophyll. In basil (*Ocimum basilicum*), the dry biomass of the roots increased linearly with increasing the V concentrations from 0 to 40 mg L^{-1} (Akoumianaki-Ioannidou, Barouchas et al. 2015). Recently, the research showed that application of 5 μM V enhanced plant development, initiated floral bud growth and speeded up flowering in pepper (*Capsicum annuum*). V at a low concentration is helpful for plant growth because it causes expansion of cells by making tissues more elastic with increased water volume (García-Jiménez, Trejo-Téllez et al. 2018). But at higher concentrations, vanadium negatively influences plant developmental functions by disrupting enzymatic behavior, expression levels and oxygen radicals (ROS) output (Reiter, Tan et al. 2015). Vanadium acts as a hazardous substance when found in high concentrations and enhances the production of ROS that induces oxidative cell damage, alters lipid membranes and hinders growth of plants by interrupting the regular metabolic functions of the plant (Vachirapatama, Jirakiattiku et al. 2011).

Vanadium toxicity has adverse effects on seed germination, as it leads to change in soil pH, which influences seed germination (Wu, Zhang et al. 2020). In addition, studies have shown that the chemical form of V(V) is more lethal, unlike V(IV) and V(III), to seed germination (Yang, Teng et al. 2011) by injuring embryonic seed tissues and thus disturbing their growth and development (Yuan, Imtiaz et al. 2020). Studies have shown effect of V on seed germination in *Setaria viridis* (Amorim, Welz et al. 2007), *Glycine max* (Yang, Wang et al. 2017), *Medicago sativa* L. (Wu, Zhang et al. 2020) and *Oryza sativa* L. (Yuan, Imtiaz et al. 2020). In addition to seed germination, toxicity V has adverse effects on plant physiology attributable to chloroplastic damage, root and shoot mortality and leaf chlorosis. In rice crop V concentrations of 0, 15, 35 and 70 mg L^{-1}, biomass (dry and fresh) and root growth decreased significantly as a result of a decline in total root volume, root tips, root fork, root surface area and root cross-section (Altaf, Diao et al. 2020). In particular, exposure of V concentration > 35 mg/l to rice seedlings decreased fresh biomass, height of shoots and roots (Yuan, Imtiaz et al. 2020). However, in alfalfa, V concentration > 0.5 mg/l significantly inhibited root growth and root vitality, while V concentration of ≤ 10 mg L^{-1} V did not influence seed germination, final survival rate and seedling height of alfalfa, and at V concentration of 50 mg L^{-1} intensified testa color and changed seed structure were noticed (Wu, Zhang et al. 2020). Furthermore, V concentration ≥ 130 ppm induced detrimental effects on chickpea plants. In chickpea plants, decreased cell viability or cell death was witnessed when plants were exposed to higher V concentration (Imtiaz, Ashraf et al. 2018). Reduction of root biomass by vanadium has been found in *Brassica juncea* L. (Imtiaz, Ashraf et al. 2018), *Brassica napus* (Gokul, Cyster et al. 2018), and *Mentha × villosa* (Barouchas, Akoumianaki-Ioannidou et al. 2019); *Medicago sativa* L. (Gan, Chen et al. 2020).

Excessive V concentration hampers the photosynthesis, which is a vital process for plant growth and development. V at high concentration disrupts the enzymatic activities, carbon dioxide production, nutrient uptake and electron transport chain and causes lipid peroxidation and decrease in chlorophyll content (a and b), carotenoids which can increase the permeability of cell membranes (Chen, Li et al. 2016; Imtiaz, Ashraf et al. 2018). In a recent study, in *Ipomoea aquatica* Forsk, it was observed that V stress leads to decrease in chlorophyll synthesis due to decrease in magnesium and iron uptake, which are needed for chlorophyll synthesis (Chen, Li et al. 2016). In contrast to this, in soyabean, chlorophyll synthesis increased at high soil V concentration of 100–500 ppm; this can be due to ability of soyabean plants to tolerate high levels of V concentration (Gan, Chen et al. 2020). Immoderate levels of V restrict the absorption and translocation of essential minerals and thus disturbs the nutrient metabolism in the soil-plant system (Aihemaiti, Jiang et al. 2019). Excessive V accumulation allows V to be competitively bound to common carriers of mineral elements or to replace nutrient element ions from their respective binding sites within the plant cell (Aihemaiti, Jiang et al. 2019). In *Mentha pulegium* L. at 40 mg/L of V, there was decreased uptake of essential nutrients like calcium, magnesium, manganese and potassium (Akoumianaki-Ioannidou, Barouchas et al. 2015). Also, in *Setaria viridis*, there was decreased uptake of iron, zinc and magnesium at 55.17 mg L^{-1} of V (Aihemaiti, Jiang et al. 2019)

Furthermore, at higher concentration of V, reactive oxygen species (ROS) like hydrogen peroxides, free radicals and superoxides are produced in supra-optimal concentrations in plant organelles, viz. chloroplasts, mitochondria and endoplasmic reticulum. These ROS inhibit the electron transport chain, leading to reduced plant growth and even complete plant death (Baken, Larsson et al. 2012). In chickpea seedlings, ROS production induced by high V concentration resulted in oxidative damage of cell and reduced protein content (Imtiaz, Ashraf et al. 2018). Likewise, in *Brassica napus*, oxidative damage and reduced chlorophyll content was witnessed at toxic levels of V (Gokul, Cyster et al. 2018). In addition to ROS production, high concentration of V also causes lipid peroxidation in chickpea seedlings (Imtiaz, Ashraf et al. 2018) which results in hampered growth of plant.

10.6 Phytoremediation

In essence, phytoremediation refers to the use of plants and associated soil bacteria to mitigate ecologically polluting or harmful effects. Phytoremediation is commonly accepted as an environmental preservation technology that is cost effective and is considered an alternative to methods of engineering that are typically more detrimental to the soil (Jeelani, Yang et al. 2017). Phytoremediation can be used to extract heavy metals from polluted soil as an important method. Phytoremediation is an efficient and cost-effective way to extract heavy metals from wastewater (Williams 2002; Jeelani, Yang et al. 2017; Noor, Daud et al. 2017). To achieve appropriate levels of pollutants in the atmosphere, phytoremediation of polluted areas should preferably not exceed one decade. However, phytoremediation is confined to the plant rootzone. This technology also has restricted applications where contaminant concentrations damage plants. For distinct conditions and types of pollutants, phytoremediation methods are available. These include various procedures, such as stabilization or degradation in situ and pollutant removal (i.e., volatilization or extraction), as shown in the Figure 10.3. It involves different technologies (Table 10.1) such as phytodegradation, phytovolatilization, phytostabilization and phytoextraction.

Specific plants were used in many experiments involving wetlands to extract metals from wastewater and have been shown to be effective in removing 50–99 percent of arsenic, cadmium, polycyclic hydrocarbons and vanadium (Sahu 2014; Jeelani, Yang et al. 2017; Wu, Zhang et al. 2020).

There are over 400 plant species listed as hyperaccumulators, in addition to periodic phytoremediation (Zhang, Peng et al. 2007; Sun, Wang et al. 2013; Sun, Li et al. 2019). Hyperaccumulators are plants in which the shoots of the plant extract extraordinarily high quantities of metals, these concentrations being excessive compared to other species (Rascio and Navari-Izzo 2011). Three simple hyperaccumulator distinguishers are enhanced metal uptake rate, rapid root to shoot metal allocation and the ability to store metals in the shoots (Rascio and Navari-Izzo 2011). In particular, *Acorus calamus* is a promising candidate for phytoremediation and a potential hyperaccumulator because of its robust root system, high adaptability and extensive biomass (Sun, Wang et al. 2013; Chen, Li et al. 2016). Plants that hyperaccumulate heavy metals are commonly used to remediate polluted heavy metal ecosystems, with the aid of rhizosphere microorganisms that are both ecofriendly and

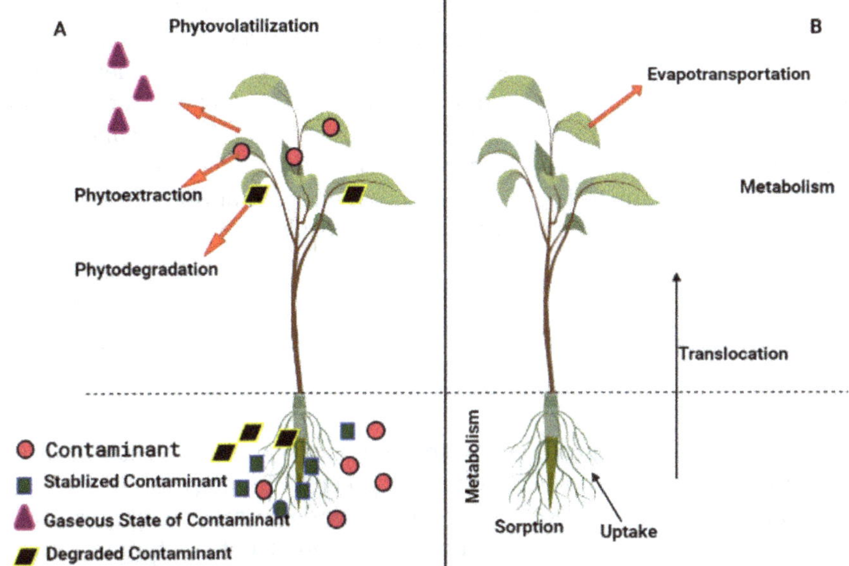

FIGURE 10.3
Schematic model of the various phytoremediation technologies, including contaminant removal and containment.

TABLE 10.1

Different Methods Involved in Remediation Process

1. **Phytostabilization:**	Phytostabilization helps to preserve and stop further dispersal of pollutants in the soil. In the roots or inside the rhizosphere, pollutants may stabilize.
2. **Photodegradation:**	Phytodegradation involves the direct degradation of organic pollutants by releasing of enzymes from the roots or by metabolic operations inside plant tissues.
3. **Phytovolatilization:**	For phytovolatilization, the absorption of contaminants and their transformation into a gaseous form through plant roots and their release into the atmosphere are important.
4. **Phytoextraction:**	Phytoextraction exploits plants' ability to absorb toxins in aboveground, harvestable biomass. To minimize toxin soil concentration, this method requires repeated biomass harvesting.

cost-effective compared to physical and chemical approaches. Nevertheless, only a few studies on the phytoremediation of vanadium-related native macrophyte contamination (Khan, Kazi et al. 2011; Qian, Gallagher et al. 2014) are required to explore more macrophyte candidates for their possible use for in situ phytoremediation. Hydroponic experiments found that both *P. australis* and *C. demersum* are best to remove vanadium. In phytoremediation, Bermuda grass, *P. australis* and *quadrifolia* achieved greater heavy metal removal by combining terrestrial plants and aquatic macrophytes (Jiang, Xing et al. 2018). Even at micrograms per liter, *demersum* was able to effectively absorb heavy metals, serving as a potential candidate for low-level heavy metal contamination phytoremediation, particularly in vanadium contaminated sites. Strong metals are a significant threat to biodiversity and human health. Phytoremediation is a newly developed technique that, using plants and associated soil microbes, provides a cost-effective approach to reducing contaminant content or toxic effects in the atmosphere.

10.7 Bioremediation

Sustainable development is part of the conservation and promotion of the environment and the successful pursuit of green technology for the management of a wide variety of anthropogenically polluted marine and terrestrial ecosystems. Bioremediation is an increasingly prevalent solution to conventional waste treatment approaches and media that can eliminate pollution by the normal microbial behavior of different microbial strain consortia (Juwarkar, Singh et al. 2010). Environmental problems are dynamic and also unique to time and space. There are frequent shifts in the existence and complexity of problems. New challenges arise, and there is a constant need for new technologies. Biotechnology has tremendous potential to fulfill global needs and optimize defense, conservation and management (Azadi and Ho 2010). Although some implementations are direct applications of biotechnology (Dowling and Doty 2009), several are for ecological mitigation, pollution control and waste reduction.

Autotrophic microorganisms oxidize elemental sulfur S(0) and zero valent iron Fe(0) and also use bicarbonate as a carbon by-product to form volatile fatty acids (VFAs) (Zhang, Jiang et al. 2018). The main substrates for V(V) reducers are V(V) and V(IV) reductions. V(V) can be extracted using two distinct bio-reduction pathways connected to the membrane (Wang, Liu et al. 2008). Intracellular reducing agents reduce vanadate oxyanion and pump it as a vanadyl cell in the bio-reduction process. The vanadate reductase enzyme reduces V(V) to V(VI) in the cell membrane (Carpentier, De Smet et al. 2005).

Soil conditioners and indigenous macrophytes may mitigate in situ vanadium-polluted water and soil. The use of various interactions (fulvic and humic acids, calcium oxide, ferrous sulfate, hydro-oxide, iron and aluminum (hydro-) manure) helps in the elimination of bio-availability, toxicity and vanadium absorption through the complexion, precipitation and reduction of the process (Larsson, Baken et al. 2015; Zou, Li et al. 2019; Zou, Xiang et al. 2019). The use of microbial consortia to degrade and detoxify environmental toxins (Margesin, Hämmerle et al. 2007; Singh, Kang et al. 2008; Zhao and Poh 2008) is another promising, rapidly evolving method among these recent developments, based on the emerging Green Chemistry and Green Engineering theory.

10.8 Conclusion and Future Prospectus

In conclusion, V at low concentration imparts plant growth and development; however, at elevated concentration, it has negative impacts on plant growth and seed germination by hampering chlorophyll production by overproduction of ROS and lipid peroxidation. V has a stimulating impact on plant growth and development at various phases of development. V has the potential to be a beneficial factor that helps increase agricultural crop production. Physiological, biochemical and genomic methods will help to elucidate novel modes of action and encourage the widespread use of V in plants, as well as its future applications in bioremediation and biofortification. More research is required to determine the concentration threshold for V hyperaccumulators in plant shoots. There are actually only a few plant species known to be able to accumulate V in the shoot tissues. Screening of possible V hyperaccumulators is desired. The comprehensive dynamics of biochemical valence shifts and V delivery at the subcellular level are worth exploring to better understand the processes by which plants defend themselves from V stress. Further research to elucidate the molecular mechanism of vanadium toxicity and stress tolerance is required.

References

Adriano, D. C. (2001). *Bioavailability of trace metals. Trace elements in terrestrial environments.* Springer: 61–89.

Aihemaiti, A., Y. Gao, Y. Meng, X. Chen, J. Liu, H. Xiang, Y. Xu and J. Jiang (2020). "Review of plant-vanadium physiological interactions, bioaccumulation, and bioremediation of vanadium-contaminated sites." *Science of the Total Environment* **712**: 135637.

Aihemaiti, A., J. Jiang, Y. Gao, Y. Meng, Q. Zou, M. Yang, Y. Xu, S. Han, W. Yan and T. Tuerhong (2019). "The effect of vanadium on essential element uptake of Setaria viridis' seedlings." *Journal of Environmental Management* **237**: 399–407.

Akoumianaki-Ioannidou, A., P. E. Barouchas, E. Ilia, A. Kyramariou and N. K. Moustakas (2016). "Effect of vanadium on dry matter and nutrient concentration in sweet basil (*'Ocimum basilicum'* L.)." *Australian Journal of Crop Science* **10**(2): 199–206.

Akoumianaki-Ioannidou, A., P. E. Barouchas, A. Kyramariou, E. Ilia and N. K. Moustakas (2015). "Effect of vanadium on dry matter and nutrient concentration in pennyroyal (*Mentha pulegium* L.)." *Bulletin UASVM Horticulture* **72**(2): 295–298.

Allan, D. L. and W. M. Jarrell (1989). "Proton and copper adsorption to maize and soybean root cell walls." *Plant Physiology* **89**(3): 823–832.

Altaf, M. M., X.-P. Diao, A. Ur Rehman, M. Imtiaz, A. Shakoor, M. A. Altaf, H. Younis, P. Fu and M. U. Ghani (2020). "Effect of vanadium on growth, photosynthesis, reactive oxygen species, antioxidant enzymes, and cell death of rice." *Journal of Soil Science and Plant Nutrition* **20**(4): 2643–2656.

Amorim, F. A., B. Welz, A. C. Costa, F. G. Lepri, M. G. R. Vale and S. L. Ferreira (2007). "Determination of vanadium in petroleum and petroleum products using atomic spectrometric techniques." *Talanta* **72**(2): 349–359.

Azadi, H. and P. Ho (2010). "Genetically modified and organic crops in developing countries: A review of options for food security." *Biotechnology Advances* **28**(1): 160–168.

Baken, S., M. A. Larsson, J. P. Gustafsson, F. Cubadda and E. Smolders (2012). "Ageing of vanadium in soils and consequences for bioavailability." *European Journal of Soil Science* **63**(6): 839–847.

Barceloux, D. G. and D. Barceloux (1999). "Vanadium." *Journal of Toxicology: Clinical Toxicology* **37**(2): 265–278.

Barouchas, P. E., A. Akoumianaki-Ioannidou, A. Liopa-Tsakalidi and N. K. Moustakas (2019). "Effects of vanadium and nickel on morphological characteristics and on vanadium and nickel uptake by shoots of mojito (Mentha× villosa) and lavender (*Lavandula anqustifolia*)." *Notulae Botanicae Horti Agrobotanici Cluj-Napoca* **47**(2): 487–492.

Bertrand, D. and V. Lee (1950). "Survey of contemporary knowledge of biogeochemistry: 2. The biogeochemistry of vanadium." *Bulletin of the American Museum of Natural History* **94**: 403–456.

Bradl, H. B. (2004). "Adsorption of heavy metal ions on soils and soils constituents." *Journal of Colloid and Interface Science* **277**(1): 1–18.

Carpentier, W., L. De Smet, J. Van Beeumen and A. Brigé (2005). "Respiration and growth of Shewanella oneidensis MR-1 using vanadate as the sole electron acceptor." *Journal of Bacteriology* **187**(10): 3293–3301.

Chen, L., J.-R. Liu, J. Gao, W.-F. Hu and J.-Y. Yang (2020). "Vanadium in Soil-Plant System: Source, Fate, Toxicity, and Bioremediation." *Journal of Hazardous Materials*: 124200.

Chen, T., T. Q. Li and J. Y. Yang (2016). "Damage suffered by swamp morning glory (*Ipomoea aquatica* Forsk) exposed to vanadium (V)." *Environmental Toxicology and Chemistry* **35**(3): 695–701.

Del Carpio, E., L. Hernández, C. Ciangherotti, V. V. Coa, L. Jiménez, V. Lubes and G. Lubes (2018). "Vanadium: History, chemistry, interactions with α-amino acids and potential therapeutic applications." *Coordination Chemistry Reviews* **372**: 117–140.

Dowling, D. N. and S. L. Doty (2009). "Improving phytoremediation through biotechnology." *Current Opinion in Biotechnology* **20**(2): 204–206.

Drava, G., L. Cornara, P. Giordani and V. Minganti (2019). "Trace elements in *Plantago lanceolata* L., a plant used for herbal and food preparations: New data and literature review." *Environmental Science and Pollution Research* **26**(3): 2305–2313.

Dutta, S. K. and D. R. Lodhari (2018). *Extraction of nuclear and non-ferrous metals.* Springer: 149–154.

Fang, G., W. Wu, C. Liu, D. D. Dionysiou, Y. Deng and D. Zhou (2017). "Activation of persulfate with vanadium species for PCBs degradation: A mechanistic study." *Applied Catalysis B: Environmental* **202**: 1–11.

Gan, C.-d., T. Chen and J.-y. Yang (2020). "Remediation of vanadium contaminated soil by alfalfa (*Medicago sativa* L.) combined with vanadium-resistant bacterial strain." *Environmental Technology & Innovation* **20**: 101090.

García-Jiménez, A., L. I. Trejo-Téllez, D. Guillén-Sánchez and F. C. Gómez-Merino (2018). "Vanadium stimulates pepper plant growth and flowering, increases concentrations of amino acids, sugars and chlorophylls, and modifies nutrient concentrations." *PLoS One* **13**(8): e0201908.

GCVIR (2014–2017). "(Global and China vanadium industry report)." *Research in China* 1–85.

Ghosh, M. and S. Singh (2005). "A review on phytoremediation of heavy metals and utilization of it's by products." *Asian Journal on Energy and Environment* **6**(4): 18.

Gokul, A., L. Cyster and M. Keyster (2018). "Efficient superoxide scavenging and metal immobilization in roots determines the level of tolerance to Vanadium stress in two contrasting *Brassica napus* genotypes." *South African Journal of Botany* **119**: 17–27.

Hall, J. á. (2002). "Cellular mechanisms for heavy metal detoxification and tolerance." *Journal of Experimental Botany* **53**(366): 1–11.

Hou, M., M. Li, X. Yang and R. Pan (2019). "Responses of nonprotein thiols to stress of vanadium and mercury in maize (*Zea mays* L.) seedlings." *Bulletin of Environmental Contamination and Toxicology* **102**(3): 425–431.

Imtiaz, M., M. Ashraf, M. S. Rizwan, M. A. Nawaz, M. Rizwan, S. Mehmood, B. Yousaf, Y. Yuan, A. Ditta and M. A. Mumtaz (2018). "Vanadium toxicity in chickpea (*Cicer arietinum* L.) grown in red soil: Effects on cell death, ROS and antioxidative systems." *Ecotoxicology and Environmental Safety* **158**: 139–144.

Imtiaz, M., M. S. Rizwan, S. Xiong, H. Li, M. Ashraf, S. M. Shahzad, M. Shahzad, M. Rizwan and S. Tu (2015). "Vanadium, recent advancements and research prospects: A review." *Environment International* **80**: 79–88.

Jeelani, N., W. Yang, L. Xu, Y. Qiao, S. Anand X. Leng (2017). "Phytoremediation potential of Acorus calamus in soils co-contaminated with cadmium and polycyclic aromatic hydrocarbons." *Scientific Reports* **7**(1): 1–9.

Jiang, B., Y. Xing, B. Zhang, R. Cai, D. Zhang and G. Sun (2018). "Effective phytoremediation of low-level heavy metals by native macrophytes in a vanadium mining area, China." *Environmental Science and Pollution Research* **25**(31): 31272–31282.

Juwarkar, A. A., S. K. Singh and A. Mudhoo (2010). "A comprehensive overview of elements in bioremediation." *Reviews in Environmental Science and Bio/Technology* **9**(3): 215–288.

Khan, S., T. G. Kazi, H. I. Afridi, N. F. Kolachi, N. Ullah and K. Dev (2013). "Speciation of vanadium in coal mining, industrial, and agricultural soil samples using different extractants and heating systems." *Journal of AOAC International* **96**(1): 186–189.

Khan, S., T. G. Kazi, N. F. Kolachi, J. A. Baig, H. I. Afridi, A. Q. Shah, S. Kumar and F. Shah (2011). "Hazardous impact and translocation of vanadium (V) species from soil to different vegetables and grasses grown in the vicinity of thermal power plant." *Journal of Hazardous Materials* **190**(1–3): 738–743.

Kubes, J., M. Skalicky, L. Tumova, J. Martin, V. Hejnak and J. Martinkova (2019). "Vanadium elicitation of *Trifolium pratense* L. cell culture and possible pathways of produced isoflavones transport across the plasma membrane." *Plant Cell Reports* **38**(5): 657–671.

Kumar, V., S. AlMomin, A. Al-Shatti, H. Al-Aqeel, F. Al-Salameen, A. Shajan and S. Nair (2019). "Enhancement of heavy metal tolerance and accumulation efficiency by expressing Arabidopsis ATP sulfurylase gene in alfalfa." *International Journal of Phytoremediation* **21**(11): 1112–1121.

Larsson, M. A., S. Baken, E. Smolders, F. Cubadda and J. P. Gustafsson (2015). "Vanadium bioavailability in soils amended with blast furnace slag." *Journal of Hazardous Materials* **296**: 158–165.

Marcano, L., I. Carruyo, Y. Fernández, X. Montiel and Z. Torrealba (2006). "Determination of vanadium accumulation in onion root cells (*Allium cepa* L.) and its correlation with toxicity." *Biocell* **30**(2): 259.

Margesin, R., M. Hämmerle and D. Tscherko (2007). "Microbial activity and community composition during bioremediation of diesel-oil-contaminated soil: Effects of hydrocarbon concentration, fertilizers, and incubation time." *Microbial Ecology* **53**(2): 259–269.

Noor, N., A. Daud, M. Amansyah and A. Mallongi (2017). "Affectivity dose of *Acorus calamus* (Sweet flag) to reduce the ammonia in hospital wastewater." *Journal of Environmental Science and Technology* **10**: 139–146.

Pilon-Smits, E. (2005). "Phytoremediation." *Annual Review of Plant Biology* **56**: 15–39.

Qian, Y., F. J. Gallagher, H. Feng, M. Wu and Q. Zhu (2014). "Vanadium uptake and translocation in dominant plant species on an urban coastal brownfield site." *Science of the Total Environment* **476**: 696–704.

Rascio, N. and F. Navari-Izzo (2011). "Heavy metal hyperaccumulating plants: How and why do they do it? And what makes them so interesting?" *Plant Science* **180**(2): 169–181.

Rauser, W. E. (1999). "Structure and function of metal chelators produced by plants." *Cell Biochemistry and Biophysics* **31**(1): 19–48.

Reijonen, I. (2017). "Chemical bioavailability of chromium and vanadium species in soil: Risk assessment of the use of steel industry slags as liming materials." Doctoral Dissertation, University of Helsinki, YEB.

Reiter, R. J., D.-X. Tan, Z. Zhou, M. H. C. Cruz, L. Fuentes-Broto and A. Galano (2015). "Phytomelatonin: Assisting plants to survive and thrive." *Molecules* **20**(4): 7396–7437.

Roychoudhury, A. (2020). "Vanadium Uptake and Toxicity in Plants." *SF Journal of Agricultural and Crop Management* **1**(2): 1010.

Sahu, O. (2014). "Reduction of heavy metals from waste water by wetland." *International Letters of Natural Sciences* **12**: 35–43.

Shaheen, S. M., D. S. Alessi, F. M. Tack, Y. S. Ok, K.-H. Kim, J. P. Gustafsson, D. L. Sparks and J. Rinklebe (2019). "Redox chemistry of vanadium in soils and sediments: Interactions with colloidal materials, mobilization, speciation, and relevant environmental implications-A review." *Advances in Colloid and Interface Science* **265**: 1–13.

Shi, Y. X., V. Mangal and C. Guéguen (2016). "Influence of dissolved organic matter on dissolved vanadium speciation in the Churchill River estuary (Manitoba, Canada)." *Chemosphere* **154**: 367–374.

Singh, B. (1971). "Effect of vanadium on the growth, yield and chemical composition of maize (*Zea mays* L.)." *Plant and Soil* **34**: 209–213.

Singh, S., S. H. Kang, A. Mulchandani and W. Chen (2008). "Bioremediation: Environmental clean-up through pathway engineering." *Current Opinion in Biotechnology* **19**(5): 437–444.

Sun, H., Z. Wang, P. Gao and P. Liu (2013). "Selection of aquatic plants for phytoremediation of heavy metal in electroplate wastewater." *Acta Physiologiae Plantarum* **35**(2): 355–364.

Sun, W., X. Li, J. Padilla, T. A. Elbana and H. M. Selim (2019). "The influence of phosphate on the adsorption-desorption kinetics of vanadium in an acidic soil." *Journal of Environmental Quality* **48**(3): 686–693.

Teng, Y., J. Yang, Z. Sun, J. Wang, R. Zuo and J. Zheng (2011). "Environmental vanadium distribution, mobility and bioaccumulation in different land-use Districts in Panzhihua Region, SW China." *Environmental Monitoring and Assessment* **176**(1–4): 605–620.

Thakur, S., L. Singh, Z. Ab Wahid, M. F. Siddiqui, S. M. Atnaw and M. F. M. Din (2016). "Plant-driven removal of heavy metals from soil: Uptake, translocation, tolerance mechanism, challenges, and future perspectives." *Environmental Monitoring and Assessment* **188**(4): 206.

Tian, L.-Y., J.-Y. Yang, C. Alewell and J.-H. Huang (2014). "Speciation of vanadium in Chinese cabbage (*Brassica rapa* L.) and soils in response to different levels of vanadium in soils and cabbage growth." *Chemosphere* **111**: 89–95.

Tian, L.-Y., J.-Y. Yang and J.-H. Huang (2015). "Uptake and speciation of vanadium in the rhizosphere soils of rape (*Brassica juncea* L.)." *Environmental Science and Pollution Research* **22**(12): 9215–9223.

Tripathi, D., V. Mani and R. P. Pal (2018). "Vanadium in biosphere and its role in biological processes." *Biological Trace Element Research* **186**(1): 52–67.

Vachirapatama, N., Y. Jirakiattiku, G. Dicinoski, A. T. Townsend and P. R. Haddad (2011). "Effect of vanadium on plant growth and its accumulation in plant tissues." *Songklanakarin Journal of Science and Technology* **33**(3): 255–261.

Wang, X., Y. Liu, G. Zeng, L. Chai, X. Song, Z. Min and X. Xiao (2008). "Subcellular distribution and chemical forms of cadmium in *Bechmeria nivea* (L.) Gaud." *Environmental and Experimental Botany* **62**(3): 389–395.

Weis, J. S. and P. Weis (2004). "Metal uptake, transport and release by wetland plants: Implications for phytoremediation and restoration." *Environment International* **30**(5): 685–700.

Welch, R. M. (1973). "Vanadium uptake by plants: Absorption kinetics and the effects of pH, metabolic inhibitors, and other anions and cations." *Plant Physiology* **51**(5): 828–832.

Williams, J. B. (2002). "Phytoremediation in wetland ecosystems: Progress, problems, and potential." *Critical Reviews in Plant Sciences* **21**(6): 607–635.

Wu, Z.-Z., Y.-X. Zhang, J.-Y. Yang, Y. Zhou and C.-Q. Wang (2020). "Effect of vanadium on testa, seed germination, and subsequent seedling growth of alfalfa (*Medicago sativa* L.)." *Journal of Plant Growth Regulation* **40**(29): 1566–1578.

Yang, J., Y. Teng, J. Wang and J. Li (2011). "Vanadium uptake by alfalfa grown in V-Cd-contaminated soil by pot experiment." *Biological Trace Element Research* **142**(3): 787–795.

Yang, J., M. Wang, Y. Jia, M. Gou and J. Zeyer (2017). "Toxicity of vanadium in soil on soybean at different growth stages." *Environmental Pollution* **231**: 48–58.

Yuan, Y., M. Imtiaz, M. Rizwan, X. Dong and S. Tu (2020). "Effect of vanadium on germination, growth and activities of amylase and antioxidant enzymes in genotypes of rice." *International Journal of Environmental Science and Technology* **17**(1): 383–394.

Zhang, W., J. Jiang, K. Li, T. Li, D. A. Li and J. Wang (2018). "Amendment of vanadium-contaminated soil with soil conditioners: A study based on pot experiments with canola plants (*Brassica campestris* L.)." *International Journal of Phytoremediation* **20**(5): 454–461.

Zhang, X.-B., L. Peng, Y.-S. Yang and W.-R. Chen (2007). "Phytoremediation of urban wastewater by model wetlands with ornamental hydrophytes." *Journal of Environmental Sciences* **19**(8): 902–909.

Zhao, B. and C. L. Poh (2008). "Insights into environmental bioremediation by microorganisms through functional genomics and proteomics." *Proteomics* **8**(4): 874–881.

Zou, Q., D. A. Li, J. Jiang, A. Aihemaiti, Y. Gao, N. Liu and J. Liu (2019). "Geochemical simulation of the stabilization process of vanadium-contaminated soil remediated with calcium oxide and ferrous sulfate." *Ecotoxicology and Environmental Safety* **174**: 498–505.

Zou, Q., H. Xiang, J. Jiang, D. Li, A. Aihemaiti, F. Yan and N. Liu (2019). "Vanadium and chromium-contaminated soil remediation using VFAs derived from food waste as soil washing agents: A case study." *Journal of Environmental Management* **232**: 895–901.

11

Microbial Community Responses and Bioremediation

Baogang Zhang, Han Zhang and Jinxi He

School of Water Resources and Environment, MOE Key Laboratory of Groundwater Circulation and Environmental Evolution, China University of Geosciences (Beijing)

CONTENTS

11.1 Introduction 223
11.2 Microbial Community in Vanadium Mining and Smelting Areas 224
11.3 Spatiotemporal Dynamics of the Microbial Community 231
11.4 Microbial Remediation Concept and Mechanisms 239
11.5 Functional Microbes and Electron Donors 242
11.6 Influences of Co-Existing Substances 246
11.7 Strengthening Strategy for Microbial Remediation 254
References 260

11.1 Introduction

Vanadium is a transition metal that presents in the active site of vanadium nitrogenase, which is ubiquitously found in the soil environment (Bellenger et al., 2014). The mineral provides metabolic function critical to nutrition processing in higher life forms (French and Jones, 1993). Vanadium in high valent state can be mobilized as result of variation in geochemistry and enter the subsurface environment, subsequently reshaping the microbial community in soil. On the other hand, as an essential component of soil, microorganisms played an important role in soil ecosystem, exhibiting ability to rapidly mutate and evolve in earth and adapt living in a wide range of environmental conditions. The more toxic V(V) can be biologically converted into less bioavailable V(IV) in a microbial catalyzed process through which vanadium-reducing microbes utilize vanadium to facilitate their metabolic mechanisms. Therefore, the versatility of microbes to reduce V(V) makes

DOI: 10.1201/9781003173274-11

bioremediation a promising practice that can be applied in a vanadium contaminated environment. This chapter aims to describe and summarize microbial community responses to vanadium pollution induced by mining and smelting operations. The process, mechanisms and optimization strategies of vanadium bioremediation are also presented.

11.2 Microbial Community in Vanadium Mining and Smelting Areas

The presence of vanadium in soil, water and the atmosphere occurs because of intensive mining and smelting activities in regions with abundant vanadium deposits. For example, Panzhihua, China, is rich in vanadium titanomagnetite resource, accounting for 64% of the total vanadium supply nationwide (Teng et al., 2011). In places that are heavily contaminated with vanadium, microorganisms always play important roles in the biogeochemical cycle of vanadium in different environmental matrices (soil, water and atmosphere). Recently, microbial community responses to vanadium in these matrices with spatiotemporal dynamics have been concerning. Cao et al. collected topsoil samples (up to 20 cm depth) from five different vanadium production sites in Panzhihua including a concentrator (CO), waste dump (WD), mining plant (MP), tailing reservoir (TR) and smelter (SM) (Cao et al., 2017). In the studies, the microbial community structure in soil samples was investigated by molecular biology techniques. Actinobacteria, Bacteroidetes and Proteobacteria are the most abundant bacterial phyla identified in samples from MP and WD. Bacteroidetes, Proteobacteria and Firmicutes are highly enriched in samples of CO, SM and TR. Bacteroidetes is more sensitive to vanadium toxicity, with decreased relative abundance in samples containing elevated vanadium content. As displayed in Figure 11.1, microbial community composition at the genus level is revealed. In MP soil originating from site (denoted as MP-0), *Geobacter* and *Comamonas* are the two heavy metal-reducing genera with great relative abundance. *Geobacter metallireducens* can respire vanadium to facilitate its growth, which can reduce vanadate (Ortiz-Bernad et al., 2004). The relative abundance of *Stenotrophomonas* and *Azospira* is high in WD-0 soil. Previous studies relate *Stenotrophomonas sp.* to bioconversion of Cu(II) and Cr(VI) (Ghosh et al., 2015). *Azospira oryzae* reduces toxic Se into harmless element Se (Hunter, 2007). *Anaeromyxobacter* and *Clostridium* are found with elevated abundance in CO-0. It is reported that *Anaeromyxobacter* converts soluble selenium to precipitated selenium (0) (He and Yao, 2010), while *Clostridium* reduces U(VI) to insoluble U(IV) (Vecchia et al., 2010). *Zoogloea,* known for its contribution in removal of Cd, Zn and Cr from simulated industrial wastewater, is found in quantity in SM-0 (Solisio et al., 1998). TR-0 features *Klebsiella* and *Acinetobacter; Klebsiella sp.* is a

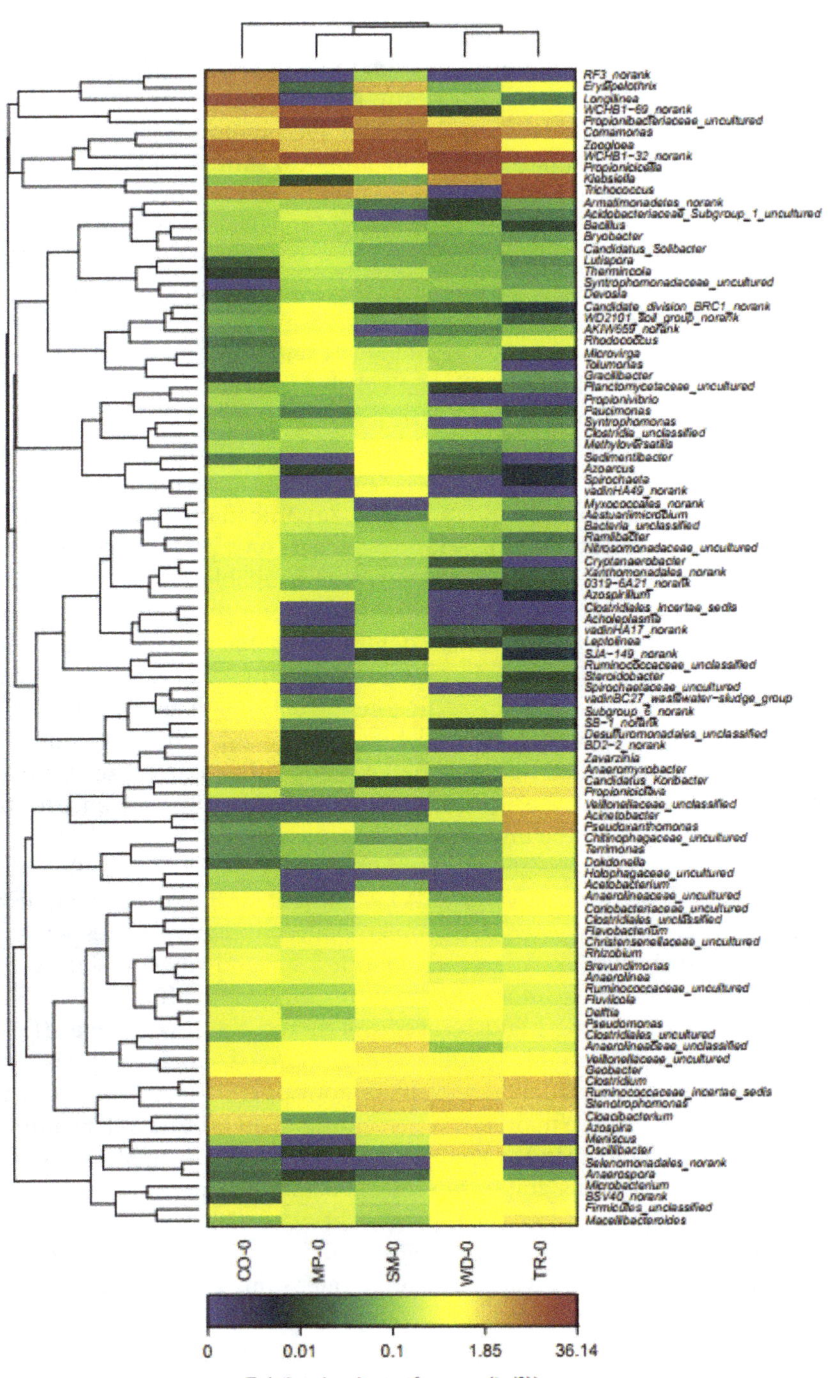

FIGURE 11.1
Hierarchical cluster analysis of bacterial communities in the original soil samples.
Source: Cao et al., 2017.

reducer for Fe(III) and ammonium oxide (Su et al., 2016), and *Acinetobacter sp HK-1* is responsible for chromium fixation in treatment of Cr(VI) containing wastewater (Zhang, Lu et al., 2014). Fungi also exhibits important ecological functions in a microbial community. At the phylum level, Ascomycota and Ciliophora predominate in the soil samples, similar to previous studies at heavy metal processing sites (Taylor et al., 2006; Zhang, Dong et al., 2014). At the genus level, *Aspergillus Niger* and *Penicillium simplicissimum* from Ascomycota are of higher abundance, both of which demonstrate excellent vanadium recovery ability (Mirazimi et al., 2015; Rasoulnia and Mousavi, 2016). The composition within the microbial community varies depending on sampling locations; however, bacterial genera with good tolerance to heavy metal were present in all sites with relatively higher abundance, possibly due to the vanadium presence.

The interaction between geochemical parameters and the microbial community is studied with statistical tools. According to result from a redundancy analysis (RDA) shown in Figure 11.2, organic matter, available sulfur and available phosphorus incur significant effects on a bacterial community. TR-0 and MP-0 show positive correlation with available sulfur and organic matter, indicating their essential roles in modulating the microbial community. Available phosphorus is a major determinant that shapes the microbial community in SM-0 and CO-0. In addition, vanadium can affect phosphate metabolism due to sharing an analogous structure (Olness et al., 2005). Vanadium also regulates the bioavailability of sulfur by forming coordination bonds with sulfur, consequentially exerting influence on the microbial community (Taylor et al., 2006). This outcome provides solid evidence for the impact of vanadium on the microbial community.

The mobilization of topsoil vanadium may also contribute to pollution in surface water (SW), groundwater (GW), sediment (SD) and other environmental matrices. To investigate the microbial community in response to vanadium distribution, Zhang et al. collected groundwater, surface water and sediment samples from a vanadium mining plant (MP), smelter (SM), concentrator (CO), tailing pond reservoir (TR) and the Jinsha River (JS) in Panzhihua (Zhang, Wang et al., 2019). The abundance of microorganisms in groundwater is relatively low due to its oligotrophic nature. Proteobacteria and Bacteroidetes are the main phyla presenting in all groundwater samples, both of which participate in V(V) reduction (Zhang et al., 2015). At genus level, *Novosphingobium*, with its ability to reduce cobalt, dominates in all groundwater samples obtained from MP (Raghu et al., 2008). *Flavobacterium* and *Acinetobacter* are heavy metal reducers detected in high abundance (Kumar et al., 2011). *Sphingobium* prevails in groundwater sample from TR, exhibiting high tolerance to toxic heavy metals (Wang et al., 2013).

The microbial community composition was also characterized in surface water and sediment. At the phylum level, the relative abundance of Proteobacteria and Actinobacteria in surface waters is greater than that in sediment. Metal-related genera are also observed. *Synechococcus* presents

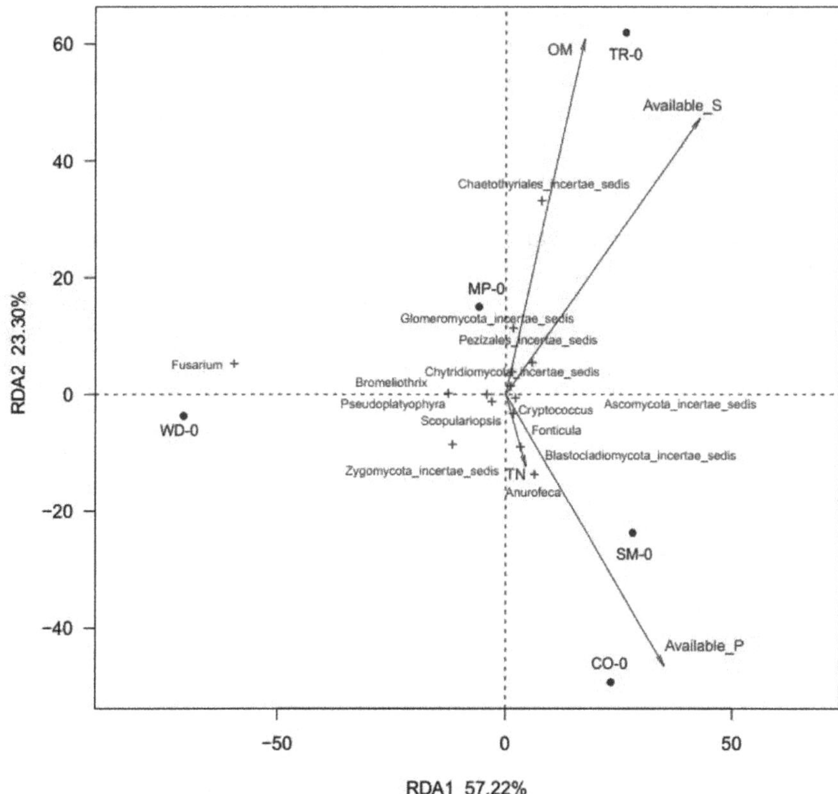

FIGURE 11.2
Redundancy analysis of the effect of environmental factors on bacterial communities (OTUs on 97% similarity level).
Source: Cao et al., 2017.

only in surface water from TR in high relative abundance. The study confirms its resistant to copper-induced toxicity (Polti et al., 2014). *Exiguobacterium* is abundant in sediment samples of JS, while *Cyanobacteria* is only detected in sediment samples from TR. Both genera are associated with reduction of Cr(VI) and Fe(III) (Mohapatra et al., 2017; Xu et al., 2016). In comparison, vanadium imposes less toxicity than chromium and copper. For genera showing microbial tolerance to other heavy metals, the likelihood of them exhibiting the same resistance to vanadium is high (Aihemaiti et al., 2018). Moreover, microbial abundance in TR surface water is lower than that in groundwater samples. This observation can be explained by direct contact between surface water and tailings entailing greater toxic effects on the microbial community. Surface water exhibits greater diversity microorganisms in term of the Shannon index compared to sediment.

Nonmetric multidimensional scaling (NMDS) assists the visualization of similarity and difference within the microbial community. As seen in Figure 11.3a, surface soil, groundwater and subsoil samples form separate clusters presenting distinctive community compositions. However, similarity in community composition is observed in surface water and sediment samples from JS, possibly attributable to impact from similar geochemical conditions. Geochemical parameters are the major driver to regulate the composition pattern within the microbial community, as demonstrated in Figure 11.3b. Vanadium is positively correlated with *Trichococcus*, while it is also vital to abundance of *Geobacter* and *Comamonas*. Figure 11.3c presents the microbial response to geochemical parameters in aqueous matrix, where pH, TOC, and NO_3^- regulate the occurrence of metal-associated bacteria. In addition,

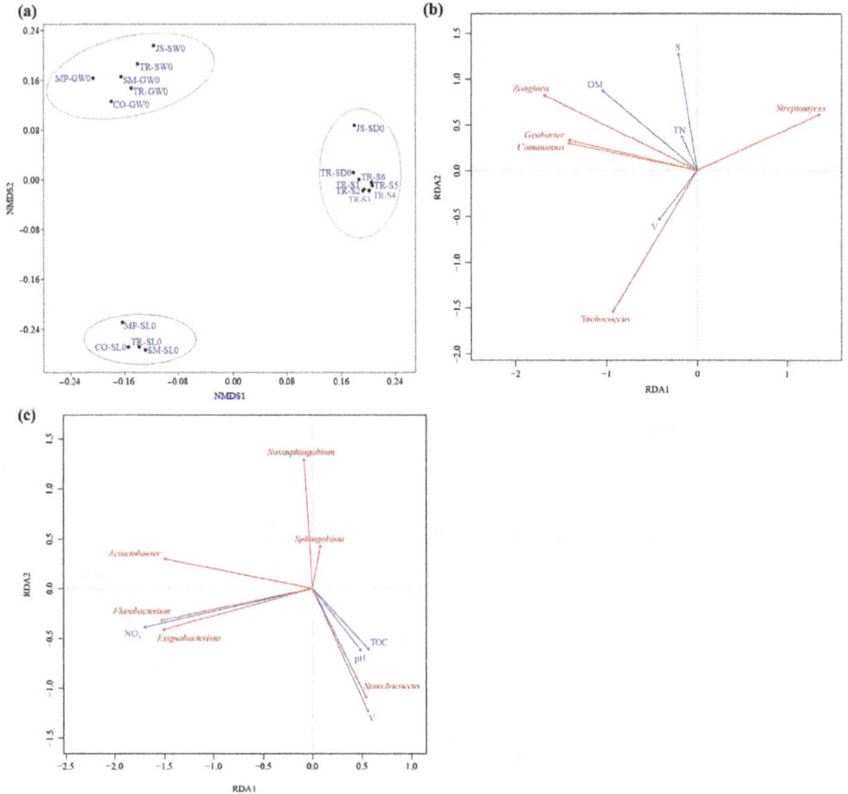

FIGURE 11.3
NMDS plots and RDA of microbial communities. (a) NMDS plots for all original samples; RDA for (b) solid samples and (c) aqueous samples.
Source: Zhang, Wang et al., 2019.

Synechococcus is greatly affected by vanadium. These results showcase the function of vanadium in reshaping microbial communities.

In the smelting process, the facility flue releases a substantial amount of toxic gas containing various types of metals into the air, triggering the interaction between microorganisms and heavy metal constituents (Li et al., 2020). Due to the seasonal cycle of atmospheric conditions, aerosol composition varies greatly. Seasonal sampling of aerosols around a vanadium smelter is carried out in Panzhihua to study the evolution of microorganisms in aerosols and their tolerance to heavy metals (Wang, Zhang, Wang et al., 2020). This microbial community may originate from soil through atmospheric movement. The microbial community is studied at the class level. As seen in Figure 11.4a, Bacilli, Alphaproteobacteria, Betaproteobacteria and Gammaproteobacteria predominantly exist in the aerosol sample. However, the composition varies seasonally. The abundance of Alphaproteobacteria is elevated to 77.1% in the spring, while Bacilli reaches 6.4% in the summer. At genus level, heavy metal tolerant taxa are displayed in Figure 11.4b; V(V) reducers comprising *Acinetobacter, Brevundimonas* and *Pseudomonas* are present in all seasons. Several more tolerant genera prevail in summer, including *Bacillus*, exhibiting tolerance to Zn, As, Cu and Ni, *Geobacter*, known to be a V(V) reducer, and *Thaurea*, a bio-agent for dissimilative reduction of Fe(III) (Miao et al., 2015; Smith et al., 2014; Zhang, Wang et al., 2019). The accumulations of metal-tolerant/-reducing microbes is probably due to the selection pressure induced by vanadium during their interaction in the atmospheric zone.

The microbial community on vanadium mine tailing and adjacent soil in Chenxi County, Hunan Province, China, has also been studied (Sun et al., 2020). Figure 11.5a depicts the drastic difference in microbial community structure between soil and tailings, with a lower value of Shannon index in tailings than that in the nearby soil. NMDS analysis also reveals clusters formed separately between soil and tailings samples, as shown in Figure 11.5b. Figure 11.5c presents the microbial composition at the family level. *Burkholeriacae* is detected with relative abundance of 12.2 ± 2.1% in mine tailings and 6.3 ± 3.2% in soil, respectively; *Solimonadaceae*, even though absent in soil, is significantly enriched in tailing samples, accounting for 16.0 ± 3.4%. *Polaromonas*, belonging to the *Burkholderiaceae* family, plays important roles in V(V) reduction. Retrieved metagenome-assembled genomes lay out the evidence to unravel the genes encoded by *Polaromonas spp.*, highlighting the abundance of cymA, omcA and narG, which are conductive to V(V) reduction. The study demonstrates the impact force driving the microbial evolution within the community structures largely due to the vanadium stress.

The effects of vanadium and other geochemical parameters on the microbial community are robust factors that shape the microbial community in different environmental matrices. Studies clarify the impact by unraveling the microbial community structure, with emphasis on microorganisms with

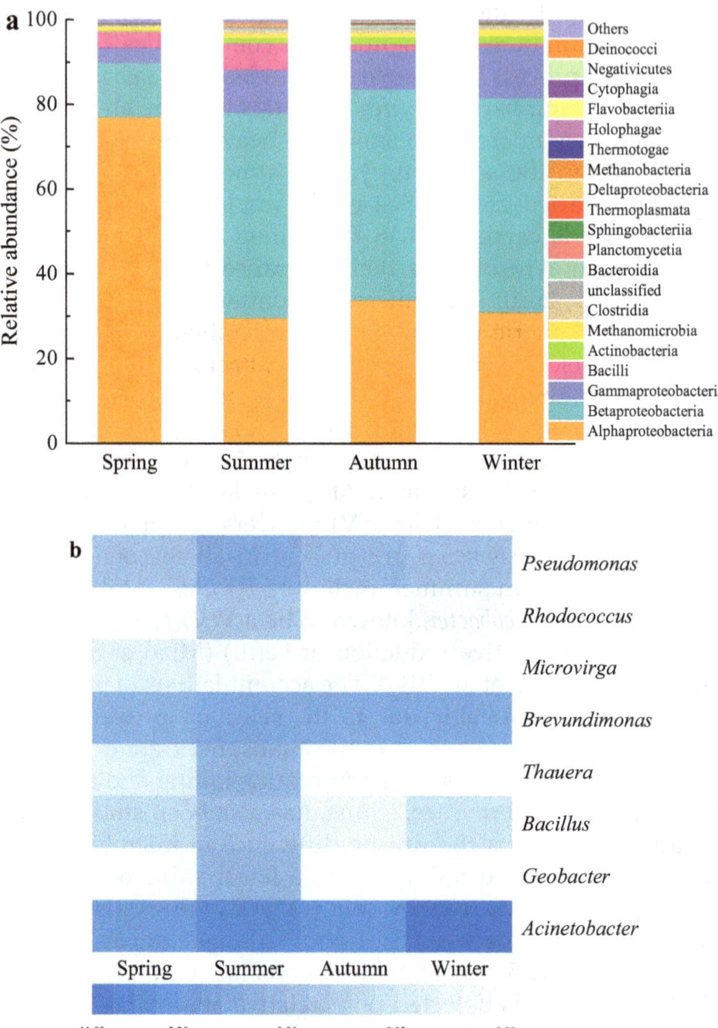

FIGURE 11.4
Microbial community dynamics in bioaerosols from vanadium smelter in Panzhihua, China, revealed by high-throughput Illumina sequencing at (a) class and (b) genus levels.
Source: Wang, Zhang, Wang et al., 2020.

greater adaptability to vanadium. The bioavailability of vanadium also varies as a result of a series of geochemical processes, leading to further evolution of the microbial community. The enriched vanadium reducing microbes imply that bioremediation of the vanadium-polluted environment can be conducted with indigenous microorganisms.

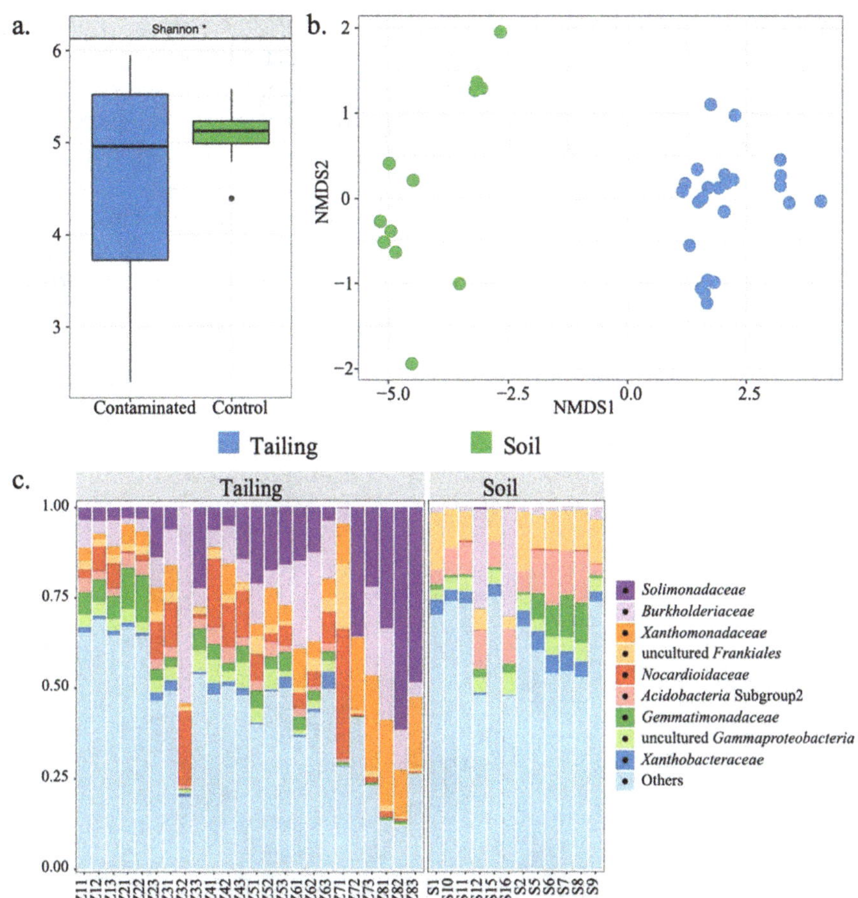

FIGURE 11.5
Characterization of the bacterial community in the V tailings (24 samples) and the adjacent soils (13 samples). (a) Difference in alpha diversity as measured by the Shannon index (* indicates $p < 0.05$); (b) Bray–Curtis distance plotted as an NMDS plot; and (c) bacterial community structures in the V tailings and the adjacent soils.
Source: Sun et al., 2020.

11.3 Spatiotemporal Dynamics of the Microbial Community

The vanadium-bearing dust released from the smelter or other flue sources is discharged into the atmosphere and subsequently deposited into soil (Xiao et al., 2015; Schlesinger et al., 2017), resulting in an overaccumulation of vanadium content in the soil. Long-term exposure to heavy metals entails a decline in microbial diversity and a change in composition within the microbial community, promoting metal-tolerant bacteria to dominate (Beattie et al., 2018; Zhao et al., 2019). With the microbes playing an essential part in bioremediation of vanadium (Yelton et al., 2013; Chen et al., 2021; Aihemaiti et al., 2020), understanding the spatiotemporal dynamics of indigenous microbial communities can provide useful information for planning a bioremediation strategy.

Wang et al. studied the soil microbial community in the vicinity of the vanadium smelter in Panzhihua (Wang, Zhang, Li et al., 2020). Topsoil (0–10 cm) and subsoil (0–170 cm) samples were collected from four directions at varying distance (10–2,000 m) relative to the smelter plant. The distribution of vanadium content and index value for potential ecological risk were calculated and displayed, as shown in Figure 11.6. For topsoil, vanadium concentration decreases at locations farther away from the smelter in all directions (north, east, south, west), among which northern area is less impacted by vanadium, with concentration up to 1,051 mg/kg.

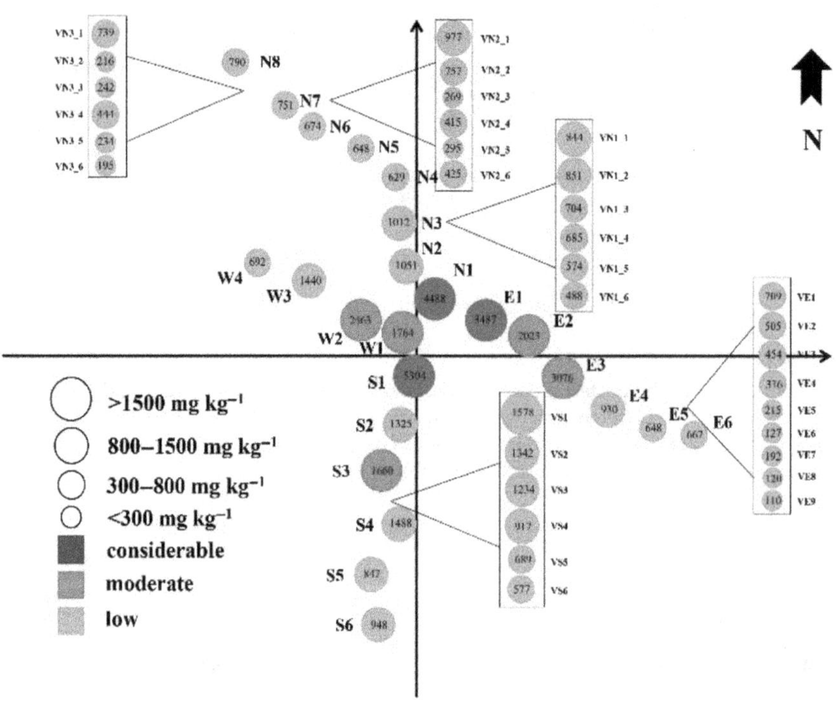

FIGURE 11.6
Vanadium distributions and potential ecological risk in surface and profile soils in Panzhihua, China. Different size circles show different vanadium concentration, and various colors stand for different ecological risks.

Source: Wang, Zhang, Li et al., 2020.

This surface vanadium distribution was in line with the finding where deposition of metal bearing dust into the soil is subject to the influence of locally dominant wind direction (Wang, Zhang, Wang et al., 2020; Dundar, 2006). Meanwhile, vanadium content in subsoil decreased along depth gradient, over which soil geochemistry including pH, redox, organic matters and presence of other metals varied accordingly (Frohne et al., 2015; Luo et al., 2017). The microbial community evolved distinctively in response to spatial variation in vanadium and other geochemistry. At the phylum level, Proteobacteria (4.29–41.85%), Actinobacteria (16.27–50.57%) and Chloroflexi (7.9–33.38%) predominated in the topsoil community. The relative abundance of Actinobacteria increased significantly in deeper soil zone, while a decreasing trend was observed for Firmicutes along depth gradient. The overall top abundance genera identified in the soil include *Bacillus* (6.61 ± 9.82%), *Unclassified-c-Acidobacteria* (5.58 ± 3.32%), *Unclassified-f-Anaerolineaceae* (3.73 ± 2.30%) and *Unclassified-o-Gaiellales* (3.68 ± 4.65%). However, the distribution pattern of microbial community differed between topsoil and subsoil, with even more enrichment for *Unclassfied-c-Acidobacteria* (2.64–12.90%) and *Unclassified-f-Anaerolineaceae* (0.58 ± 10.03%) in topsoil, while *Bacillus* (4.55–37.13%) and *Unclassified-o-Gaiellales* (1.39 ± 16.67%) are elevated in subsoil.

To explore the microbial response to variation in the level of vanadium content, mathematical modeling is performed to delineate the relationship between vanadium concentration and the relative abundance of the vanadium reducing bacteria (VRB) assemblage, comprising *Bacillus, Clostridium, Comamonadaceae, Geobacter* and *Pseudomonas*. As seen in Figure 11.7, growth of VRB assemblage can be stimulated by presence of a moderate level of vanadium content, as their relation can be better depicted by Gaussian equation ($R^2 = 0.653$, $p < 0.001$). In comparison, high vanadium concentration imposes inhibitory effects on microbial diversity and richness ascribed to higher toxicity. In a soil ecosystem, different VRB may compete with each other for limited resource to support their growth. There are also other factors imposing either stimulant or inhibitory effects on individual VRB taxa, including co-contaminants and nutrients such as total nitrogen (TN) and available phosphate (AP), which are positively correlated with *Bacillus* and *Closdrium* in the study. Tests are performed under an incubation environment in which the microbial community is under long-term vanadium exposure at either high or low pollutant loading. A Principal Coordinate Analysis (PCoA) (not shown here) implies distinctive evolution processes in which the microbial community suffers greater disruption under a higher contamination level, attaining a significant decrease in microbial diversity. In comparison, less impact results under low level exposure. These results also coincide with findings using original soil samples, where moderate level of vanadium result in higher values for microbial diversity and richness. This finding provides detailed insight into the impacts of vanadium on the microbial community.

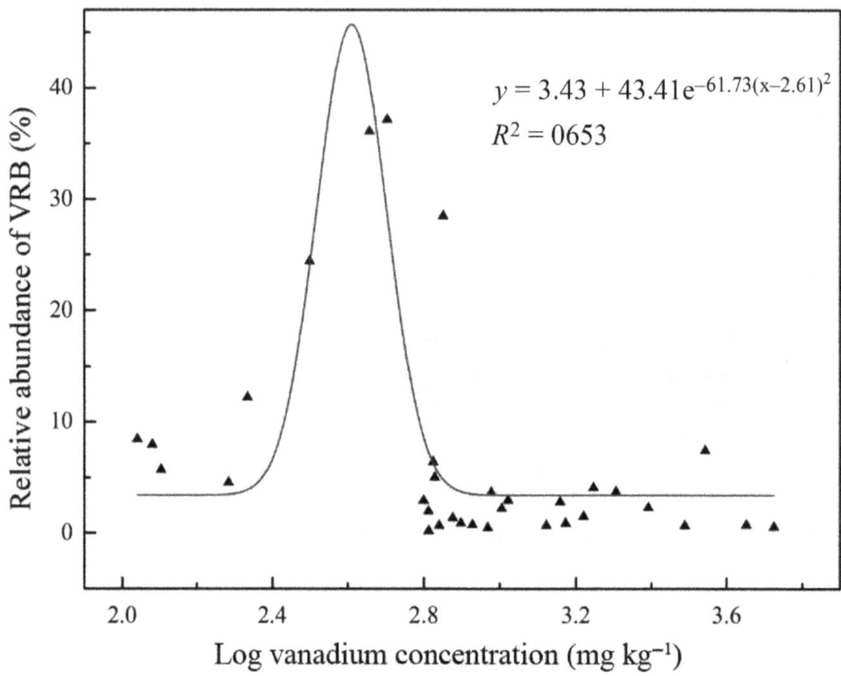

FIGURE 11.7
The relationships between the log value of vanadium concentration and the abundance of VRB assemblage in soils.
Source: Wang, Zhang, Li et al., 2020.

Zhang et al. reveal the interaction between vanadium and the microbial community driven by spatiotemporal factors in the vanadium smelter in Panzhihua, China (Zhang, Zhang et al., 2020). Similar to the previous study, topsoil (0–10 cm) and subsoil (0–110 cm) samples were collected in all direction (north, east, south, west) at varying distance (10–800 m) relative to the Panzhihua smelter plant. However, the sampling events take place in four successive seasons (spring, summer, fall and winter) at same sampling locations. Figure 11.8 shows the significant spatiotemporal variation in vanadium distribution for both topsoil and subsoil. Topsoil retains the most vanadium content throughout the seasons, while subsoil vanadium peaks during summer, fall and winter, with a decreasing trend along the depth gradient.

As for the microbial community, in general, vanadium slightly inhibits the diversity of topsoil community. Diversity also decreases along the sampling depth. The highest diversity is obtained in spring and significantly reduced during the rainfall season (summer, autumn and winter), which is consistent with excessive accumulation of subsurface vanadium during this period. This pattern can be explained by a drought–rewetting process in which

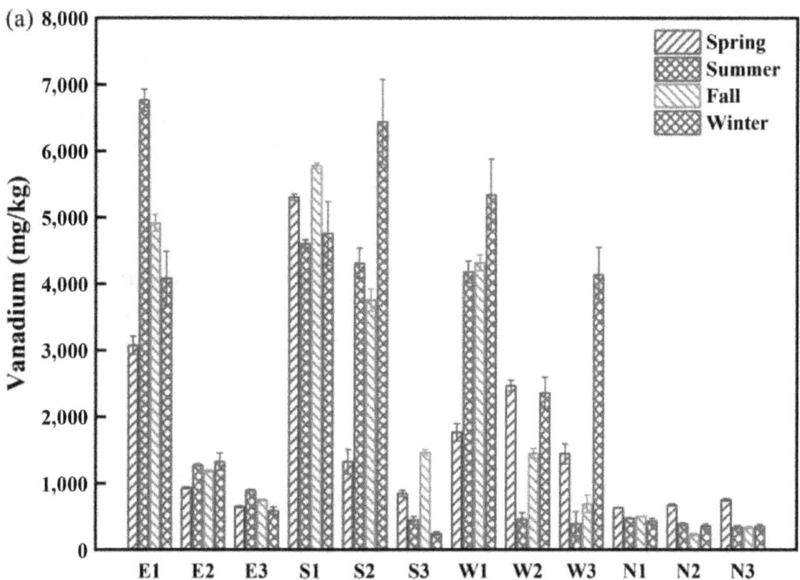

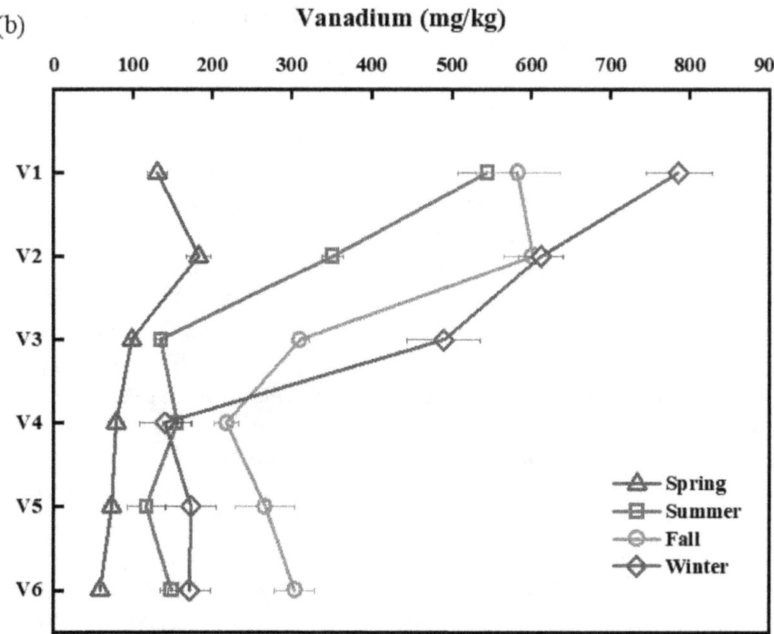

FIGURE 11.8
Spatiotemporal variations in vanadium concentrations for (a) topsoil and (b) subsoil sampled at smelter site.

Source: Zhang, Zhang et al., 2020.

geochemical parameters (pH, redox potential) governing the vanadium mobility vary drastically, leading to vanadium remobilization in the subsurface (Yang et al., 2014; Huang et al., 2015). As a result, the spatial distribution of the dominant phyla changes over time. In topsoil, Actinobacteria predominates in summer (27.3 ± 5.81%), fall (33.1 ± 5.56%) and winter (28.9 ± 5.5%), while Proteobacteria prevails in spring (35.2 ± 2.61%). In the subsoil, Proteobacteria becomes the most abundant in spring (34.5 ± 1.61%) and fall (34.5 ± 2.25%) but is surpassed by Actinobacteria in summer (38.4 ± 3.01%) and winter (41.9 ± 1.74%). Key genera are identified, including *Rhodospirillaceae, Lysobacter, Ramlibacter, Geobacter, Streptomyces* and *Bacillus* (collectively called MTG), with the ability associated with vanadium bioreduction and metal tolerance reported previously (Xu et al., 2015; Kaplan et al., 2019; Ortiz-Bernad et al., 2004; Polti et al., 2014; Zhang, Wang et al., 2019). In topsoil, according to an RDA analysis provided in Figure 11.9, most MTG negatively correlates with vanadium due to exceedingly severe toxicity presented by a substantial amount of topsoil vanadium content. However, a distinctive pattern is observed in the subsoil; for example, vanadium reducers *Bacillus* and *Geobacter* exhibit significantly positive correlation with vanadium content. Compared to topsoil, subsoil is contaminated with much lower amount of vanadium, which may stimulate the growth of certain V(V) reducers. Other geochemical parameters also drive the microbial response patterns. For example, available phosphorus (AP) is important to the immobilization of V(V), serving as a mediator for metals accumulation in biomass (Gadd, 2004). Organic matter is vital here to provide a carbon source for microbial metabolism (Yelton et al., 2013).

To further explore the relationship between MTG and other genera, a co-occurrence network analysis is carried out, as presented in Figure 11.10. Topological properties differ significantly between the topsoil and subsoil network. Fewer MTG are identified within the topsoil network, presenting less ecological linkage with other genera. Topsoil with elevated vanadium levels plays down the growth of MTG but promotes the ecological roles played by other taxa showing better adaptation, which is commonly possessed by taxa ubiquitously found in soils (Yin et al., 2015). In comparison, MTG in subsoil is more abundant, exhibiting greater interspecific interaction with other taxa, such as genera responsible for carbon and nitrogen cycling, which are vital to support metabolic mechanism of MTG. By comparing to the randomly generated network, both networks show nonrandom co-occurrence patterns, indicating a deterministic process in which microbial interaction is driven by habitat filtering effects ascribed to spatial and temporal variation in geochemistry. Revealing the microbial interrelationships benefits the development of targeted strategies to promote V(V) reducers for successful bioremediation.

In a soil ecosystem, microorganisms are very sensitive and inclined to form facilitative association in dealing with environmental stressors (Bar-Massada, 2015; Jiao et al., 2016; Li et al., 2017). Variation in vanadium concentration strongly affects the composition and succession of the microbial community

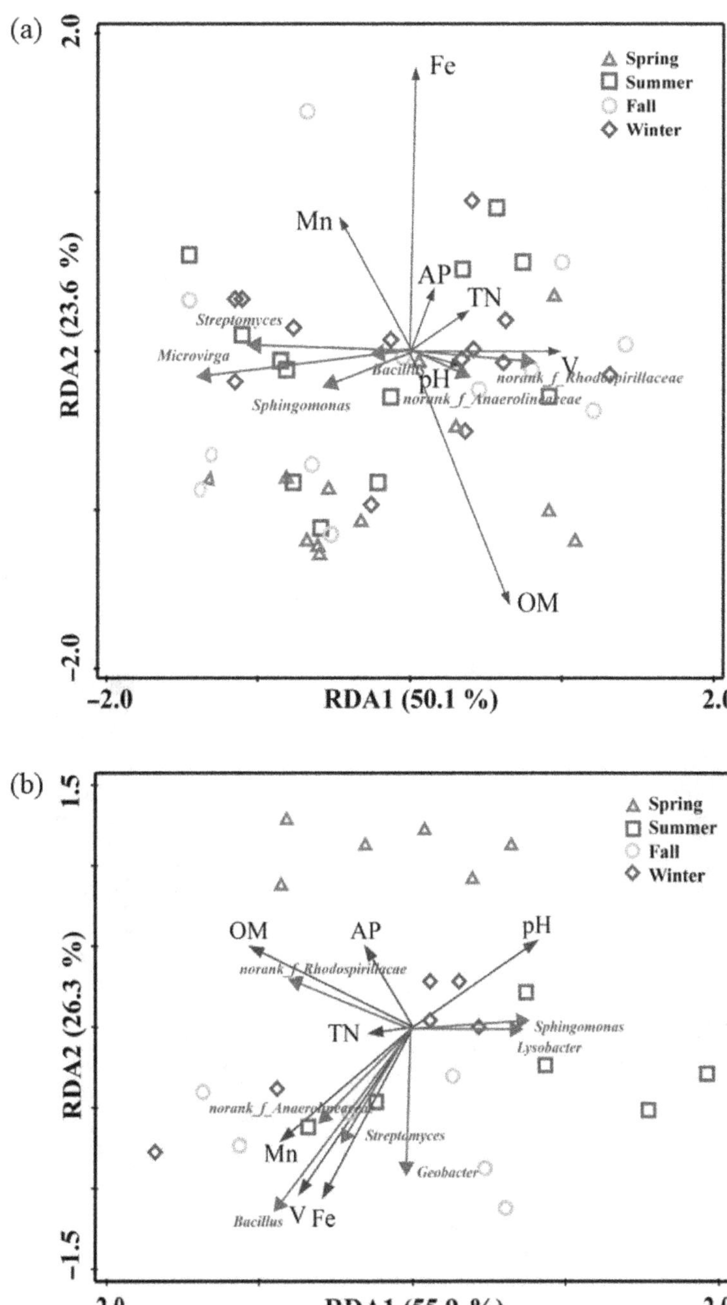

FIGURE 11.9
Redundancy analysis (RDA) based on the Bray-Curtis method to evaluate the influence of environmental variables on microbial community in (a) topsoil and (b) subsoil.
Source: Zhang, Zhang et al., 2020.

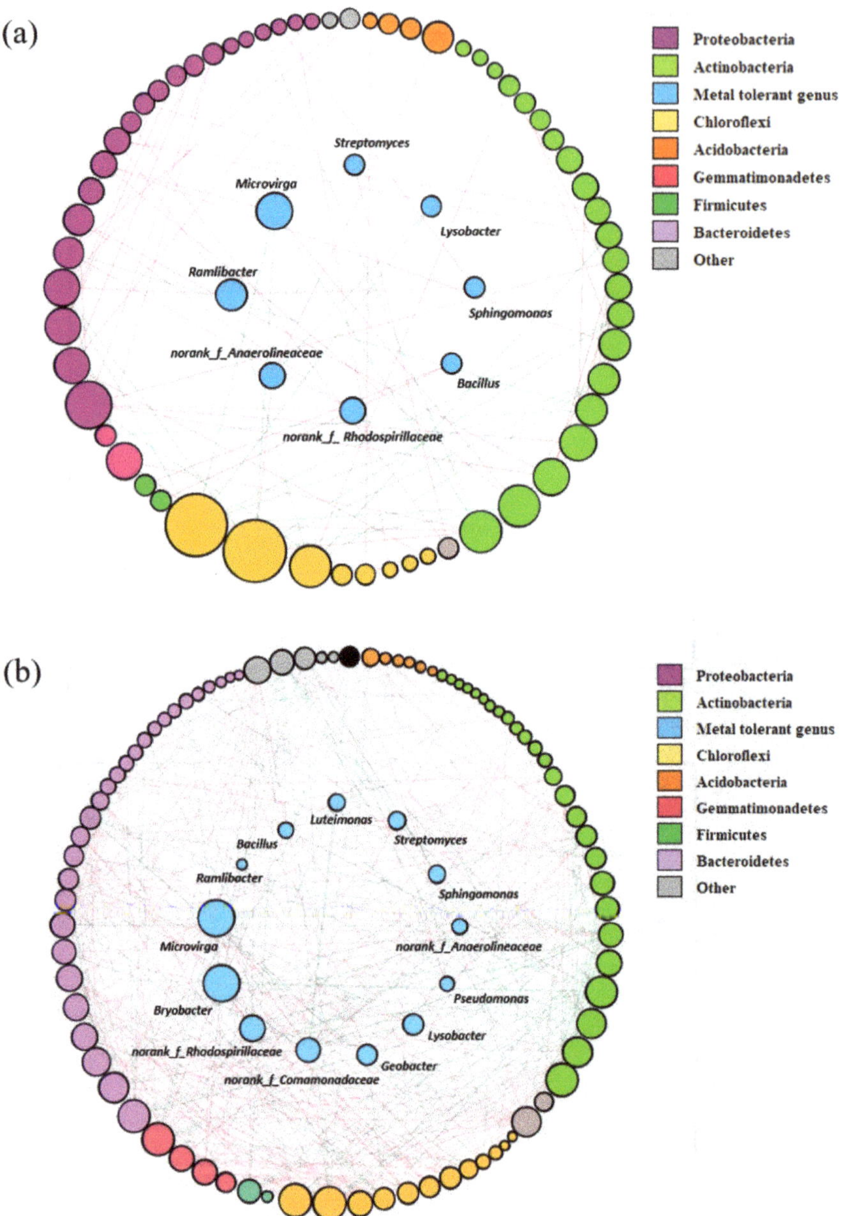

FIGURE 11.10
Co-occurrence network analysis of interspecific interaction among bacterial genera in topsoil and subsoil system.
Source: Zhang, Zhang et al., 2020.

in the soil. The interspecific interaction among taxa is largely regulated by the cyclic change in local geochemistry following spatiotemporal gradients. The microbial community studies have laid solid ground to further explore the microbial behaviors under changing environments, which is vital for decision making on managing vanadium contaminated sites. Furthermore, understanding the principles from spatiotemporal dynamics of the microbial community helps guide site bioremediation and management.

11.4 Microbial Remediation Concept and Mechanisms

The reduction of V(V) to less toxic V(IV) mediated by microorganisms is one of the most important process in the biogeochemical cycle of vanadium (Huang et al., 2015). Bioremediation incorporates the process where microorganisms are employed to remove V(V) from a contaminated environment by inducing microbial reduction (Lovley and Coates, 1997). Bioremediation is a more promising solution for vanadium removal due to its high efficiency, low-cost and green nature (Lovley and Coates, 1997; Yelton et al., 2013).

The bioprocess of V(V) reduction to V(IV) is governed by either electron transfer mediated by vanadium respiring bacteria or vanadium binding to reductase of other terminal electron acceptors, involving both extracellular and intracellular pathways.

Extracellular polymeric substances (EPS) are composed of a mixture of polysaccharides, mucopolysaccharides and proteins localized on bacterial cell surfaces with direct binding to substrate (Liu et al., 2002; Sutherland, 2001). EPS plays an essential role in mediating electron transfer between microorganisms and electron acceptors. V(V) can form binding with EPS through functional groups such as hydroxyl, acetamido and amino groups, offering protection for microbial cells in which normal metabolic functions are taking place, including ATP production, V(V) reduction and TCA cycles (Lai et al., 2018). EPS may also trigger the reaction to consume reactive oxygen species produced under high vanadium exposure, alleviating the oxidative stress that causes damage and alternation to amino acids and protein structures (Han et al., 2017).

V(V) can also be extracellularly reduced by cytochrome c (Myers et al., 2004). Cytochromes c such as OmcA and MtrC are the integral components of extracellular electron transfer pathways (Seeliger et al., 1998; Carpentier et al., 2005). Whole-genome sequencing analysis identifies 39 cytochromes c in dissimilatory metal-reducer *Shewanella,* including CymA, MtrA, MtrB, MtrC and OmcA, anchored on either inner or outer membrane to facilitate electron transfer (Heidelberg et al., 2002). CymA, the inner membrane reductases, is believed to mediate the electron release from the inner membrane quinone/quinol pool; electrons subsequently travel across the periplasm

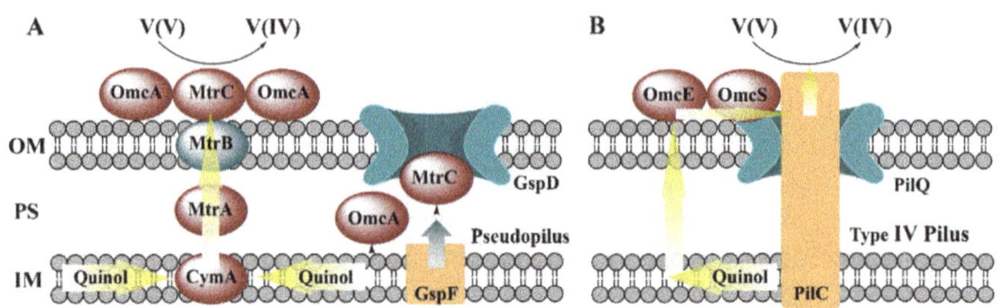

FIGURE 11.11
Proposed models depicting electron transfer pathways for *S. oneidensis* MR-1 (A) and *G. sulfurreducens* (B) (modified from Fig. 1 in Shi et al., 2007).

reductases, over which MtrA interacts with transmembrane protein MtrB embedded in the outer membrane that is necessary for transporting extracellular electron to insoluble acceptors (Beliaev and Saffarini, 1998); eventually, the electrons arrive at MtrC and OmcA, serving as terminal reductases for reduction ofV(V) to V(IV) on the cell surface (Shi et al., 2007). The whole process is depicted in Figure 11.11. Moreover, the absence of OmcA results in 62% loss of removal rate for *Shewanella oneidensis* MR-1 during V(V) bioreduction (Myers et al., 2004). Gene knock-out studies on *Geobacter sulfurreducens* highlights the monoheme cytochrome c *OmcF*, underpinning the extracellular electron transfer in the reduction of metal oxy-hydroxides (Teixeira et al., 2018). In addition, *Soehngenia and Anaerolinea* also show an ability for V(V) reduction by mediating extracellular reduction through cytochrome c (Zhang, Cheng et al., 2019).

V(IV) precipitate is obtained in the cytoplasm of *Shewanella loihica* PV-4, implying the occurrence of intracellular reduction (Wang et al., 2017). Vanadate shares similar structure to some inorganic ions, such as phosphate (Bowman, 1983). Studies verify that both V(V) and V(IV) can enter the cell using radio-labeled vanadium (^{48}V), possibly because of ion promiscuity in which vanadate is misidentified and taken up by cells through the phosphate transport system (Willsky et al., 1984; Olness et al., 2005). Nicotinamide adenine dinucleotide (NADH) is upregulated in the V(V) bioreduction, indicating the occurrence of the intracellular pathway (Zhang, Cheng et al., 2019), where NADH serves as an electron donor in the cytoplasm (He et al., 2011). The contribution of functional genes encoding for nitrate/nitrite reductase for V(V) bioreduction has been established in previous studies (Xu et al., 2015). Studies confirm that nitrate dissimilatory reduction is able to utilize nitrate stored intracellularly (Kamp et al., 2015). In addition, previous studies also examine the nitrate reductases, which are either membrane-bounded or soluble in periplasm, locating the presence of vanadium in the periplasmic enzyme (Antipov et al., 1998). The result indicates a potential pathway where intracellularly stored vanadate is processed by nitrate reductase to achieve

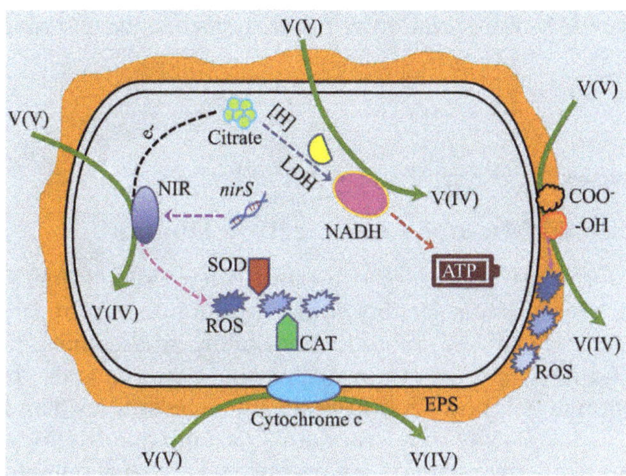

FIGURE 11.12
Proposed mechanisms of V(V) reduction by *L. raffinolactis* with citrate as an electron donor.
Source: Zhang et al., 2021.

bioreduction. A similar result is also obtained in a pure culture test utilizing *Lactococcus raffinolactis* to perform vanadium (V) bioreduction. The system achieves 86.5% removal efficiency during a 10-d operation with citrate acting as the electron donor for 50 mg L^{-1} initial V(V) (Zhang et al., 2021). An ultrathin section (50–60 nm) of the cells is retrieved. The result of a line scan expresses higher intracellular vanadium intensity, confirming the occurrence of an intracellular bioreduction pathway. The metabolic pathway of *Lactococcus raffinolactis* is displayed in Figure 11.12. Among denitrification genes (*nirS, napA* and *narG*), *nirS* stands out with the highest relative abundance, indicating the dominance of nitrite reduction pathway in V(V) bioreduction by *Lactococcus raffinolactis*. In addition, nitrite reductase activity also coincides with the *nirS* gene, exhibiting significant improvement in activity.

Extracellular and intracellular V(V) reduction pathways can also be inferred from mixed culture studies. In the CH_4-driven V(V) reduction bioprocess, the relative abundance of genes encoding an ABC transporter is increased from 0.38% to 0.87% (Zhang, Jiang et al., 2020), implying that V(V) enters into the cytoplasm through membrane-bound protein for intracellular detoxification. Significant upregulation of cytochrome c, NADH and the nitrite reductase gene *nirS* is enriched in the mixture culture system utilizing ethanol as an organic electron donor (Zhang, Cheng et al., 2019). With natural mackinawite as an electron donor, V(V) can also be biotically reduced effectively (He et al., 2020). Nitrate reductase encoded by *napA* gene proliferates in this bioprocess. This differs from biosystems for V(V) reduction with organic electron donors. Intensive occurrence of redox reaction involving autotrophic processes may be conducive to the upregulation of the periplasmic nitrate reductase NapA (Li et al., 2012).

11.5 Functional Microbes and Electron Donors

Functional microbes catalyze the bioconversion process for V(V), which requires an appropriate selection of electron donors. To improve the efficiency and efficacy of V(V) bioreduction, many studies have targeted the discovery and development of novel specialist bacterial taxa, coupled with electron donors to support their growth. Previous studies demonstrate the capability of a wide array of microorganisms, both autotrophic and heterotrophic, comprising the bacterial, eukaryotic and archaeal domains to reduce V(V) to V(IV). The implementation of microorganisms for bioremediation can be grouped into two approaches: pure culture and mixed culture, as listed in Table 11.1 and Table 11.2.

A pure culture capable of V(V) bioreduction is explored for mechanism investigation and remediation augmentation. In early studies, vanadate reduction was initially reported by employing *Pseudomonas vanadiumreductans* and *Pseudomonas isachenkovii* (Lyalikova and Yurkova, 1991). *Geobacter metallireducens* was first used in V(V) bioreduction coupled to acetate oxidation, achieving complete V(V) removal within six days (Ortiz-Bernad et al., 2004). *Shewanella oneidensis* MR-1 was also cultured to conduct V(V) bioreduction, which was used as the only electron acceptor for the strain to grow on, resulting in 99% of V(V) reduction in seven days (Carpentier et al., 2005). *Shewanella loihica* PV-4 has been reported to have the ability to reduce V(V) both inside and outside the cell under anaerobic conditions (Wang et al., 2017). Several *Pseudomonas aeruginosa* strains such as A17, A03 and C25a have also been demonstrated to reduce up to 38–60% V(V) under micro aerobic conditions (Ianieva and Smirnova, 2014). *Mesophilic* and *thermophilic* are archaea strains that can achieve V(V) bioreduction utilizing organic compounds or H_2 as electron donors (Zhang, Dong et al., 2014). More functional pure cultures are expected to be isolated from diverse habitats for scientific and remediation purposes.

Microorganisms in mixed cultures are widely applied in bioremediation of heavy metal due to its metabolic diversity, exhibiting more resilience and greater adaptation to external stress, which otherwise imposes inhibitory effects on a single bacterial strain. In addition, mixed cultures are more readily available than pure cultures, mainly utilizing bacteria living indigenously in the treatment area, making it more suitable for a practical scenario. By employing immobilized mixed anaerobic sludge, V(V) is mainly transformed into V(IV) forming insoluble precipitates, attaining 87% removal efficiency in 12 hours of operation (Zhang et al., 2015). *Lactococcus* and *Spirochaeta*

TABLE 11.1

V(V) Bioreduction Process Utilizing Pure Culture

Microbial Category			Initial V(V) mg/L	Electron Donor mg/L	Removal Performance			References
Phylum	Genus	Species			Time d	Efficiency %	Rate mg/L/d	
Proteobacteria	*Pseudomonas*	*Pseudomonas isachenkovii*	6000	1000.0 mg/L lactate	/	/	/	Lyalikova and Yurkova, 1991
		Pseudomonas aeruginosa	1000	10000.0 mg/L citrate	26	38.2–60.6	14.7–23.3	Ianieva and Smirnova, 2014
	Shewanella	*Shewanella oneidensis* MR-1	255	4500.0 mg/L lactate	6.7	99	37.8	Carpentier et al., 2005
		Shewanella oneidensis MR-1	114.8	1350.0 mg/L lactate	1	73.3	84.1	Myers et al., 2004
		Shewanella loihica PV-4	50.6	1152.0 mg/L citrate	27	71.3	1.3	Wang et al., 2017
	Halomonas	*Haloalkaliphilic Halomonas Strain* Mono	500	2800.0 mg/L acetate	5	100	100	Antipov and Khijniak, 2016
		Enterobacter cloacae EV-SA01	102	880.0 mg/L pyruvate	0.3	52	158.9	van Marwijk et al., 2009
	Acidocella	*Acidocella aromatica* PFBC	51	1800.0 mg/L fructose	14.6	70	2.5	Okibe et al., 2016
	Geobacter	*Geobacter metallireducens*	51	600.0 mg/L acetate	6	100	8.5	Ortiz-Bernad et al., 2004
Euryarchaeota	*Methanos-arcinacaea*	*Methanosarcina mazei*	510	0.5 mg/L methanol	30	100	92.9	Zhang, Dong et al., 2014
				1800.0 mg/L acetate			56.2	
	Methanother-mobacter	*Methanother-mobacter thermautotrophicus*	255	80:20 = H_2/ CO_2	30	100	15.89	
Firmicutes	*Lactococcus*	*Lactococcus raf f inolactis*	50	300 mg/L citrate	10	86.5 ± 2.17	4.32 ± 0.28	Zhang et al., 2021

TABLE 11.2
V(V) Bioreduction Process Utilizing Mixed Culture

Electron Donor Type	Donor Concentration mg/L	Initial V(V) mg/L	Enriched Genus	Removal Performance Time d	Efficiency %	Rate mg/L/d	References
acetate	640	51	*Geobacter, Anaerolineaceae*	3	97.0 ± 1.0	15.8 ± 0.3	Wang et al., 2018
ethanol	425	10	*Pseudomonas, Soehngenia, Anaerolinea*	7	96.0 ± 1.8	1.37 ± 0.03	Zhang, Wang et al., 2019
acetate	64	51	*Geobacter, Longilinea, Syntrophobacter, Spirochaeta, Anaerolinea*	3	92.8 ± 1.0	12.37± 0.85	Liu et al., 2017
sawdust	7500	75	*Bacteroidetes vadinHA17 norank, Anaerolineaceae norank*	10	0.903	6.5	Hao et al., 2021
woodchip-S(0)	60	10, 50	*Deltaproteobacteria, Geobacter*	135	68.5–98.2	9.74–14.4	Li and Zhang, 2020
acetate; glucose; citrate; lactate; soluble starch	800	75	*Actinobacteria, Chlorobaculum, Proteiniphilum*	0.5	65–76	97.5–114	Liu et al., 2016
methanol	800	10	*Delta proteobacteria, Anaerolineae*	7	100	1.43 ± 0.03	Shi, Zhang et al., 2020
acetate	60	50	*Desulfuromonas, Syntrophobacter*	2	83.9 ± 3.84	20.88 ± 0.96	Wang et al., 2020
S(0)	25 g/L						
S(0)	200 g/L	10, 50	*Geobacter, Bacteroidetes-vadinHA17*	276	88.3–99.2	0.082–0.26	Shi Zhang et al., 2020
FeS	400 g/L	10	*Thiobacillus*	150	100	10	He et al., 2021
S(0)	25 g/L	50	*Ignavibacterium, Chlorobia*	5	97.5 ± 1.2	9.75 ± 0.12	Zhang et al., 2018
Fe(0)	25 g/L		*Geobacter*		86.6 ± 2.5	8.66 ± 0.25	
CH4	86.8 mmol/ m2-d	2, 5, 10	*Methylophilus*	113	73, 73–100, 63	109, 267, 420	Lai et al., 2018
acetate	750	75	*Bacillus, Thauera*	3	65.2 ± 1.9–98.7 ± 3.6	16.32–23.28	Zhang, Wang et al., 2019
		10	*Novosphingobium, Acinetobacter, Sphingobium*	0.5	78.0 ± 3.5–88.3 ± 3.7	14.16–17.746	
acetate	384	183	*Polaromonas*	50	/	1.8	Sun et al., 2020

are enriched during the bioprocess from 0.01% to 59.36% and 1.54% to 4.12%, respectively. Studies have related both genera to dissimilatory reduction of heavy metal (Sintubin et al., 2009; Martins et al., 2010), implying that they are responsible for V(V) reduction in this bioprocess. One advantage of using a microbial consortium during bioremediation is that different microbes can coordinate with one another for diversification of metabolic pathways to prevent the inhibition incurred on a single bio-route. A synergistic relationship is examined extensively in bioreactor studies employing a culture mixture for vanadium (V) bioreduction (Zhang, Cheng et al., 2019; Li and Zhang, 2020), where oxidizers utilize electron donors to synthesize organic metabolites that can be used for heterotrophic V(V) reducers. For example, *Paludibacter*, belonging to fermenting bacteria, significantly increased from 1.34% to 8.46%, which can ferment complex organic matter into products of acetic, butyric, lactic acid and CO_2/H_2, while the produced lactic acid can provide nutrients to assist *Lactococcus* in removal of vanadium (V) in a mixed culture system (Zhang et al., 2013).

Electron donors play an essential role in assuring the success of bioremediation. Studies are performed to explore the variants of electron donors that can be applied to stimulate the bioprocess for V(V) bioreduction. V(V) loaded bioreactors are added with five commonly found soluble organic carbon sources, comprising acetate, lactate, citrate, glucose and soluble starch, among which acetate results in the highest removal efficiency, up to 75.6% during a 12-hour cycle (Liu et al., 2016). As shown in Figure 11.13, with greater molecular weight of the organic compounds, less V(V) removal can be achieved, whereas compounds with a simple structure are more easily utilized by microorganisms. The study also found a positive correlation between removal efficiency and the initial dosage of electron donor. However, the improvement in removal efficiency reaches a threshold if exceedingly high dosage is applied, because methanogenesis will compete with dissimilatory reduction process at a higher initial dosage level. More electron donors are employed in the soluble form, including methanol, ethanol and glucose, achieving various levels of bioreduction for V(V) (Hao et al., 2018; Antipov and Khijniak, 2016; Okibe et al., 2016).

Insoluble organic matter is also easily acquired as electron donors. Wood chips, rice husks and sawdust can be recycled from agricultural operation and exhibit great potential for environmental application. In contrast to soluble reagents, insoluble organic donors can be continuously released in relatively low concentration into the designated treatment zones, inducing lasting effects of biostimulation. Sawdust originating from pine is employed as the sole carbon source to support the microbial vanadium (V) removal, achieving removal efficiency of 90.3% in nutrient solution within ten days (Hao et al., 2021). In column experiments, bio-columns are inoculated with domesticated sludge (BS), ore mining soil (BP) and farm soil (BU), respectively. The removal efficiency of microbial Figure 11.14 provides the time history of V(V) removal with sawdust addition, attaining removal efficiency

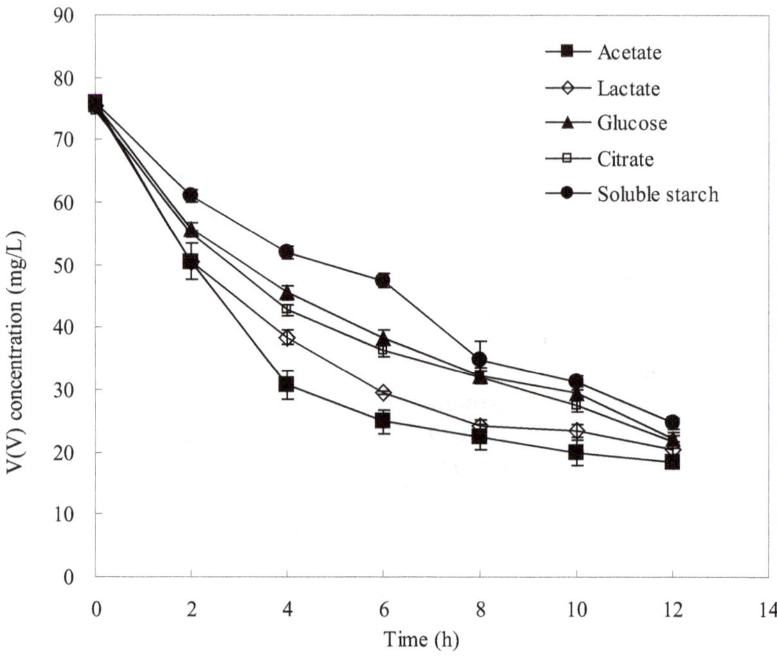

FIGURE 11.13
Time histories of V(V) concentrations in the bioreactors with five kinds of carbon sources during the 12 h operation.
Source: Liu et al., 2016.

of 58.7%, 54.8% and 38.4%, respectively. The study observed the persistent release of dissolved organic carbon (DOC) throughout operation stages, with initial DOC level up to 208.2 mg/L after 1.5 days. However, there is a decline in removal efficiency in later operation stages due to the exhausted bioavailable organic source and short contact time of the reactor.

To ensure effectiveness and cost savings during remediation operations, more types of electron donors as well as their efficacy with functional microbes should be examined. A novel design on the nutrient package, especially variants of electron donors, is also required.

11.6 Influences of Co-Existing Substances

Vanadium exists in geological environments under the influence of complexity in geochemistry. Figure 11.15 illustrates the interactive pattern in which substances comprising gases, organic compounds, nutrients and

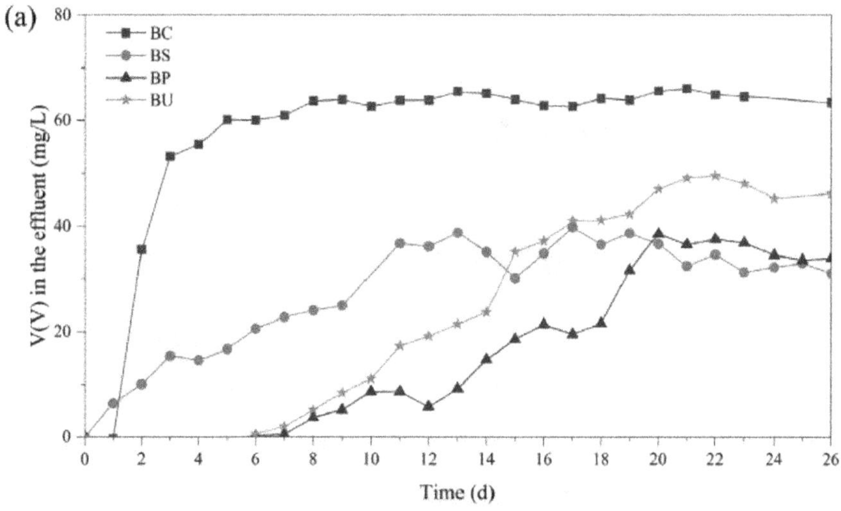

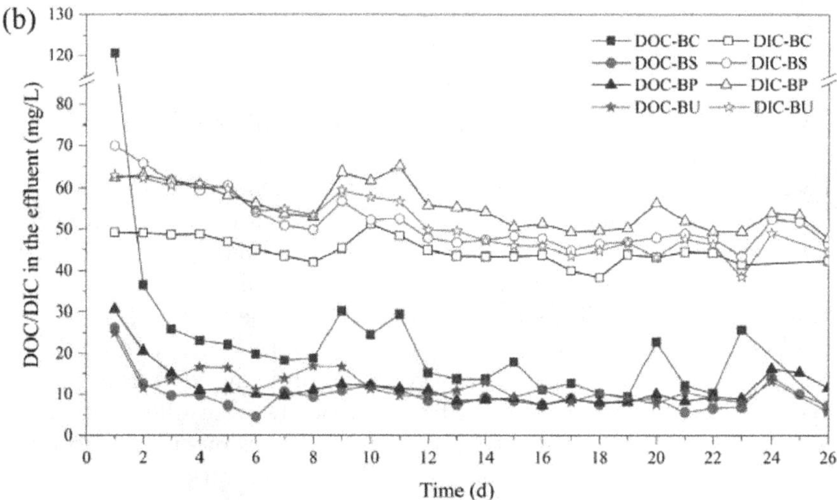

FIGURE.11.14
V(V) (a) and DOC/DIC concentrations (b) in the effluents of column bioreactors. (BC: bioreactor without inoculation; BS: bioreactor inoculated with sludge; BP: bioreactor inoculated with domesticated soil samples from V ore mining area (petroleum associated minerals); BU: bioreactor inoculated with domesticated farmland soil samples near V ore mining area (uvanite)).
Source: Hao et al., 2021.

other metal that co-occur with vanadium take part in the biogeochemical cycle of vanadium, eventually deciding the fate and transport of vanadium in the environmental matrix. Study of co-existing substances and their influence on vanadium biotransformation is important not only for

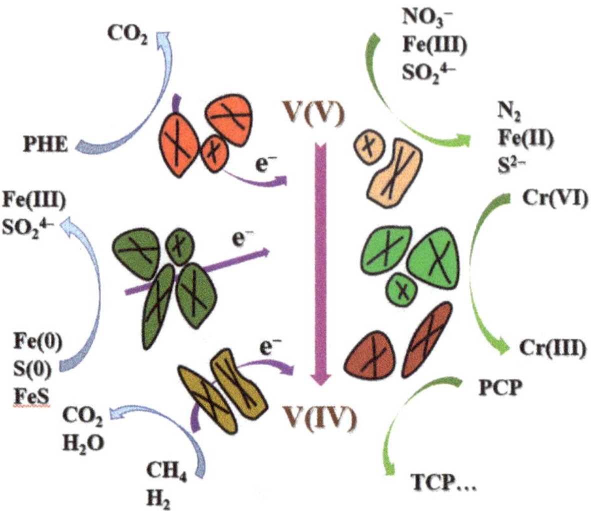

FIGURE 11.15
Interaction pathway between V(V) and other co-existing substances.

tracking the environmental risks but also for optimizing the bioprocess for vanadium (V) reduction.

Hydrogen (H_2) and methane (CH_4) are non-toxic gases commonly found in groundwater aquifers. H_2 is dissolved in groundwater with concentration in the magnitude of several millimoles per liters (Berta et al., 2018). CH_4 can be detected in groundwater aquifers, usually at low concentration < 10 μg/L (Bell et al., 2017). Both H_2 and CH_4 can be used to stimulate the microbial growth in bioremediation application. In a recent study, H_2 was used as an electron donor in the bioprocess of reducing V(V) from contaminated groundwater (Jiang et al., 2018). Bioreactors were filled with sludge obtained from an upflow anaerobic sludge blanket as inoculum and loaded with V(V) containing groundwater. Under near-neutral pH with 50 mg/L initial loading of V(V), 95.5% removal efficiency was achieved in a 7-d operation. Autohydrogenotrophic denitrifiers *Dechloromonas* and *hydrogenophaga* were detected at elevated abundance during the bioprocess, which may also contribute to V(V) reduction. In another study, Zhang et al. prepared bioreactors filled with V(V) contaminated synthetic groundwater and injected CH_4 as the sole electron donor (Zhang, Jiang et al., 2020). All reactors were cultivated for three months under anaerobic condition to ensure depletion of organic residues, and 95.8 ± 3.1% removal of 1 mM V(V) were achieved within a seven-day cycle. *Methylomonas* prevailed within the microbial community to drive the direct vanadate reduction coupling to methane oxidation. *Methylomonas*'s association to membrane-bound nitrate reductase had previously been determined (Xu et al., 2015). *Methanobacterium* is responsible for reverse methanogenesis in which VFA intermediates were obtained from anaerobic CH_4 oxidation (Yu et al., 2017). Heterotroph V(V) reducers can utilize the organic metabolites to reduce vanadate. These reducing gases can contribute to natural attenuation of V(V) through bioreduction by indigenous microorganisms.

Mackinawite is naturally occurring in aquifers, providing Fe^{2+} and S^{2-}, both of which can facilitate the anaerobic bioreduction for V(V). He et al. simulated V(V) contaminated groundwater flowing through the mackinawite-filled column for 150 days, achieving complete V(V) elimination at effluent sample with 10 mg/L initial loading (He et al., 2021). Figure 11.16a presents the breakthrough curve for column operation, where 144 PV and 30 PV are required to reach complete breakthrough for the biotic and abiotic column, respectively. Therefore, the biotic contribution for V(V) removal in the column is 76.4 ± 1.01%, while a 23.5 ± 1.25% reduction of V(V) is attributed to abiotic removal. Moreover, the rate of breakthrough is calculated based on the relationship between pore volume and column length, as shown in Figure 11.16b. The biotic column has a longer efficacy of V(V) removal, with migration of V(V) front at 0.69 ± 0.12 cm/PV, much lower than 2.78 ± 0.04 cm/PV in the abiotic column. Both *Thiobacillus* and *Sulfuricurvum* are present throughout the bioprocess, responsible for autotrophic oxidation of S^{2-} and Fe^{2+}, leading to VFA production. Heterotrophic reducers *Pseudomonas* and *Spirochaeta* are enriched in an upward trend in the bioprocess, utilizing VFA residuals to achieve synergistic V(V) removal. The study delivers strong implication that V(V) can be attenuated with the presence of naturally bearing mineral.

The combustion processes of fossil fuel are responsible for discharging vanadium and polycyclic aromatic hydrocarbon into the environment, leading to another representative scenario for co-contamination of V(V) and phenanthrene (PHE) (Dominguez et al., 2019). Under anaerobic conditions, batch bioreactors and a column experiment are established to explore the synchronous removal of V(V) and PHE (Shi, Cui et al., 2020). Under co-presence mode, complete V(V) is accomplished in a bioreactor, while removal efficiency of PHE reaches 82.0 ± 0.8%. Figure 11.17 shows the time-history of pollutant removal in a column setting for a dual contaminant system. In addition, PHE decomposers are more sensitive to increases in toxicity as the result of elevation in influent of both V(V) and PHE. Pronounced elevation of relative abundance of *Mycobacterium* and *Clostridium* is detected in bioprocess, in which methanol is used as a co-metabolic substrate to facilitate PHE biodegradation. Bioconversion of PHE also supplies extra electrons to heterotrophic V(V) reducers, coupling with direct methanol utilization, to aid vanadium (V) bioreduction. Therefore, biodetoxification of V(V) and PHE can be successfully achieved via bioremediation.

There are common electron acceptors such as NO_3^-, Fe^{3+}, SO_4^{2-} and CO_2 in subsurface environments. A previous study conducted an assessment of

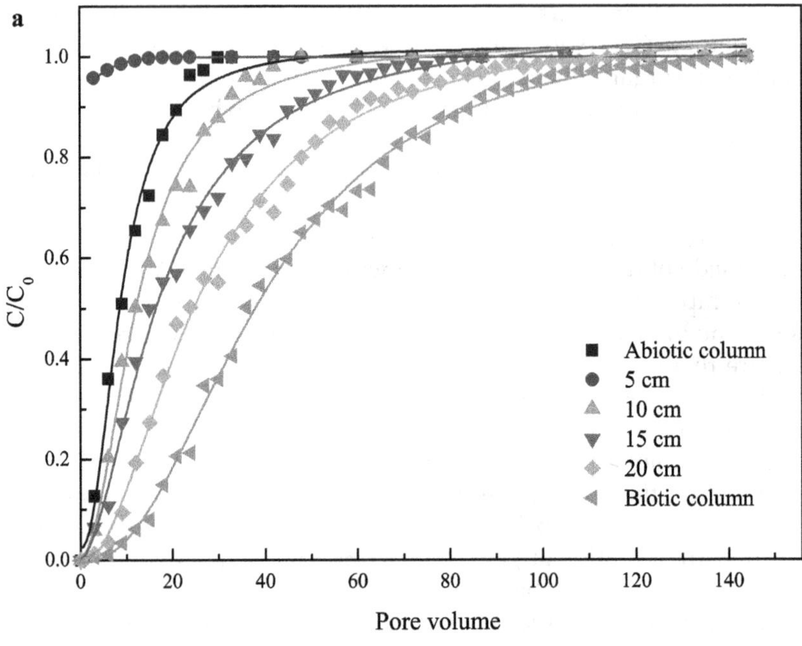

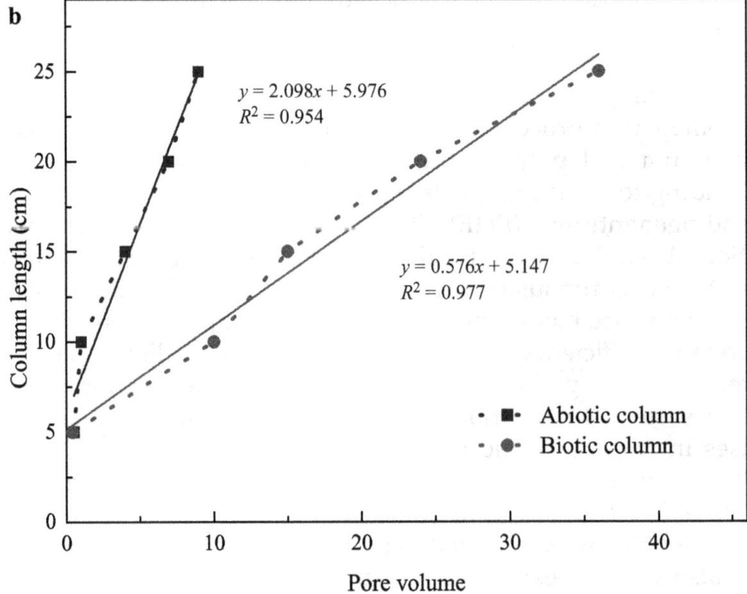

FIGURE 11.16
V(V) breakthrough curves of biotic and abiotic columns. (a) V(V) concentration in the effluent of both columns and in different heights of biotic column; (b) linear regression equation of V(V) front migration.

Source: He et al., 2021.

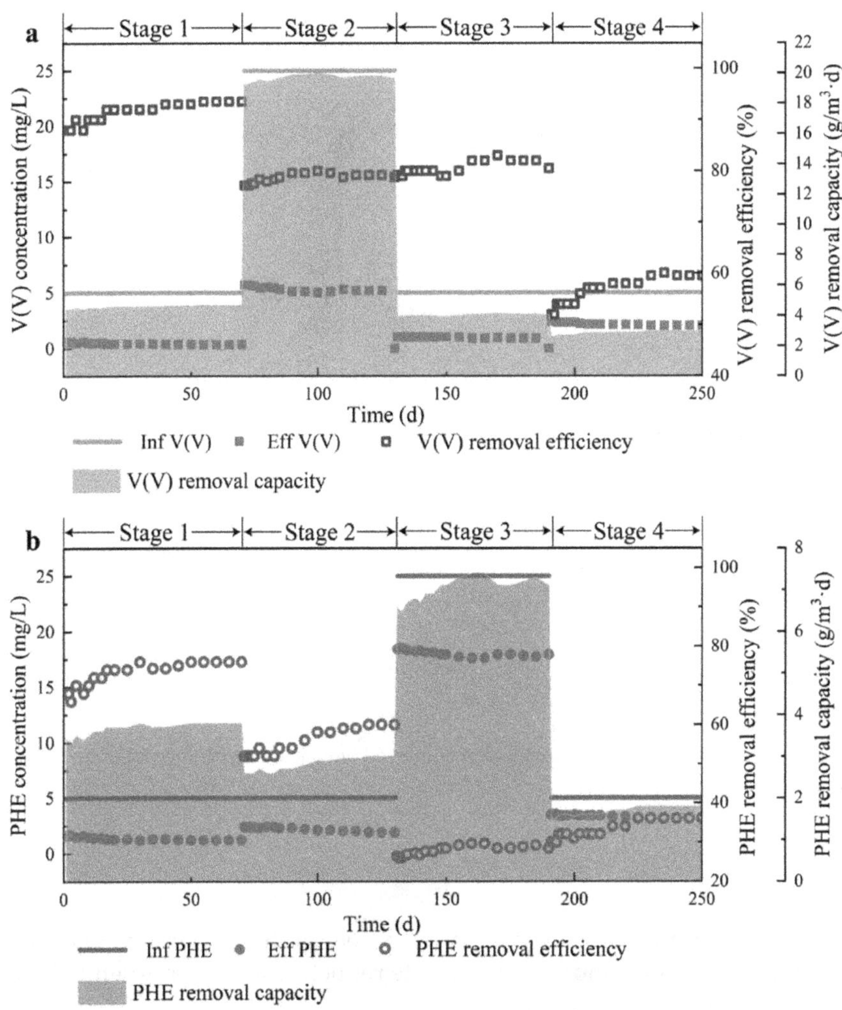

FIGURE 11.17
Time profiles of (a) vanadate [V(V)] and (b) phenanthrene (PHE) concentration corresponding to removal efficiency and capacity in the biological column during a 250-d operation.
Source: Shi, Zhang et al., 2020.

the impact of common electron acceptors on bioreduction of V(V) (Liu et al., 2017). It was found that Fe^{3+}, SO_4^{2-} and CO_2 adversely affect V(V) reduction at varying extents, possibly because these higher redox potential substances compete for electrons with V(V). In comparison, NO_{3-} first inhibits then promotes V(V) immobilization. With its presence, 90 ± 0.4% V(V) removal efficiency can be achieved. In addition, NO_{3-} increases the diversity of the microbial community, whereas Fe^{3+}, SO_4^{2-} and CO_2 cause a decline in the

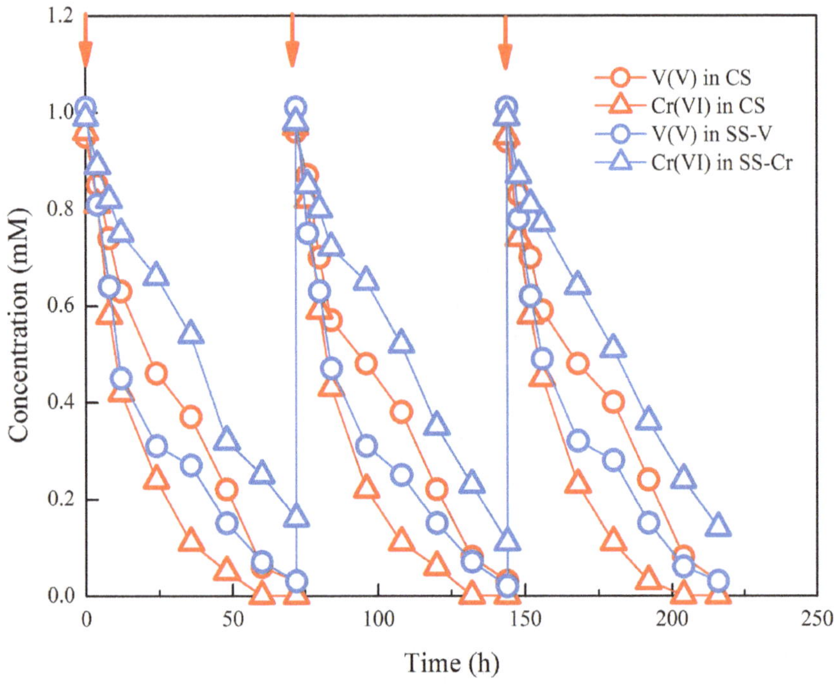

FIGURE 11.18
Time courses of V(V) and Cr(VI) removal in employed bioreactors in three consecutive cycles.
Source: Wang et al., 2018.

diversity and evenness. The observation agrees with the previous finding of vanadium (V) forms binding with nitrate reductase and subsequently being reduced (Antipov et al., 2000).

In many industrial sites, it is common to have scenarios in which vanadium and other high valance pollutants including heavy metal and organics co-present in an environmental matrix, such as chromium and chlorophenol (Zhang, Wang et al., 2019). In research conducted by Wang et al., bioreactors were supplied with V(V) and chromium (VI) [Cr(VI)] simultaneously(Wang et al., 2018). Figure 11.18 exhibits the time history of contaminant removal. With an initial concentration of 1 mM for both metals, after a 72 h operation, the removal efficiency of V(V) and Cr(VI) are 97.0 ± 1.0% and 99.1 ± 0.7%, respectively. Cr(VI) bio-reduction is enhanced while V(V) detoxification is inhibited, compared with a biosystem with a single metal. *Geobacter* and *Anaerolineaceae* predominate the V(V) only system, while *Thauera* and *Syntrophobacter* are more abundant in Cr(VI) contaminated system. In addition, *Spirochaeta* and *Spirochaetaceae* are determined to be favored by a dual

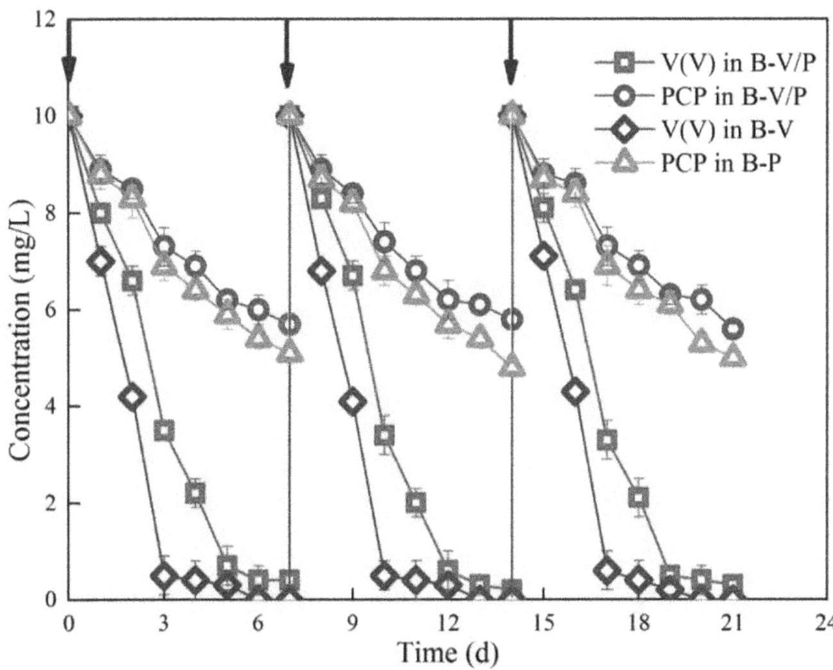

FIGURE 11.19
Time histories of V(V) and PCP concentrations over three consecutive operation cycles in different bioreactors.
Source: Zhang, Cheng et al., 2019.

contaminant system. In a biosystem with co-contamination of V(V) and pentachlorophenol (PCP), V(V) and PCP are synchronously bioreduced, with removal efficiency of 96.0 ± 1.8% and 43.4 ± 4.6%, respectively (Zhang, Cheng et al., 2019). However, PCP acts as a co-existing electron acceptor to compete with vanadium for electrons to achieve dechlorination, compromising the removal efficiency of V(V), for which complete removal is accomplished in a PCP absent system, as seen in Figure 11.19. Unlike Cr(VI), mentioned previously, PCP removal is also suppressed remarkably as a result of vanadium presence, exhibiting about an 8.3% drop in removal efficiency in the dual contaminant system compared with a PCP only system. These results indicate that these high valence substances can be removed synchronously with V(V) through microbial activities, further proving the efficacy of bioremediation.

Co-presence of other chemical compounds certainly affects the bioreduction of V(V) in the natural environment. Further study is needed to testify to the performance persistency of V(V) bioreduction in coping with a complexity of geochemical conditions. However, studies find that the interaction between vanadium and other chemical substances may promote the production of some metabolites or simultaneous removal of other pollutants, which may unlock new possibilities to enhance the efficacy of bioremediation.

11.7 Strengthening Strategy for Microbial Remediation

The success of V(V) bioreduction depends on smooth cooperation between microbial taxa as well as electron donors. *Geobacter metallireducens* (Ortiz-Bernad et al., 2004), *Shewanella oneidensis* (Myers et al., 2004), *Methanosarcina mazei* and *Methanothermobacter thermautotrophi* (Zhang, Dong et al., 2014) are specialists in reducing V(V). Most of them are heterotrophic, consuming organic substrates in the bioprocess (Zhang et al., 2015). However, the content of available organic matter intrinsic to the soil may become scarce at greater depth. Therefore, several strengthening strategies for microbial remediation are developed (Table 11.3). At the Rifle site near to the mining tail where groundwater is contaminated with vanadium, acetate is injected directly into the subsurface for biostimulation. Complete removal in the test area is achieved within a 23 month timespan (Yelton et al., 2013). In addition, MNP counts are performed on vanadium reducing cells in the column inoculated with original sediment. With acetate addition, 2.4×10^5 cells/g and 1.1×10^6 cells/g are obtained in sediment with high vanadium content and low vanadium content, respectively, both of which are higher than biomass in the control. This field study proves the effectiveness of bioremediation for V(V)-contaminated environment.

Introduction of AN external source of soluble organic matter requires high volume injection, possibly inducing secondary pollution and biofouling in the treatment zone (Mamais et al., 2016). Therefore, inorganic electron donors gain increasingly more attention in order to strengthen the performance and practicability of biostimulation. Elemental sulfur [S(0)] and zero-valent iron [Fe(0)] can be used as electron donors in the V(V) bioreducing process, where autotrophs are able to synthesize organic compounds with either S(0) or Fe(0) as the sole electron donor, providing energy for heterotrophic V(V) reducers (Zhang et al., 2018). In bioreactor test, after a two-month period of cultivation with 5 g S(0) (B-S) and 5 g Fe(0) (B-Fe) added as electron donors in the reactors, removal efficiency of V(V) reached $97.5 \pm 1.2\%$ and $88.6 \pm 2.5\%$ in a 120 hours cycle for B-S and B-Fe, respectively, displayed in Figure 11.20. These results are comparable to V(V) removals with organic electron donors (Wang et al., 2018). Although Fe(0) can abiotically reduce V(V), the bioprocess improves the longevity of Fe(0) through alleviating its passivation. The 16s rRNA analysis result highlights the enrichment of *Spirochaeta* and *Geobacter*, both of which are reported as functional species for reducing V(V) (Ortiz-Bernad

TABLE 11.3

Strengthening Strategies for Improving V(V) Bioremediation Efficiency

	Additives	Dosage	Pollution System	Initial V(V) mg/L	Microorganisms	Removal Performance Time d	Efficiency %	Rate mg/L/d	References
Biostimulation	S(0)	5g	V(V)	50	*Syntrophobacter, Spirochaeta GeobacterDesulfurella*	2	83.9	21.0	Wang et al., 2021
	S(0)	100g	V(V)+Cr(VI)	10	*Sulfuricurvum Geobacter*	0.25	88.3	35.3	Shi et al., 2020
	Fe(0)	5g	V(V)	50	*GeobacterFerritrophicum Caldithrix*	5	86.6	8.7	Zhang et al., 2018
	S(0)+H2		V(V)+Nitrate	30	*Bacillus Bacteroidia Epsilonproteobacteria*	5	95.04	5.7	Chen et al., 2018
	H2	3L	V(V)	85	*Dechloromonas Hydrogenophaga*	7	91	11.1	Jiang et al., 2018
bioaugmentation	Mixed microorganisms	50ml	V(V)	75	*Acetobacterium Oscillibacter*	0.5	87	130.5	Zhang et al., 2015
	Lactococcus raffinolactis		V(V)	50	*Lactococcus raffinolactis*	10	86.5	4.3	Zhang et al., 2021
	Shewanella loihica PV-4		V(V)+Cr(VI)	50.6	*Shewanella loihica PV-4*	27	71.3	1.3	Wang et al., 2017
electrical stimulation	Bioelectricity	529 mW/m2	V(V)	200	*Dysgonomonas Klebsiella*	7	60.7	17.3	Qiu et al., 2017
	Bioelectricity	543.4 mW/m3	V(V)	75	*Macellibacteroides Enterobacter Lactococcus*	0.5	93.6	140.4	Hao et al., 2015
	Bioelectricity	419 mW/m4	V(V)	75	*Deltaproteobacteria Bacteroidetes Spirochaetes*	0.5	76.8	115.2	Zhang et al., 2015

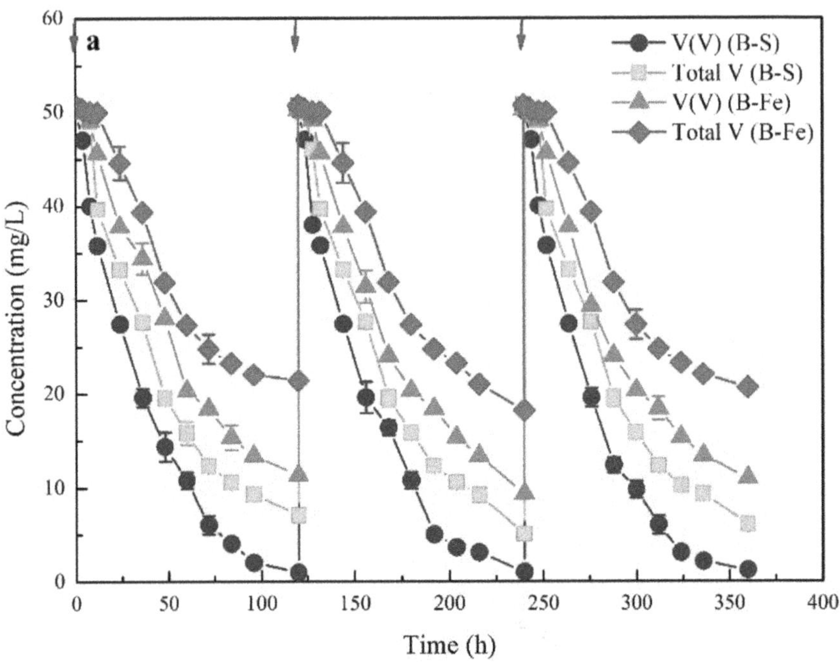

FIGURE 11.20
Time-course concentrations of V(V) and total V in three consecutive operating cycles.
Source: Zhang et al., 2018.

et al., 2004; Hao et al., 2015). The archaeal community also experience variation in community. For examples, *Methanosarcina* is dominant in B-S, which utilizes hydrogen as a substrate to reduce V(V) (Liu et al., 2011) In B-Fe, the accumulation of *Caldithrix* is determined, with potential to employ molecular hydrogen for nitrate reduction (Miroshnichenko et al., 2003). Synergy between autotrophs and heterotrophs played a major role in the reduction of V(V) by S(0) and Fe(0).

Furthermore, inorganic and organic substances can be used together in a V(V) bioreducing process. Wang et al. prepared bioreactors designated for simulation of mixotrophic (S(0) and acetate, denoted as MB), heterotrophic (acetate, denoted as HB) and autotrophic conditions (S(0), denoted as AB) (Wang et al., 2021); 83.9 ± 3.84% vanadium (V) was removed in MB within 48 h, which is higher than that obtained in HB (47.2 ± 2.59%) and AB (35.5 ± 1.71%). Figure 11.21a exhibits slower TOC utilization in MB, with a 0.46 ± 0.01 mg/L·d consumption rate, implying greater cost effectiveness. Figure 11.21b distinguishes the sulfate accumulation derived from oxidation of S(0). The sulfate accumulation is significantly reduced in MB, leaving only 48.0 ± 2.95 mg/L at the end of reaction cycle compared to 93.0 ± 3.95 mg/L in AB,

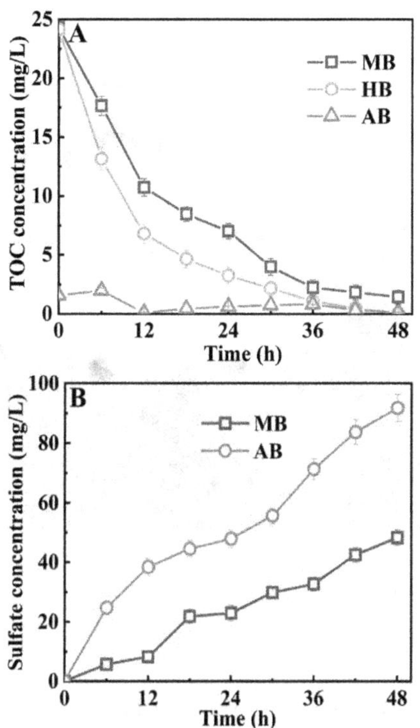

FIGURE 11.21
Fates of the added electron donors during microbial V(V) reduction in a typical cycle. (A) TOC variations in MB and HB; (B) sulfate accumulations in MB and AB.
Source: Wang et al., 2021.

suggesting less sulfate accumulation. In another study, Li et al. constructed a biological permeable reaction barrier (bio-PRB) filled with woodchips and S(0) as dual electron donors to treat V(V) contaminated groundwater (Li and Zhang, 2020). Figure 11.22 presents the time history of the vanadium reduction process during the 135-day operation; 98.2 ± 0.32% vanadium (V) removal efficiency was attained, with the maximum removal rate of 38.1 ± 1.17 mg/L·d. Flow rate, initial vanadium loading and presence of co-contaminants also play roles in regulating the efficacy of the bioprocess. During the bioprocess, cellulose in the woodchip undergoes hydrolysis before donating electrons for heterotrophic V(V) reduction. Presence of co-donor S(0) also enhances the bioreduction efficiency in comparison to results obtained by a woodchip-only system. The woodchip-S(0) stimulate bioprocess incorporating both heterotrophic and autotrophic process, which take place simultaneously in the mixotrophic condition, with functional genera including *Geobacter, Bacteroides, Sulfuricurvum* and *Thiobacillus* identified that are conducive to the process.

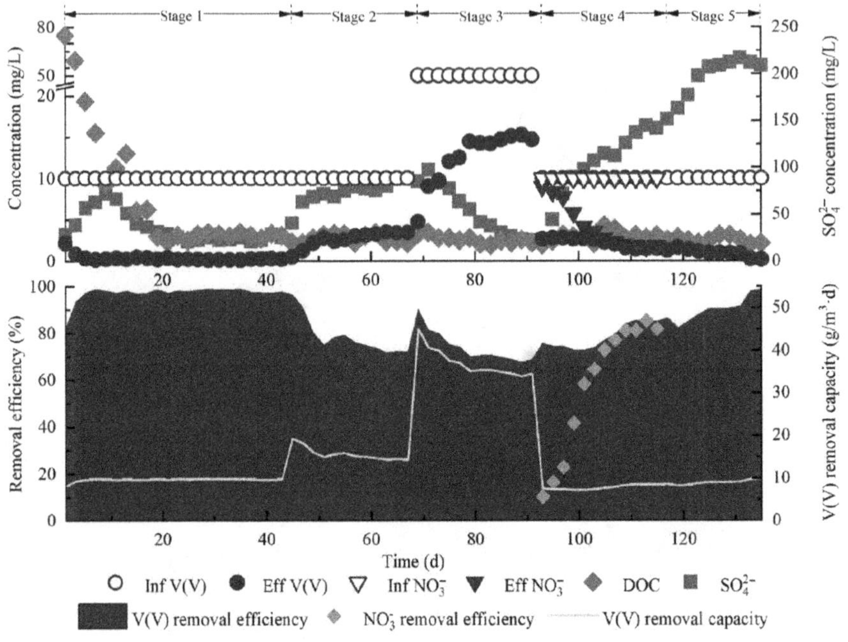

FIGURE 11.22

Time-course of vanadium (V) and nitrate removals with DOC and sulfate variation in woodchip-S(0) packed column during 135-d operation.

Source: Li and Zhang, 2020.

Bioremediation performance can be augmented utilizing functional microbial strains. Bioaugmentation is also explored to improve the efficiency of V(V)-polluted environment. Zhang et al. for the first time reports on the *Lactococcus raffinolactis*, a gram-positive strain expressing ability to reduce V(V) in sediment soil (Zhang et al., 2021). As shown in Figure 11.23, the variation in removal efficiency is largely attributed to additional dosage, with 5% concentration attaining the greatest removal efficiency of 86.5 ± 0.28%. This result proves that addition of functional V(V) reducers promotes bioremediation efficiency. More efforts should be made to optimize the bioaugmentation process for maintaining high activities and preventing the loss of functional strains.

Applying a voltage within an appropriate range can promote the reproduction and activity of bacteria. Hao et al. introduces bioelectricity from a single chamber microbial fuel cell (MFC) into a bioelectrical reactor (BER), displayed in Figure 11.24 (Hao et al., 2015). The MFC is filled with electrolyte components, with plain carbon and carbon fiber serving as materials for cathode and anode, respectively. The BER is seeded with domesticated sludge containing V(V) and electrolyte solution and connecting with MFC

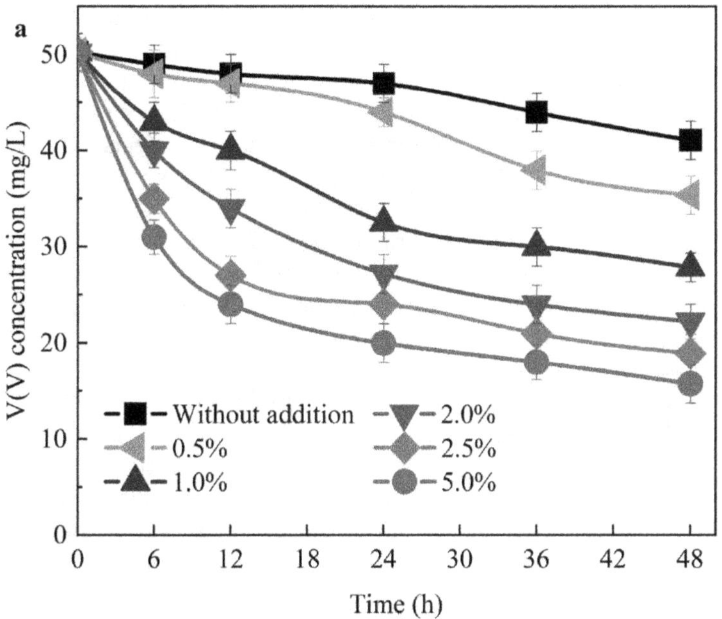

FIGURE 11.23
V(V) variations with time under different addition percentage of *Lactococcus raffinolactis*.
Source: Zhang et al., 2021.

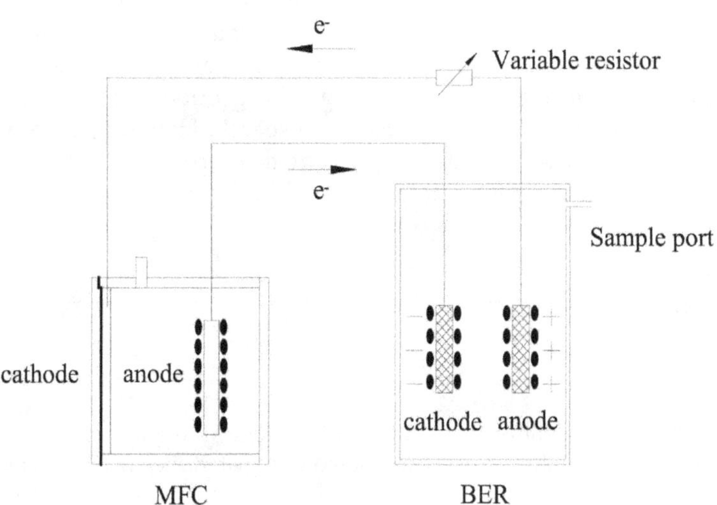

FIGURE 11.24
Experimental apparatus employed in the present study.
Source: Hao et al., 2015.

through copper wire. Glucose is used as the organic substrate in MFC to power the generation of bioelectricity, resulting in the maximum power density of 543.4 mW/m^3. With initial V(V) loading of 75 mg/L, the BER can make the removal efficiency of V(V) as high as 93.6% after 12 hours under 600 mV voltage. Calculations indicate limited electron-supply effects for V(V) reduction from external circuit. However, weak current is sufficient to promote the microbial activities. *Enterobacter, Spirochaeta* and *Lactococcus* are metal reducers presenting in high abundance in a BER (van Marwijk et al., 2009; Martin et al., 2010; Sintubin et al., 2009). Fermentative bacteria prevail in BER ascribed to glucose substrate, which is metabolized into micromolecular substances such as H_2, sugar and organic acid for vanadium reducers. In addition, pronounced enrichment of *Enterobacter* is identified in BER, facilitating the electron transfer to electron acceptors including a solid electrode due to its electrochemically active nature (Rezaei et al., 2009). With advances in novel materials for MFC components, electrochemical stimulation can be of great use to improve the bioreduction of V(V), especially with bioelectricity generated from an MFC-like configuration; energy can be harvested efficiently to save the operation cost, which will introduce great benefit in the practical application.

This chapter describes microbial responses to vanadium and summarizes microbially mediated processes for bioremediation of a V(V) polluted environment. Microbial communities in various environmental media are examined. Functional microbes are identified from both in-situ and lab studies. Many determinants are examined, including selection of the electron donor, presence of co-contaminants and co-occurring geochemical conditions. The results provide more insight into the metabolic process of V(V) bioreduction. Mechanism pathways are also characterized in view of removal performance, the microbial community, biological markers and the content of biogeochemical intermediates, providing solid ground to optimize the bioremediation process and advance future research. However, there is still a major lack of on-site studies dealing with real complexity of site conditions, which will require more effort to be realized in the future.

References

Aihemaiti, A., Gao, Y.C., Meng, Y., Chen, X.J., Liu, J.W., Xiang, H.L., Xu, Y.W., Jiang, J.G., 2020. Review of plant-vanadium physiological interactions, bioaccumulation, and bioremediation of vanadium-contaminated sites. *Sci Total Environ.* 712, 135637.

Aihemaiti, A., Jiang, J., Li, D., Liu, N., Yang, M., Meng, Y., Zou, Q., 2018. The interactions of metal concentrations and soil properties on toxic metal accumulation of native plants in vanadium mining area. *Int. Biodeterior. Biodegrad.* 222(15), 216–226.

Antipov, A.N., Khijniak, T.V., 2016. Vanadate reduction under alkaline conditions by haloalkaliphilic *Halomonas* strains. *Microbiology*. 85, 658–663.

Antipov, N.A., Lyalikova, N.N., Khijniak, V.T., L'vov, P.N., 1998. Molybdenum-free nitrate reductases from vanadate-reducing bacteria. *FEBS Lett*. 441, 257–260.

Antipov, A.N., Lyalikova, N.N., L'vov, N.P., 2000. Vanadium-binding protein excreted by vanadate-reducing bacteria. *Iubmb Life*. 49, 137–141.

Bar-Massada, 2015. Complex relationships between species niches and environmental heterogeneity affect species co-occurrence patterns in modelled and real communities. *Proc. R. Soc. B*. 282.

Beattie, R.E., Henke, W., Campa, M.F., Hazen, T.C., McAliley, L.R., Campbella, J.H., 2018. Variation in microbial community structure correlates with heavy-metal contamination in soils decades after mining ceased. *Soil Biol. Biochem*. 126, 57–63.

Beliaev, A.S., Saffarini, D. A., 1998. *Shewanella putrefaciens* mtrB encodes an outer membrane protein required for Fe(III) and Mn(IV) reduction. *J. Bacteriol*. 180, 6292–6297.

Bell, R.A., Darling, W.G., Ward, R.S., Basava-Reddi, L., Halwa, L., Manamsa, K., Ó Dochartaigh, B.E., 2017. A baseline survey of dissolved methane in aquifers of Great Britain. *Sci Total Environ*. 602, 1803–1813.

Bellenger, J.P., Xu, Y., Zhang, X., Morel, F.M.M., Kraepiel, A.M., 2014. Possible contribution of alternative nitrogenases to nitrogen fixation by asymbiotic N2-fixing bacteria in soils. *Soil Biol. Biochem*. 69, 413–420.

Berta, M., Dethlefsen, F., Ebert, M., Schäfer, D., Dahmke, A, 2018. Geochemical effects of millimolar hydrogen concentrations in groundwater—an experimental study in the context of subsurface hydrogen storage. *Environ. Sci. Technol*. 52(8), 4937–4949.

Bowman, B., 1983. Vanadate uptake in Neurospora crassa occurs via phosphate transport system II. *J. Bacteriol*. 153(1), 286–291.

Cao, X., Diao, M., Zhang, B., Liu, H., Wang, S., Yan, M., 2017. Spatial distribution of vanadium and microbial community responses in surface soil of Panzhihua mining and smelting area, China. *Chemosphere*. 183, 9–17.

Carpentier, W., De Smet, L., Van Beeumen, J., Brige, A., 2005. Respiration and Growth of *Shewanella oneidensis* MR-1 using vanadate as the sole electron acceptor. *J. Bacteriol*. 187(10), 3293–3301.

Chen, D., Xiao, Z., Wang, H., Yang, K., 2018. Toxic effects of vanadium (V) on a combined autotrophic denitrification system using sulfur and hydrogen as electron donors. *Bioresour. Technol*. 264, 319–326.

Chen, L., Liu, J.R., Hu, W.F., Gao, J., Yang, J.Y., 2021. Vanadium in soil-plant system: Source, fate, toxicity, and bioremediation. *J. Hazard. Mater*. 405, 124200.

Dominguez, C.M., Romero, A., Santos, A., 2019. Selective removal of chlorinated organic compounds from lindane wastes by combination of nonionic surfactant soil flushing and Fenton oxidation. *Chem. Eng. J*. 376, 120009.

Dundar, M.S., 2006. Vanadium concentrations in settled outdoor dust particles. *Environ. Monit. Assess*. 123, 345–350.

French, R.J., Jones, P.J.H., 1993. Role of vanadium in nutrition: Metabolism, essentiality and dietary considerations. *Life Sci*. 52, 339–346.

Frohne, T., Diaz-Bone, R.A., Laing, G.D., Rinklebe, J., 2015. Impact of systematic change of redox potential on the leaching of Ba, Cr, Sr, and V from a riverine soil into water. *J. Soils Sediments*. 15, 623–633.

Gadd, G.M., 2004. Microbial influence on metal mobility and application for bioremediation. *Geoderma*. 122, 109–119.

Ghosh, A., Sinha, K., Das, P., 2015. Optimization of reduction of copper using *stenotrophomonas maltophilia* PD2 biomass and artificial neural network modeling. *Environ. Eng. Manage. J.* 14, 37–44.

Han, X.M., Wang, Z.W., Chen, M., Zhang, X.R., Tang, C.Y., Wu, Z.C., 2017. Acute responses of microorganisms from membrane bioreactors in the presence of NaOCl: Protective mechanisms of extracellular polymeric substances. *Environ. Sci. Technol.* 51, 3233–3241.

Hao, L.T., Liu, Y., Chen, N., Hao, X., Zhang, B., Feng, C., 2021. Microbial removal of vanadium(V) from groundwater by sawdust used as a sole carbon source. *Sci Total Environ*. 751, 142161.

Hao, L.T., Zhang, B.G., Feng, C.P., Zhang, Z.Y., Lei, Z.F., Shimizu, K., Cao, X.L., Liu, H., Liu, H.P., 2018. Microbial vanadium (V) reduction in groundwater with different soils from vanadium ore mining areas. *Chemosphere*. 202, 272–279.

Hao, L.T., Zhang, B., Tian, C., Liu, Y., Shi, C., Cheng, M., Feng, C., 2015. Enhanced microbial reduction of vanadium (V) in groundwater with bioelectricity from microbial fuel cells. *J. Power Sources*. 287, 43–49.

He, C., Zhang, B.G., Lu, J.P., Qiu, R., 2021. A newly discovered function of nitrate reductase in chemoautotrophic vanadate transformation by natural mackinawite in aquifer. *Water Res*. 189, 116664.

He, M.Y., Li, X.Y., Liu, H.L., Miller, S.J., Wang, G.J., Rensing, C., 2011. Characterization and genomic analysis of a highly chromate resistant and reducing bacterial strain *Lysinibacillus fusiformis* ZC1. *J. Hazard. Mater.* 185, 682–688.

He, Q., Yao, K., 2010. Microbial reduction of selenium oxyanions by Anaeromyxobacter dehalogenans. *Bioresour. Technol.* 101, 3760–3764.

Heidelberg, J.F., Paulsen, I.T., Nelson, K.E., Gaidos, E.J., Nelson, W.C., Read, T.D., Eisen, J.A., Seshadri, R., Ward, N., Methe, B., Clayton, R.A., Meyer, T., Tsapin, A., Scott, J., Beanan, M., Brinkac, L., Daugherty, S., DeBoy, R.T., Dodson, R.J., Durkin, A.S., Haft, D.H., Kolonay, J.F., Madupu, R., Peterson, J.D., Umayam, L.A., White, O., Wolf, A.M., Vamathevan, J., Weidman, J., Impraim, M., Lee, K., Berry, K., Lee, C., Mueller, J., Khouri, H., Gill, J., Utterback, T.R., McDonald, L.A., Feldblyum, T.V., Smith, H.O., Venter, J.C., Nealson, K.H., Fraser, C.M., 2002. Genome sequence of the dissimilatory metal ion-reducing bacterium *Shewanella oneidensis*. *Nat. Biotechnol.* 20(11), 1118–1123.

Huang, J.H., Huang, Evans, L., Glasauer, S., 2015. Vanadium: Global (bio)geochemistry. *Chem. Geol.* 417, 68–89.

Hunter, W., 2007. An Azospira oryzae (syn Dechlorosoma suillum) strain that reduces selenate and selenite to elemental red selenium. *Curr. Microbiol.* 54, 376–381.

Ianieva, O.D., Smirnova, G.F., 2014. Vanadate reduction by Pseudomonas aeruginosa strains. *Mikrobiol. Zh.* 76 31–35.

Jiang, Y.F., Zhang, B.G., He, C., Shi, J.X., 2018. Synchronous microbial vanadium (V) reduction and denitrification in groundwater using hydrogen as the sole electron donor. *Water Res*. 141, 289–296.

Jiao, S., Liu, Z., Lin, Y., Yang, J., Chen, W., Wei, G., 2016. Bacterial communities in oil contaminated soils: Biogeography and co-occurrence patterns. *Soil Biol. Biochem*. 98, 64–73.

Kamp. A., Høgslund, S., Risgaard-Petersen, N., Stief, P., 2015. Nitrate storage and dissimilatory nitrate reduction by eukaryotic microbes. *Front. Microbiol.* 6, 1492.

Kaplan, H., Ratering, S., Felix-Henningsen, P., Schnell, S., 2019. Stability of in situ immobilization of trace metals with different amendments revealed by microbial 13 C-labelled wheat root decomposition and efflux-mediated metal resistance of soil bacteria. *Sci. Total Environ.* 659, 1082–1089.

Kumar, A., Bisht, B. S., Joshi, V., 2011. Bioremediation potential of three acclimated bacteria with reference to heavy metal removal from waste. *Int. J. Environ. Sci. Technol.* 2(2), 896–908.

Lai, C.Y., Dong, Q.Y., Chen, J.X., Zhu, Q.S., Yang, X., Chen, W.D., Zhao, H.P., Zhu, L., 2018. Role of extracellular polymeric substances in a methane based membrane biofilm reactor reducing vanadate. *Environ Sci Technol.* 52, 10680–10688.

Li, J., Zhang, B., 2020. Woodchip-sulfur packed biological permeable reactive barrier for mixotrophic vanadium (V) detoxification in groundwater. *Sci. China Technol. Sci.* 63, 2283–2291.

Li, W., Liu, L., Xu, L., Zhang, J., Yuan, Q., Ding, X., et al., 2020. Overview of primary biological aerosol particles from a Chinese boreal forest: Insight into morphology, size, and mixing state at microscopic scale. *Sci. Total Environ.* 719, 137520.

Li, X., Meng, D., Li, J., Yin, H., Liu, H., Liu, X., Cheng, C., Xiao, Y., Liu, Z., Yan, M., 2017. Response of soil microbial communities and microbial interactions to long-term heavy metal contamination. *Environ. Pollut.* 231, 908–917.

Li, Y., Katzmann, E., Borg, S., Schüler, D., 2012. The periplasmic nitrate reductase Nap is required for anaerobic growth and involved in redox control of magnetite biomineralization in Magnetospirillum gryphiswaldense. *J. Bacteriol.* 194, 4847–4856.

Liu, D., Dong, H., Bishop, E.M., Wang, H.M., Agrawal, A., Tritschler, S., Eberl, D., Xie, S., 2011. Reduction of structural Fe(III) in nontronite by methanogen Methanosarcina barkeri. *Geochim. Cosmochim.* Acta. 75, 1057–1071.

Liu, H., Fang, H.H.P., 2002. Characterization of electrostatic binding sites of extracellular polymers by linear programming analysis of titration data. *Biotechnol Bioeng.* 80(7), 806–811.

Liu, H., Zhang, B.G., Xing, Y., Hao, L., 2016. Behavior of dissolved organic carbon sources on the microbial reduction and precipitation of vanadium (V) in groundwater. *RSC Adv.* 6, 97253–97258.

Liu, H., Zhang, B.G., Yuan, H.Y., Cheng, Y.T., Wang, S., He, Z., 2017. Microbial reduction of vanadium (V) in groundwater: Interactions with coexisting common electron acceptors and analysis of microbial community. *Environ. Pollut.* 231, 1362–1369.

Lovley, D.R., Coates, J.D., 1997. Bioremediation of metal contamination. *Curr. Opin. Biotechnol.* 8(3), 285–289.

Luo, X.U., Yu, L., Wang, C.Z., Yin, X.Q., Mosa, A., Lv, J.L., Sun, H.M., 2017. Sorption of vanadium (V) onto natural soil colloids under various solution pH and ionic strength conditions. *Chemosphere.* 169, 609–617.

Lyalikova, N.N., Yurkova, N.A., 1991. Role of microorganisms in vanadium concentration and dispersion. *Geomicrobiol. J.* 10, 15–26.

Mamais, D., Noutsopoulos, C., Kavallari, I., Nyktari, E., Kaldis, A., Panousi, E., Nikitopoulos, G., Antoniou, K., Nasioka, M., 2016. Biological groundwater treatment for chromium removal at low hexavalent chromium concentrations. *Chemosphere.* 152, 238–244.

Martins, M., Faleiro, M.L., Da Costa, A.M.R., Chaves, S., Tenreiro, R., Matos, A.P, Costa, M.C., 2010. Mechanism of uranium (VI) removal by two anaerobic bacterial communities. *J. Hazard. Mater.* 184, 1–3.

Miao, Y., Liao, R., Zhang, X., Wang, Y., Wang, Z., Shi, P., et al., 2015. Metagenomic insights into Cr (VI) effect on microbial communities and functional genes of an expanded granular sludge bed reactor treating high-nitrate wastewater. *Water Res.* 76, 43–52.

Mirazimi, S., Abbasalipour, Z., Rashchi, F., 2015. Vanadium removal from LD converter slag using bacteria and fungi. *J. Environ. Manage.* 153, 144–151.

Miroshnichenko, L.M., Kostrikina, A.N., Chernyh, A.N., Pimenov, N., Tourova, P.T., Antipov, N.A., Spring, S., Stackebrandt, E., Bonch-Osmolovskaya, A., 2003. Caldithrix abyssi gen. nov., sp. nov., a nitrate reducing, thermophilic, anaerobic bacterium isolated from a Mid-Atlantic Ridge hydrothermal vent, represents a novel bacterial lineage. *Int. J. Syst. Evol.* 53, 323–329.

Mohapatra, R., Parhi, P., Thatoi, H., Panda, C., 2017. Bioreduction of hexavalent chromium by Exiguobacterium indicum strain MW1 isolated from marine water of Paradip Port, Odisha, India. *Chem. Ecol.* 33(2), 114–130.

Myers, J. M., Antholine, W. E., Myers, C. R., 2004. Vanadium(V) Reduction by Shewanella oneidensis MR-1 requires menaquinone and cytochromes from the cytoplasmic and outer membranes. *Appl. Environ. Microb.* 70(3), 1405–1412.

Okibe, N., Maki, M., Nakayama, D., Sasaki, K., 2016. Microbial recovery of vanadium by the acidophilic bacterium, Acidocella aromatica. *Biotechnol Lett.* 38(9), 1475–1481.

Olness, A., Gesch, R., Forcella, F., Archer, D., Rinke, J., 2005. Importance of vanadium and nutrient ionic ratios on the development of hydroponically grown cuphea. *Ind. Crop. Prod.* 21, 165–171.

Ortiz-Bernad, I., Anderson, R.T., Vrionis, H., Lovley, D., 2004. Vanadium respiration by *Geobacter metallireducens*: Novel strategy for in situ removal of vanadium from groundwater. *Appl. Environ. Microbiol.* 70, 3091–3095.

Polti, M., Aparicio, J., Benimeli, C., Amoroso, M., 2014. Simultaneous bioremediation of Cr (VI) and lindane in soil by actinobacteria. *Int. Biodeterior. Biodegrad.* 88(2), 48–55.

Qiu, R., Zhang, B., Li, J., Lv, Q., Wang, S., Gu, Q., 2017. Enhanced vanadium (V) reduction and bioelectricity generation in microbial fuel cells with biocathode. *J. Power Sources* 359, 379-383.

Raghu, G., Balaji, V., Venkateswaran, G., Rodrigue, A., Maruthi Mohan, P., 2008. Bioremediation of trace cobalt from simulated spent decontamination solutions of nuclear power reactors using E. coli expressing NiCoT genes. *Appl. Microbiol. Biotechnol.* 81(3), 571–578.

Rasoulnia, P., Mousavi, S., 2016. V and Ni recovery from a vanadium-rich power plant residual ash using acid producing fungi: Aspergillus niger and Penicillium simplicissimum. *RSC. Adv.* 6, 9139–9151.

Rezaei, F., Xing, D.F., Wagner, R., Regan, J.M., Richard, T.L., Logan, B.E., 2009. Simultaneous cellulose degradation and electricity production by Enterobacter cloacae in a microbial fuel cell. *Appl. Environ. Microb.* 75, 3673–3678.

Schlesinger, W.H., Klein, E.M., Vengosh, A., 2017. Global biogeochemical cycle of vanadium. *Proc. Natl. Acad. Sci. U. S. A.* 114 (52), 11092–11100.

Seeliger, S., Cord-Ruwisch, R., Schink, B., 1998. A periplasmic and extracellular c-type cytochrome of *Geobacter sulfurreducens* acts as a ferric iron reductase and as an electron carrier to other acceptors or to partner bacteria. *J Bacteriol.* 180, 3686–3691.

Shi, C.H., Cui, Y.L., Lu, J.P., Zhang, B.G., 2020. Sulfur-based autotrophic biosystem for efficient vanadium (V) and chromium (VI) reductions in groundwater. *Chem. Eng. J.* 395, 124972.

Shi, J., Zhang, B., Cheng, Y., Peng, K., 2020. Microbial vanadate reduction coupled to co-metabolic phenanthrene biodegradation in groundwater. *Water Res.* 186, 116354.

Shi, L., Squier, T.C., Zachara, J.M., Fredrickson, J.K., 2007. Respiration of metal (hydr) oxides by Shewanella and Geobacter: A key role for multihaemc-type cytochromes. *Mol. Microbiol.* 65(1), 12–20.

Sintubin, L., De Windt, W., Dick, J., 2009. Lactic acid bacteria as reducing and capping agent for the fast and efficient production of silver nanoparticles. *Appl Microbiol Biotechnol.* 84, 741–749.

Smith, J., Tremblay, P., Shrestha, P., Snoeyenbos-West, O., Franks, A., Nevin, K., et al., 2014. Going wireless: Fe (III) oxide reduction without pili by *geobacter sulfurreducens* strain JS-1. *Appl. Environ. Microbiol.* 80 (14), 4331–4340.

Solisio, C., Lodi, A., Converti, A., Borghi, M., 1998. Cadmium, zinc and chromium(III) removal from aqueous solutions by Zoogloea ramigera. *Chem. Biochem. Eng. Q.* 12, 45–49.

Su, J., Cheng, C., Huang, T., Ma, F., Lu, J., Shao, S., 2016. Novel simultaneous Fe(III) reduction and ammonium oxidation of Klebsiella sp. FC61 under the anaerobic conditions. *RSC. Adv.* 6, 12584–12591.

Sun, X., Qiu, L., Kolton, M., Häggblom. M., Xu, R., Kong, T., et al., 2020. V^{V} Reduction by *Polaromonas* spp. in Vanadium Mine Tailings. *Environ. Sci. Technol.* 54, 14442–14454.

Sutherland, I.W., 2001. Microbial polysaccharides from gram-negative bacteria. *Int Dairy J.* 11(9), 663–674.

Taylor, M., Evans, D., Young, C., 2006. Highly-oxidised, sulfur-rich, mixed-valence vanadium(IV/V) complexes. *Chem. Commun.* 40, 4245–4246.

Teixeira, L.R., Dantas, J.M., Salgueiro, C.A., Cordas, C.M., 2018. Thermodynamic and kinetic properties of the outer membrane cytochrome *OmcF*, a key protein for extracellular electron transfer in *Geobacter sulfurreducens. BBA-Bioenergetics.* 1859(10), 1132–1137.

Teng, Y.G., Yang, J., Sun, Z.J., Wang, J.S., Zuo, R., Zheng, J.Q., 2011. Environmental vanadium distribution, mobility and bioaccumulation in different land-use districts in Panzhihua region. *SW China, Environ. Monit. Assess.* 176, 605–620.

van Marwijk, J., Opperman, D.J., Piater, L.A., van Heerden, E., 2009. Reduction of vanadium (V) by *Enterobacter cloacae* EV-SA01 isolated from a South African deep gold mine. *Biotechnol. Lett.* 31, 845–849.

Vecchia, E., Veeramani, H., Suvorova, E.I., Wigginton, N., Bargar, J., Bernier-Latmani, R., 2010. U(VI) reduction by spores of Clostridium acetobutylicum. *Res. Microbiol.* 161, 765–771.

Wang, C., Wang, F., Hong, Q., Zhang, Y., Kengara, F., Li, Z., Jiang, X., 2013. Isolation and characterization of a toxic metal tolerant Phenanthrene degrader *Sphingobium* sp. in a two liquid phase partitioning bioreactor (TPPB). *Environ. Earth Sci.* 70(4), 1765–1773.

Wang, G., Zhang, B., Li, S., Yang, M., Yin, C., 2017. Simultaneous microbial reduction of vanadium (V) and chromium (VI) by *Shewanella loihica* PV-4. *Bioresour. Technol.* 227, 353–358.

Wang, S., Zhang, B.G., Diao, M.H., Shi, J.X., Jiang, Y.F., Cheng, Y.T., Liu, H., 2018. Enhancement of synchronous bio-reductions of vanadium (V) and chromium (VI) by mixed anaerobic culture. *Environ. Pollut.* 242, 249–256.

Wang, S., Zhang, B.G., Li, T.T., Li, Z.Y., F, J., 2020. Soil vanadium(V)-reducing related bacteria drive community response to vanadium pollution from a smelting plant over multiple gradients. *Environ. Int.* 138, 105630.

Wang, Y., Zhang, B., Wang, S., Zhong, Y., 2020. Temporal dynamics of heavy metal distribution and associated microbial community in ambient aerosols from vanadium smelter. *Sci. Total Environ.* 735, 139360.

Wang, Z.L., Zhang, B.G., He, C., Shi, J.X., Wu, M.X., Guo, J.H., 2021. Sulfur-based mixotrophic vanadium (V) bio-reduction towards lower organic requirement and sulfate accumulation. *Water Res.* 189, 116655.

Willsky, G.R., White, D.A., McCabe, B.C., 1984. Metabolism of added orthovanadate to vanadyl and high-molecular-weight vanadates by *Saccharomyces cerevisiae. J. Bio. Chem.* 259(21), 13273–13281.

Xiao, X.Y., Yang, M., Guo, Z.H., Jiang, Z.C., Liu, Y.N., Cao, X., 2015. Soil vanadium pollution and microbial response characteristics from stone coal smelting district. *Trans. Nonferrous Met. Soc. China.* 25(4), 1271–1278.

Xu, N., Qiu, G., Lou, W., Price, N., Qiu, B., Jiang, H., 2016. Identification of an iron permease, cFTR1, in cyanobacteria involved in the iron reduction/reoxidation uptake pathway. *Environ. Earth Sci.* 18(12), 5005–5017.

Xu, X.Y., Xia, S.Q., Zhou, L.J., Zhang, Z.Q., Rittmann, B.E., 2015. Bioreduction of vanadium (V) in groundwater by autohydrogentrophic bacteria: Mechanisms and microorganisms. *J. Environ. Sci.* 30, 122–128.

Yang, J.Y., Tang, Y., Yang, K., Rouff, A.A., Elzinga, E.J., Huang, J.H., 2014. Leaching characteristics of vanadium in mine tailings and soils near a vanadium titanomagnetite mining site. *J. Hazard. Mater.* 264, 498–504.

Yelton, A.P., Williams, K.H., Fournelle, J., Wrighton, K.C., Handley, K.M., Banfield, J.F., 2013. Vanadate and acetate biostimulation of contaminated sediments decreases diversity, selects for specific taxa, and decreases aqueous V^5+ concentration. *Environ. Sci. Technol.* 47(12), 6500–6509.

Yin, H., Niu, J., Ren, Y., Cong, J., Zhang, Xiaoxia, Fan, F., Xiao, Y., Zhang, Xian, Deng, J., Xie, M., He, Z., Zhou, J., Liang, Y., Liu, X., 2015. An integrated insight into the response of sedimentary microbial communities to heavy metal contamination. *Sci. Rep.* 5, 14266.

Yu, H., Kashima, H., Regan, J.M., Hussain, A., Elbeshbishy, E., Lee, H.S., 2017. Kinetic study on anaerobic oxidation of methane coupled to denitrification. *Enzyme Microb. Technol.* 104, 47–55.

Zhang, B.G, Cheng, Y.T., Shi, J.X., Xing, X., Zhu, Y.L., Xu, N., Xia, J.X., Borthwick, A.G.L., 2019. Insights into interactions between vanadium (V) bio-reduction and pentachlorophenol dechlorination in synthetic groundwater. *Chem. Eng. J.* 375(1), 121965.

Zhang, B.G., Jiang, Y.F., Zuo, K.C., He, C., Ren, J.Z.Y., 2020. Microbial vanadate and nitrate reductions coupled with anaerobic methane oxidation in groundwater. *J. Hazard. Mater.* 382, 121228.

Zhang, B.G., Hao, L., Tian, C., Yuan, S., Feng, C., Ni, J., Borthwick, A., 2015. Microbial reduction and precipitation of vanadium (V) in groundwater by immobilized mixed anaerobic culture. *Bioresour. Technol.* 192, 410–417.

Zhang, B.G., Li, Y.N., Fei, Y.M., Cheng, Y.T., 2021. Novel pathway for vanadium(V) bio-detoxification by gram-positive *Lactococcus raffinolactis*. *Environ. Sci. Technol.* 55(3), 2121–2131.

Zhang, B.G., Qiu, R., Lu, L., Chen, X., He, C., Lu, J., Ren, Z., 2018. Autotrophic vanadium (V) bioreduction in groundwater by elemental sulfur and zerovalent iron. *Environ. Sci. Technol.* 52, 7434–7442.

Zhang, B.G., Wang, S., Diao, M., Fu, J., Xie, M., Shi, J., et al., 2019. Microbial community responses to vanadium distributions in mining geological environments and bioremediation assessment. *J. Geophys. Res. Biogeosci.* 124, 601–615.

Zhang, H., Lu, H., Wang, J., Zhou, J., Sui, M., 2014. Cr(VI) reduction and Cr(III) immobilisation by *Acinetobacter* sp. HK-1 with the assistance of a novel quinone/graphene oxide composite. *Environ. Sci. Technol.* 48, 12876–12885.

Zhang, H., Zhang, B.G., Wang, S., Chen, J.L., Jiang, B., Xing, Y., 2020. Spatiotemporal vanadium distribution in soils with microbial community dynamics at vanadium smelting site. *Environ. Pollut.* 265, 114782.

Zhang, J.X., Dong, H., Zhao, L., McCarrick, R., Agrawal, A., 2014. Microbial reduction and precipitation of vanadium by mesophilic and thermophilic methanogens. *Chem. Geol.* 370, 29–39.

Zhang, J.X., Zhang, Y.B., Quan, X., Chen, S., Afzal, S., 2013. Enhanced anaerobic digestion of organic contaminants containing diverse microbial population by combined microbial electrolysis cell (MEC) and anaerobic reactor under Fe (III) reducing conditions. *Bioresour. Technol.* 136, 273–280.

Zhao, X.Q., Huang, J., Lu, J., Sun, Y., 2019. Study on the influence of soil microbial community on the long-term heavy metal pollution of different land use types and depth layers in mine. *Ecotox. Environ. Safe.* 170, 218–226.

12

Vanadium in Technical Applications and Pharmaceutical Issues

Dieter Rehder
Chemistry Department, University of Hamburg, Hamburg, Germany

CONTENTS

12.1 General and Environment Aspects....269
12.2 Pharmaceutical Issues....271
  12.2.1 Scheme I....278
References....280

12.1 General and Environment Aspects

Vanadium is the 21st most abundant element on Earth (occurrence 0.019%), being found in about 150 minerals and, in the form of orthovanadate ($H_2VO_4^-$), in water reservoirs (seawater, fresh water and ground water). Sea water (such as the Atlantic Ocean) contains vanadate at an average of 35 nM; variations through biological uptake, particle scavenging, atmospheric and/or biological as well as non-biological release occur (Wang 2009). Crude naphtha can also contain vanadium—in the form of oxidovanadium(IV), coordinated to porphinogens (Barcelaux 1999), commonly referred to as vanadyl petroporphyrins (Figure 12.1). Vanadium is also present in meteorites and has been detected in the interstellar medium. The omnipresence of vanadium suggests that is has played an important role in the early stages of the evolution of life (Campitelli et al. 2020). Under slightly reducing conditions vanadate is transformed to practically insoluble $VO(OH)_2$, which is widely spread in trace amounts in soils. Several groups of living organisms accumulate and/or employ vanadium; among these are ascidians and fan worms, macroalgae such as *Ascophyllum*, the mushroom fly agaric and several bacterial strains. Commonly, the deployment and/or action of vanadium by/in living organisms is coupled to a change in its oxidation state (+V, +IV, +III). Human blood

DOI: 10.1201/9781003173274-12

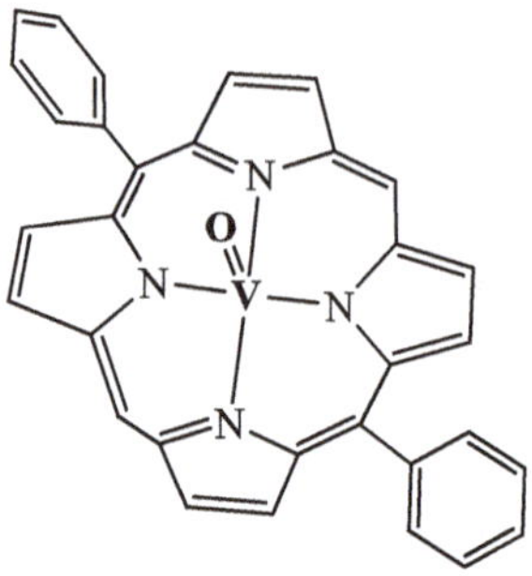

FIGURE 12.1
Oxidovanadium-diphenylporphyrin, representing one of the "vanadylporphyrins" present in crude oil (Manniko and Stoll 2019).

contains vanadate in concentrations close to 200 nM. The industrial utilization of vanadium(oxide) – for example, as a catalyst in H_2SO_4 production, as an alloying component in steel (ferrovanadium) or a redox partner in batteries and solar cells – is increasingly responsible for the distribution of vanadium(oxides) in the environment.

Industrial production of V_2O_5 by vanadium smelters brings forth emissions to the atmosphere; the V_2O_5 settles in soils, where it can convert to vanadate(V) that is taken up by (medicinal and diet-related) plants due to its similarity to phosphate (Owolabi et al. 2016). The average vanadium content in medicinal plants is about 500 μg/kg dry matter, with wild thyme having a particularly pronounced capacity to accumulate vanadium (Antal et al. 2009), a fact that may be considered of interest in the context of treating type II diabetes (vide infra). In pepper plants (such as *Capsium annuum*), vanadate was shown to increase plant growth and to accelerate flowering (García-Jeminéz et al. 2018).

The nutritional uptake and resorption of vanadium compounds in non-physiological concentrations can cause health problems: since vanadate acts as an antagonist to phosphate (Figure 12.1), vanadate will interfere with and consequently confuse or disorder the phosphate metabolism in living organisms, including DNA damage (Rodriguez-Mercado et al. 2011). Prenatal exposure to vanadium increases the risk of preterm delivery and low birth weight (Hu et al. 2017). However, the interference of vanadate with phosphate and the coordinative power of the VO^{2+} ion when interacting with biomolecules can also implicate positive effects: examples are the potentiality of vanadate – liberated from V^V coordination compounds under physiological conditions – in the treatment of diabetes (Crans et al. 2019) and cancer (Kioseouglou et al. 2015). The coordination of VO^{2+} to transferrin (Azevedo et al. 2018) and the impact of vanadium in oxidative stress (Ścibior and Kurus 2019) are additional examples for the physiological actions of vanadium. Major binders for V^V and V^{IV} in biological systems are (modified) polysaccharides (Kremer et al. 2015). The potential health benefits of vanadium and its general pharmacological capability versus the toxicological action have been addressed by, inter alia, Wong (2019), Levina and Lay (2017) and Ścibior et al. (2016); for an overview of the safety and efficacy data of vanadium, see also Ulbricht et al. (2012). Of further interest in this context is the antioxidant activity of oxidovanadium(IV) toward reactive oxygen species such as superoxide $O_2^{\bullet-}$ (Pranczk et al. 2016; Wu et al. 2016) and organic radicals (Pranczk et al. 2015), activity that can also be exploited in the context of anticancer properties of vanadium coordination compounds (Naso et al. 2016).

12.2 Pharmaceutical Issues

Vanadium compounds, or rather vanadate and/or oxidovanadium(IV) ("vanadyl") generated at physiological conditions from vanadium coordination compounds, inhibit or stimulate a variety of enzymes and are consequently also involved in immune-regulating mechanisms (Tsave et al. 2016). Vanadium coordination compounds have been shown to exhibit potentiality in the treatment of prevalent clinical pictures such as diabetes (type I and – in particular – type II), cancer and adipogenesis, as well as in viral and bacterial afflictions like pneumonia, tuberculosis, sand-fly fever and wound infection. The treatment, in particular, of human diabetes mellitus (type II), characterized – as a result of insufficient insulin reply – by hyperglycemia and disorders in the metabolism in particular of carbohydrates, has a comparatively long-standing tradition (Domingo and Gómez 2016), including clinical tests of the maltolato complex $VO(malt)_2$ (Thomson et al. 2007), **1** in Figure 12.3. Owing to (essentially minor) renal disorder with several of the probands, likely as a consequence of the vanadate-phosphate antagonism – cf. Figure 12.2 – the tests became discontinued.

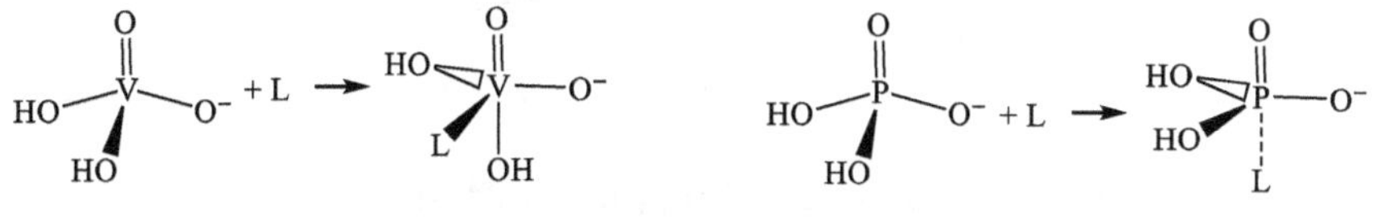

FIGURE 12.2
Interaction between vanadate and phosphate with a ligand L (an O-, N-, or S-donor). In contrast to phosphorus in phosphate, vanadium in vanadate can form stable five-coordinate complexes, while – with phosphate – the ligand L interacts non-covalently (indicated by a dashed line) and hence is in labile bonding interaction only with the central phosphorus.

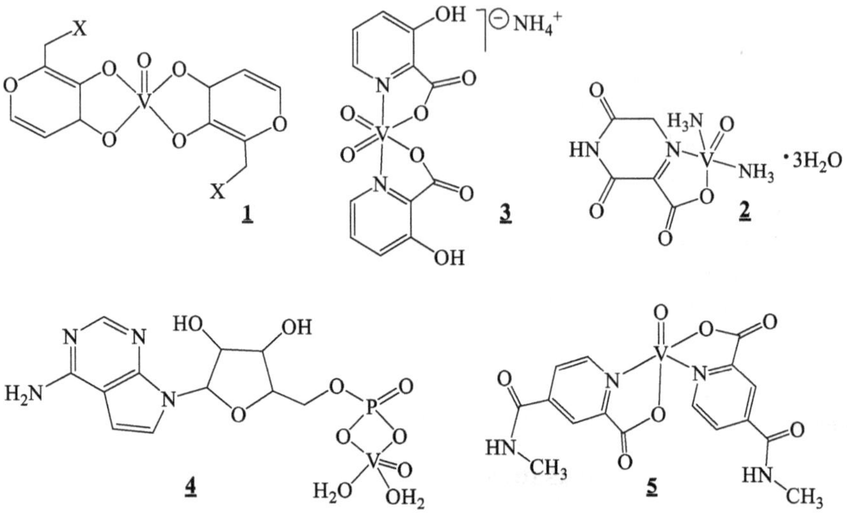

FIGURE 12.3
Selection of vanadium coordination compounds that have been employed in the treatment of diabetes (**1**: clinical tests with humans; **2** to **4**: streptocotozin induced diabetic rats; **5**: diabetic cats). These compounds significantly reduce blood glucose levels.

Sources: **1** Thomson and Orvig 2007, **2** Naglah et al. 2018, **3** Koleša-Dobravc et al. 2018, **4** Naglah et al. 2019, **5** Gätjens et al. 2003.

However, the vanadate-phosphate antagonism is also a key answer to the insulin-mimetic effect of vanadate: vanadium coordination compounds such as the classical maltolato complex (**1** in Figure 12.3), when hydrolytically and oxidatively converted to vanadate H_2VO_{4-} at physiological conditions, can enter cells via phosphate channels. Once in the intracellular environment, vanadate blocks off (deactivates) protein-tyrosine-phosphatase (PTP) via coordination to a PTP cysteinate residue (Irving and Stoker 2017; Hon et al. 2017). PTP is an enzyme that catalyzes the *de*phosphorylation of the intracellular tyrosine attached to the membrane-bound insulin receptor, a process that deactivates the insulin receptor. Blocking PTP prevents dephosphorylation and thus reactivates insulin uptake via the insulin receptor, consequently restoring intracellular glucose degradation:

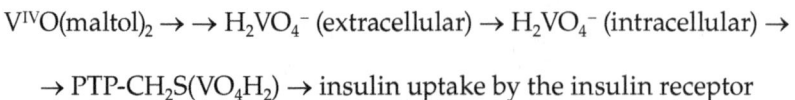

$$V^{IV}O(\text{maltol})_2 \rightarrow \rightarrow H_2VO_4^- \text{ (extracellular)} \rightarrow H_2VO_4^- \text{ (intracellular)} \rightarrow$$

$$\rightarrow \text{PTP-CH}_2\text{S}(VO_4H_2) \rightarrow \text{insulin uptake by the insulin receptor}$$

Alternatively, peroxidovanadate $H_2VO_3(O_2)-$ can inactivate PTP by oxidation of the SH group in the cysteinyl residue (Huyer et al. 1997).

The application of vanadium compounds is usually carried out intravenously, since – in the gastrointestinal tract – vanadium complexes are commonly degraded to inorganic vanadium compounds, in particular

$VO(OH)_2$, that are not easily absorbed in the gastrointestinal tract. Potential transporters for H_2VO_{4-} and VO_2^+ in the blood stream are transferrin and albumin. The scope of the efficacy of a potential antidiabetic vanadium compound is commonly tested with streptozotocin-induced diabetic rats. Somewhat surprisingly, simple vanadium compounds such as vanadate or complexes that release vanadate under physiological conditions (the anti-diabetic potentiality of vanadate is well established by now) have so far not attracted specific attention of the pharma industry (Scior et al. 2016). Actually, the ability of vanadate to reduce the blood sugar level had been reported as early as 1899 (Lyonnet et al.). Certainly, there remain concerns with respect to the off-target toxicity in the long-term use of vanadium compounds in humans. Off-target toxicity includes (i) the vanadate-phosphate antagonism and (ii) the ability of VO^{2+} to coordinate to – and thus alter and/or deactivate – the physiological function of biomolecules.

Another field directed towards the medicinal use of vanadium is the application of vanadium compounds as anticancer agents. Vanadium compounds have recently emerged as metal-based non-platinum antitumor agents, in particular in the treatment of osteosarcomas. In Figure 12.4, selected compounds are compiled that have been shown to be effective and to exert low toxicity in animal models. Metvan (compound **3**), in particular, when encapsulated into nano-structured lipid carriers for protection against degradation and for sustained release, showed a reasonably high *in vitro* effect against osteosarcoma cell lines (Cacicedo et al. 2019). Vanadate $H_2VO_4^-$, as well as the oxidovanadium(IV) compound $VO(dhp)_2$ (where dhp stands for 1,2-dimethyl-3-hydroxy-4-pyridinonate) are active against skin cancer (malignant melanoma) (Pisano et al. 2019).

The peroxide-oxidovanadium complex $[VO(O_2)edda]^+$ (edda = ethylendiamine diacetate), **1** in Figure 12.5, is active against the cancer cell lines MCF-7 (breast cancer) and A549 (adeno carcinoma; lung cancer), likely via the generation of reactive oxygen species such as superoxide $^{\bullet}O_2^-$, when applied under the mediation of Mn^{2+} and in a slightly acidic environment (Equation 12.1). Cell death is induced by apoptosis (Chen et al. 2019). Anti-cancer properties have also been reported for the non-oxidovanadium(IV) complex formed with diimidonylmethane, **2** in Figure 12.5 (Dankhoff et al. 2019). Compound **2** exhibits significant anti-proliferative activity against cancer (cell lines) such as melanoma, colon carcinoma and pancreatic carcinoma cells. The compound shows antiproliferative activity via strong interaction with/binding to DNA, resulting in cell cycle arrest (inhibition of tumor cell growth) and mitochondrial membrane damage, likely through the formation of reactive oxygen species.

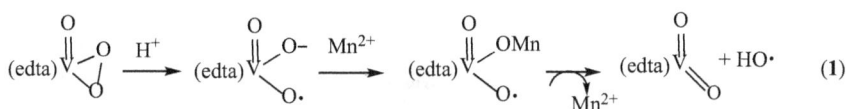

$$(edta)V(=O)(O_2) \xrightarrow{H^+} (edta)V(=O)(O^-)(O^\bullet) \xrightarrow{Mn^{2+}} (edta)V(=O)(OMn)(O^\bullet) \xrightarrow{} (edta)V(=O)(=O) + HO^\bullet \quad (\mathbf{1})$$

(Equation 12.1)

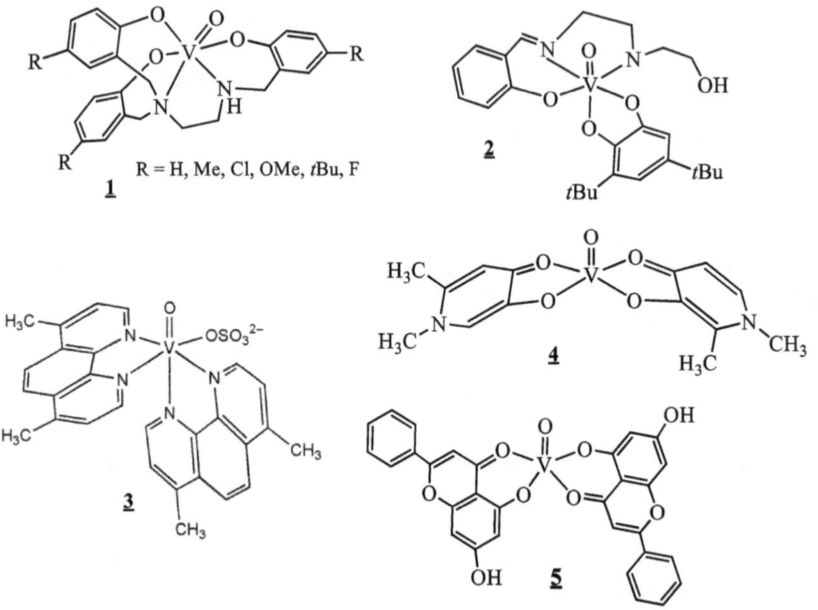

FIGURE 12.4
Vanadium-based anticancer compounds: **1** shows marked cytotoxicity towards colorectal cancer (colon HT-29) and ovarian cancer (OVCAR-3cells) (Reytman et al. 2016). For the catecholate complex **2**, no evidence of toxicity was observed in female Swiss mice (Lima et al. 2021). Metvan (**3**) is effective against bone cancer (Cacicedo et al. 2019), and the pyridinonato complex **4** causes apoptosis and cell arrest in malignant melanoma (Pisano et al. 2019). The VO-chrysin complex **5** has proved potentiality in the treatment of human osteosarcoma (Leon et al. 2016).

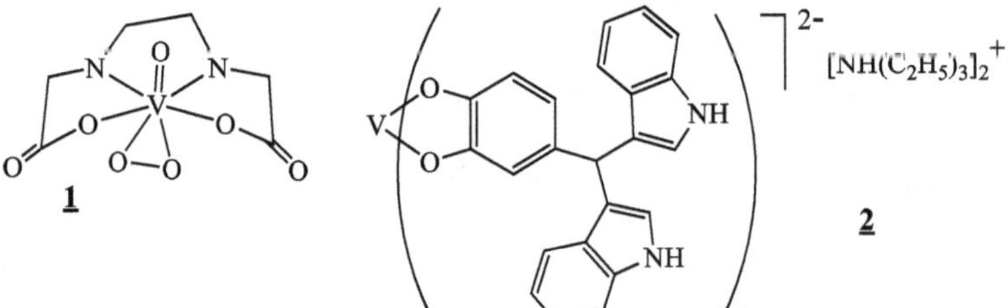

FIGURE 12.5
Two vanadium coordination compounds that have been shown to exhibit anticancer activity: The ligands are peroxide and diethylendiamine diacetate (**1**) (Chen et al. 2019) and diimidonylmethane (**2**) (Dankhoff et al. 2019).

Additional medicinal aspects include an anti-inflammatory effect (Tsave et al. 2016), the alleviation of lipid anomalies (adipogenesis) (Zhang et al. 2016) and wound healing (reorganization of connective tissues). In the latter case, the potentiality of orthovanadate $H_2VO_{4^-}$ in the healing process of incisions of the abdominal wall (laparotomy incisions) had been tested in rats with a promising outcome (Hazard et al. 2017). A potential reason for this effect is the inhibitory action of vanadate towards PTPases.

A combination of the oncolytic (non-pathogenic) rhabdovirus VSVΔ51 with the dioxidovanadium-picolinato complexes [VO_2dipic-X]- (where X = H, OH and Cl), Figure 12.5) – previously also evaluated for their antidiabetic potentiality (Willsky et al. 2011) – increases viral replication and thus promotes oncolytic virotherapy (Bergeron et al. 2019). VSVΔ51 exhibits a natural tropism for cancer cells. The active species in this application is vanadate, generated by hydrolysis of the picolinato compounds. The vanadate is likely delivered to phosphatase, that is, the active species in this vanadium-induced oncolytic virotherapy supposedly is a vanadate-phosphatase complex.

Vanadium coordination compounds have also shown promising results in the treatment of infections caused by amoeba, flagellates, bacteria and viruses. Compound **1** in Figure 12.6, a hydroxiqinoline complex of oxidovanadium(IV), displays activity against amoebiasis and cytotoxic (antitumor) activity against ovarian tumor cells (Correia et al. 2014). The oxidovanadium(IV) complex formed with the antibiotic cefuroxime (**2** in Figure 12.6) shows an antibacterial effect towards *Klebsiella pneumonia*; targeting the bacterial β-lactamases. The complex docks to the receptor protein clathrin via hydrophobic forces.

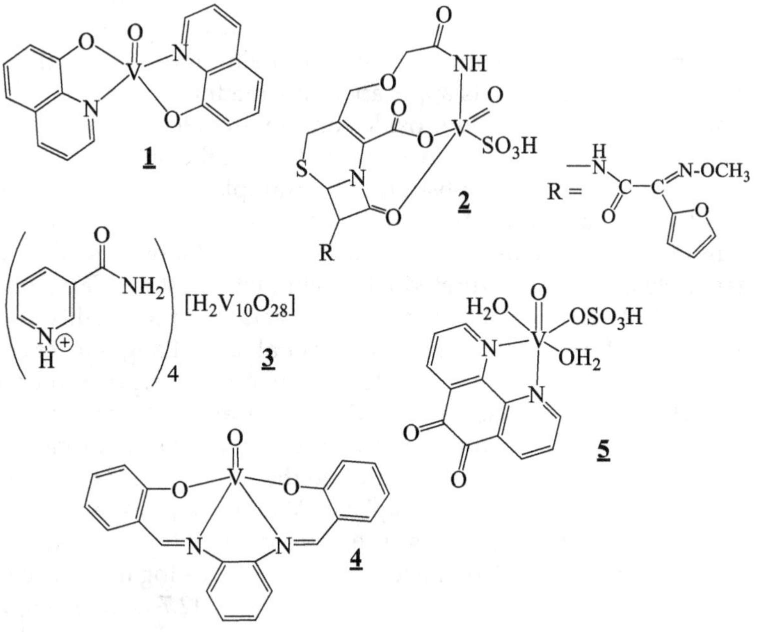

FIGURE 12.6
Vanadium coordination compounds that have been shown to exhibit activity against bacteria and viruses (and, eventually, also antitumor activity; see text). Complex **1** derives from hydrochinolin; the ligand in **2** is cefuroxime. The decavanadate **3** is effective against intestinal infections, complex **4** against leishmaniasis and complex **5** against Chagas disease.

Its antioxidant and antibiotic ability is enhanced as compared to cefuroxime itself (Datta et al. 2015). In addition to these vanadium coordination compounds, decavanadate $[H_2V_{10}O_{28}]^{4-}$ has been shown to exhibit activity against bacterial infections in the intestines (such as the gut microbia *Escherichia coli* and the protozoa *Giardia intestinalis*, causing diarrheal disease), the activity commonly depending on the nature of the pharmacologically relevant counter ion. For counter ions such as isonicotineamide, effective growth inhibition of the gut bacteria has been reported (Missina et al. 2018). Substituting one of the vanadium atoms in decavanadate for Pt(IV) or Mo(VI) hamper the growth of *Mycolicibacterium smegmatis*, a bacterium thriving in the smegma in the human genital area (Kostenkova et al. 2021). An anti-leishmanial effect (against the flagellate *Leishmania amazonensis* spread by sand flies) has been reported for the oxidovanadium salophen complex, **4** in Figure 12.6 (Almeida Machado et al. 2015). At least in part, this effect is due to the induction of the production of reactive oxygen species. The compound does not significantly damage erythrocytes. Another flagellate, *Trypanosomas cruzi* (the emastigotes of which are distributed by the bug *Triatoma infestans*), causes Chagas disease. The parasite is endemic in particular in some areas of Latin America. A perspective metal-based drug against *T. cruzi* based on vanadium is the oxidovanadium(IV) complex (**5**) formed with bidentate dinitrogen donors such as phendione (for the proposed structure of **5**, cf. Figure 12.6).

Along with its medicinal applications, vanadium is a composite in several industrial products, including its involvement in catalytically conducted processes, its use as a constituent in titanium-aluminum alloys (Popovich et al. 2015) and in steel, and its application in vanadium-based batteries and other energy devices. Vanadium oxides and hydroxides (such as V_2O_5 and $VO(OH)_2$) have widely been employed in catalysis, either as such or in combination with (coordination to) substrates. An example for the latter is mesoporous SBA-15 (a silica gel) functionalized with aminopropyl and loaded with VO^{2+} (**1** in Figure 12.7). This system catalyzes the oxidation – by tBuO$_2$H – of conjugated olefins and unsaturated fatty acid methyl esters (Campitelli and Crucianelli 2020). For recent reviews on the catalytic potentiality of vanadium in technical processes, see Sutradhar et al. 2015; Langeslay et al. 2019; for vanadium's role as a catalyst in natural (life) processes, see Rehder 2021.

High work function of mixed metal oxides, for example, by combining V_2O_5 with MoO_3 or Co_3O_4 via hydrothermal synthesis and subsequent calcination, leads to the incorporation of vanadium into the host oxides, namely the formation of $V_6Mo_xO_{40}$ (x = 6, 9) and biphasic V_2O_5-Co_3O_4, respectively. These phases provide selective materials in energy devices that can basically be employed as metallic (semi)conductors, as light-harvesting devices and light absorbers (Feste et al. 2021). Compound **2** in Figure 12.7 is an example of a catalyst that promotes the oxidative bromination of aromatic aldehydes (and thus mimics the bromoperoxidase activity of the halo-peroxidase in marine algae; vide supra). Compound **2** also exhibits catechol oxidase activity. Along with these catalytic features, an intercalating interaction with circulating tumor DNA in the peripheral blood has been noted (Majunder and Rajak 2020). Vanadium-based photoredox catalysts have been shown to be active in the environmentally relevant conversion of plastics to fuels. An example – the oxidative light-induced conversion of a polyalcohol – is provided in Equation 12.2; for the catalyst, see compound **3** in Figure 12.7 (Gazi et al. 2019). Highly active and stable vanadium catalysts based on naphthalinesufonato complexes (**4** in Figure 12.7) effectively catalyze the co-polymerization of ethylene and propylene (Hao et al. 2017). Homogenous recyclable vanadium-based catalysts, obtained through covalent binding – and thus immobilization – of vanadium coordination compounds to polystyrene have been employed in the oxidation of phenols, olefins, benzoin, cumene and organic sulfides (Maurya 2018). For an example, see compound **5** in Figure 12.7.

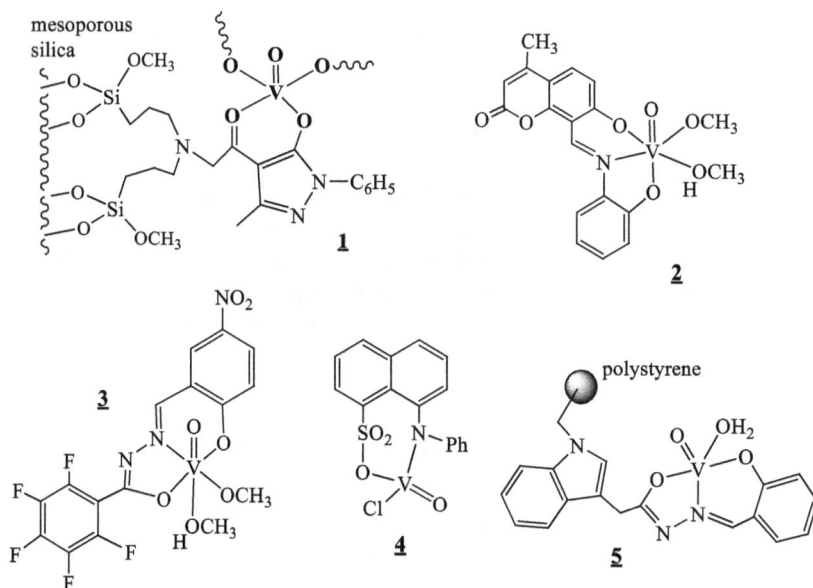

$$H_3C{-}(CH_2)_n{-}CH_2OH \xrightarrow[h\nu]{(\mathbf{3})} \left\{HCOOH + HC{\overset{O}{\underset{OCH_3}{\Large\lt}}}\right\}_n \quad (2)$$

(Equation 12.2)

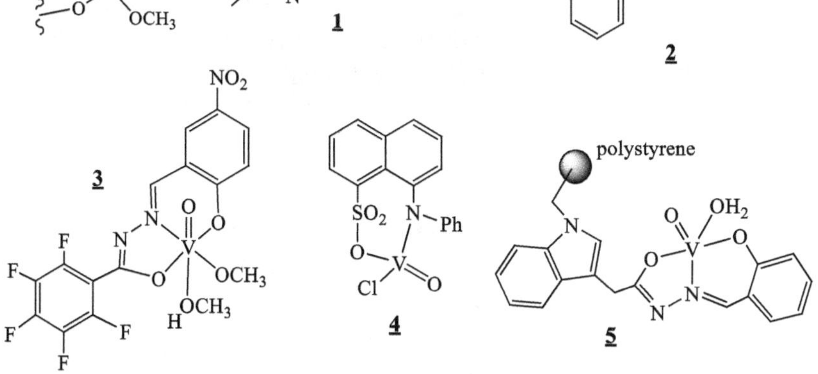

FIGURE 12.7
Examples of vanadium-based catalysts: **1** catalyzes the oxidation of olefins, **2** the oxidative bromination of aromatic aldehydes and **3** the C-C bond cleavage in lignin (models). Complex **4** is active in the co-polymerization of $H_2C{=}CH_2$ and $CH_3{-}CH{=}CH_2$, and **5** in the oxidation of various substrates; see text for details.

In the context of the detoxification of nitrogen oxides NO_x present in, for example, exhaust gases of automobiles and trucks, the vanadium-based catalyst V_2O_5-WO_3-TiO_2 has turned out to be technologically efficient. A proposed reaction mechanism is illustrated in Scheme I. Accordingly, the reaction is coupled to the activation of ammonia at the vanadium center, changing between the oxidation states V and IV:

12.2.1 Scheme I

Scheme I: The mechanism of the catalytic oxidative conversion of nitrogen oxide to nitrogen (following Marberger et al. 2016).

Vanadium-based redox catalysts such as V_xO_y-supported silica (equation (1) in Figure 12.8) and $V^{IV}O$-modified styrene nanofibers have been shown to be efficient catalysts in the oxidation of refractory sulfur compounds contained in fuels (hydro-treated crude oil and diesel), thus counteracting SO_2 emitted by combustion of organosulfur compounds present in diesel (Dembaremba et al. 2019). Dodecavanadates with [Ni/Co(ethylenediamine)$(OH)_6$] as cations have shown high-performance and low-cost potentiality in the photocatalytic reduction of CO_2 to synthesis gas ($CO + H_2$) + formic acid; equation (2) in Figure 12.8 (Yu et al. 2021). Interestingly, vanadium pentoxide particles imbedded in a matrix (paint) and applied onto (metallic) surfaces such as ships' hulls, produce – in the presence of bromide and hydrogen peroxide – hypobromous acid HOBr. HOBr in turn acts as an efficient antibacterial agent, and thus prevents biofouling when applied to, for example, ships' hulls (Natalio et al. 2012). This vanadium-catalyzed reaction is reminiscent to the bromoperoxidase activity of marine algae such as the bladder wrack *Ascophyllum nodosum*.

The technical application of vanadium is further directed towards its use in electrical devices such as batteries and solar cells. Dye-sensitized solar cells such as ruthenium-based sensitizers absorbed on nano-crystalline TiO_2 can contain the redox mediator $[V^{IV/V}O(hybeb)]^{2-/1-}$ (hybeb is the diamino-diphenolato ligand) **1** in Figure 12.9 (Apostolopoulou et al. 2015). $VO_x{\cdot}nH_2O$

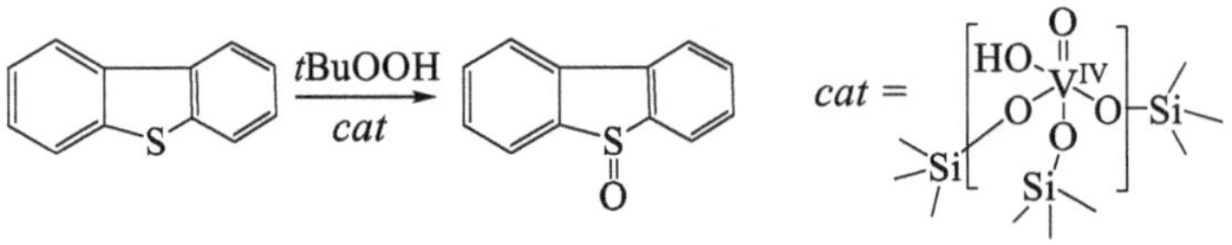

$$\text{dibenzothiophene} \xrightarrow[cat]{tBuOOH} \text{dibenzothiophene S-oxide} \qquad cat = [\text{HO}-V^{IV}(=O)(-O-Si)_3] \qquad (1)$$

$$CO_2 \xrightarrow[cat]{h\nu\ (vis)} CO + H_2 + HCOOH \quad \text{(in triethanolamine)} \qquad cat = [Co(en)_2]_6[V_{12}B_{18}O_{54}(OH)_6]\cdot 17H_2O \qquad (2)$$

FIGURE 12.8
Two reactions that demonstrate the catalysis, by vanadium compounds, of oxidative (1) and photo-induced (2) conversions.

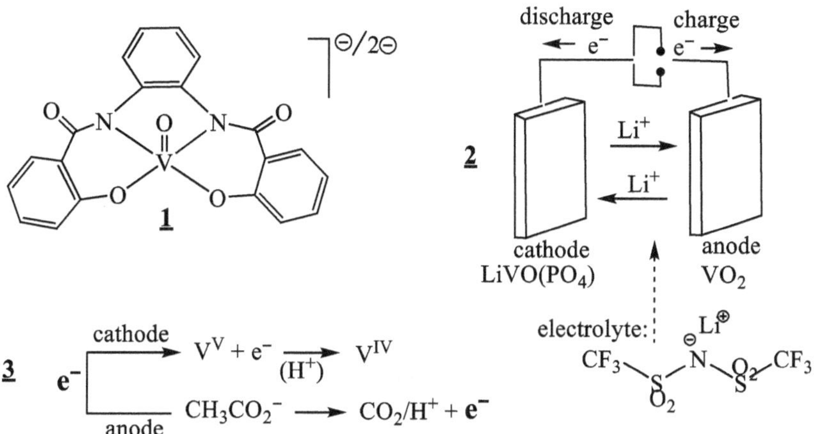

FIGURE 12.9
1: The redox mediator in a dye-sensitized solar cell. **2**: A vanadium-lithium battery for energy storage. **3**: Anodic and cathodic processes in a microbial fuel cell.

as a high-performance hole-transport layer in polymers improves the power conversion efficacy of solar cells (Cong et al. 2017). Materials with ultra-low thermal conductivity are important as thermal barrier coatings in, for example, data storage devices. The selenide Tl_3VSe_4 has been shown to be effective in this sense (Mukhopadhyay et al. 2018). Additional examples for technical applications are vanadium-doped magnetic FeCo/SiO_2 nanoparticles that effectuate increased coercivity (Desautels et al. 2016) and vanadium-graphene nano-layers that exert radiation tolerance (Kim et al. 2016).

A comparatively novel field is the applications of vanadium-based batteries (vanadium-redox batteries/vanadium-flow batteries) such as the vanadium-lithium battery **2** in Figure 12.9; **2** is suited for energy storage application, having high energy density and a long life span (Shao et al. 2018). Cathode and anode materials are vanadium-based, viz. $LiV^VO(PO_4)$ and $V^{IV}O_2$, respectively. The average operating voltage is 1.4 V, the energy density 305 Wh^{-1}. Other vanadium flow batteries are based on the $V^VO^{2+}/V^{IV}O^{2+}$ (catholyte in the cathodic region) and V^{3+}/V^{2+} (anolyte in the anionic region) redox couples, separated by an ion-conductive membrane, such as represented by a poly(vinylidene fluoride)/graphene composite (Lai et al. 2019). Microbial fuel cells have been employed to oxidize organic/inorganic matter while generating electricity (Wang et al. 2018). (Toxic) vanadate can thus be removed from vanadium-containing wastewaters via microbial reduction to VO^{2+} ($VO(OH)_2\downarrow$) in the presence of organics as the electron donor, such as acetate. The electrons are generated by bacterial consumption of the organic matter; after passage from the anode to the cathode, V^V becomes reduced to V^{IV}; **3** in Figure 12.9.

References

Antal DS, Dehelean CA, Canciu CM, Anke M. 2009. Vanadium in medicinal plants: New data on the occurrence of an element both essential and toxic to plants and man. *Anal. Univers. Oradea Fasc. Biol.* 16:5–19.

Apostolopoulou A, Vlasiou M, Tziouris PA, et al. 2015. Oxidovanadium(IV/V) complexes as new redox mediators in dye-sensitized solar cells: A combined experimental and theoretical study. *Inorg Chem* 54:3979–3988.

Azevedo CG, Correia I, dos Santos MMC, et al. 2018. Binding of vanadium to human serum-transferrin—Voltammetric and spectroscopic studies. *J Inorg Biochem* 180:211–221.

Barcelaux DG. 1999. Vanadium. *Toxicol Clin Toxicol* 37:265–278.

Benítez J, Correia I, Becco L, et al. 2013. Searching for vanadium-based prospective agents against Trypanosoma cruzi: Oxidovanadium(IV) compounds with phenanthroline derivatives as ligands. *Z Anorg Allg Chem* 639:1417–1425.

Bergeron A, Kostenkova KL, Selman M, et al. 2019. Enhancement of oncolytic virotherapy by vanadium(V) dipicolonates, *Biomet* 32:545–561.

Cacicedo ML, Ruiz MC, Scioli-Montono S, et al. 2019. Lipid nanoparticles—Metvan: Revealing a novel way to deliver a vanadium compound to bone cancer cells. *New J Chem* 43:17726–17734.

Campitelli P, Aschi M, Di Nicola C, Pettinari R, Marchetti F, Cruzianelli M. 2020. Ionic liquids vs conventional solvents as comparative study in the selective catalytic oxidations promoted by oxovanadium(IV) complexes. *Appl Cat A: General* 599:117622.

Campitelli P, Crucianelli M. 2020. On the capability of oxidovanadium(IV) derivatives to act as all-round catalytic promoters since the prebiotic world. *Molecules* 25:3073.

Chen F, Gao Z, You C, et al. 2019. Three peroxidovanadium(V) compounds mediated by transition metal cations for enhanced anticancer activity. *Dalton Trans* 48:15160–15169.

Cong H, Han D, Sun B, et al. 2017. Facile approach to preparing a vanadium oxide hydrate layer as a hole-transport layer for high-performance polymer solar cells. *ACS Appl Mater Interfaces* 9:18087–18094.

Correia I, Adão P, Roy S, et al. 2014. Hydroquinoline derived vanadium(IV and V) and copper(II) complexes as potential anti-tuberculosis and anti-tumor agents. *J Inorg Biochem* 141:83–93.

Crans DC, Henry L, Cardiff G, Posner BI. 2019. Developing vanadium as an antidiabetic or anticancer drug: A clinical and historical perspective. *Met Ions Life Sci* 19:203–230.

Dankhoff K, Ahmad A, Weber B, Biersack B, Schobert R. 2019. Anticancer properties of a new non-oxido vanadium(IV) complex with a catechol-modified 3,3′-diindolylmethane ligand. *J Inorg Biochem* 194:1–6.

Datta C, Das D, Mondal P, Chakraborty B, Sengupta M, Bhattacharjee CR. 2015. Novel water soluble neutral vanadium(V)-antibiotic complex: Antioxidant, immunomodulatory and molecular docking studies. *Eur J Med Chem* 97:214–224.

de Almeida Machado P, Mota VZ, de Lima Cavalli AC, et al. 2015. High selective antileishmanial activity of vanadium complex with stilbene derivative. *Acta Trop* 148:120–127.

Dembaremba TO, van Der Westhuizen R, Welthagen W, Ferg E, Ogunlaja AS, Tshentu ZR. 2019. Comparing the catalytic activity of silica-supported vanadium oxides and the polymer nanofiber-supported oxidovanadium(IV) complex towards oxidation of refractory organosulfur compounds in hydrotreated diesel. *Energy & Fuels* 33:7595–7603.

Desautels RD, Rowe MP, Freeland JW, Jones M, van Lierop J. 2016. Influence of vanadium-doping on the magnetism of $FeCo/SiO_2$ nanoparticles. *Dalton Trans* 45:10127–10130.

Domingo JL, Gómez M. 2016. Vanadium compounds for the treatment of human diabetes mellitus: A scientific curiosity? A review of thirty years of research. *Food Chem Toxicol* 95:137–141.

Feste PD, Crisci M, Borbon F, Tajoli F, Salerno M, Drago F, Prato M, Gross S, Gatti T, Lamberti F. 2021. Work function tuning in hydrothermally synthesized vanadium-doped MoO_3 and Co_3O_4 mesostructures for energy conversion devices. *Appl Sci* 11:2016.

García-Jeminéz A, Trejo-Téllez L, Guillén-Sánchez D, Gómez-Merino FC. 2018. Vanadium stimulated pepper plant growth and flowering increases concentrations of amino acids, sugars and chlorophylls, and modifies nutrient concentrations. *PLoS One* 13:1–20.

Gätjens J, Meier B, Kiss T, et al. 2003. A new family of insulin-mimetic vanadium complexes derived from 5-carboalkoxypicolinates. *Chem Eur J* 9:2924–2935.

Gazi S, Đodkić M, Chin KF, Ng PR, Soo HS. 2019. Visible light-driven cascade carbon-carbon bond scission for organic transformations and plastics recycling. *Adv Sci* 1902020.

Hao X, Zhang C, Li L. 2017. Use of VanadiumComplexes bearing Naphthalene-bridged Ligans as Catalyst for Copolymerization of Ethelene and Propylene. Use of Vanadium Complexes Bearin Naphthalene-bridged Ligands as Catalysts for Copolymerization of Ethylene and Propylene. *Polymers* 9:325.

Hazard SW, Zwemer CF, Mackay DR, Koduru SV, Ravnic DJ, Ehrlich HP. 2017. Topical vanadate enhances the repair of median laparotomy incision. *J Surg Res* 207: 102–107.

Hon J, Hwang MS, Charnetzki MA, et al. 2017. Kinetic characterization of the inhibition of protein tyrosine phosphatase-1B by vanadyl (VO^{2+}) chelates. *J Biol Inorg Chem* 22:1267–1279.

Hu J, Xiu W, Pan X, et al. 2017. Association of adverse birth outcomes with prenatal exposure to vanadium: A population-based cohort study. *The Lancet Planetary Health* 1:230–241.

Huyer G, Kelly J, Moffat J, et al. 1997. Mechanism of inhibition of protein-tyrosine phosphatases by vanadate and pervanadate. *J Biol Chem* 272:835–851.

Irving E, Stoker AW. 2017. Vanadium compounds as PTP inhibitors. *Molecules* 22:2269.

Kim Y, Baek J, Kim S, et al. 2016. Radiation resistant vanadium-graphene nanolayered composite. *Sci Rep* 6:24785.

Kioseouglou E, Petanidis S, Gabriel C, Salifoglou A. 2015. The chemistry and biology of vanadium compounds in cancer therapeutics. *Coord Chem Rev* 301–302:87–105.

Koleša-Dobravc T, Maejima K, Yoshikawa Y, Meden A, Yasui H, Perdih F. 2018. Bis(picolinato) complexes of vanadium and zinc as potential antidiabetic agents: Synthesis, structural elucidation and in vitro insulin-mimetic activity study. *New J Chem* 42:3619–3632.

Kostenkova K, Arhouma Z, Postal K, et al. 2021. Pt^{IV} and Mo^{VI}-substituted decavanadates inhibit the growth of *Mycobacterium smegmatis*. *J Inorg Biochem* 217:111356.

Kremer LE, McLeod AI, Aitken JB, Levina A, Lay PA. 2015. Vanadium(V) and –(IV) complexes of anionic polysaccharides: Controlled release, pharmaceutical formulations, and models of vanadium biotransformation products. *J Inorg Biochem* 147:227–234.

Lai Y, Wan L, Wang B. 2019. PVDF/graphene composite nanoporous membranes for vanadium flow batteries. *Membranes* 9:89.

Langeslay RR, Kaphan DM, Marshall CL, Stair PC, Sattelberger AP, Delferro M. 2019. Catalytic applications of vanadium: A mechanistic perspective. *Chem Rev* 119:2128–2192.

Leon IE, Cadavid-Vargas JF, Resasco A, Maschi F, Ayala MA, Carbone C, Etcheverry B. 2016. In vitro and in vivo antitumor effects of the VO-chrysin complex on a new three-dimensional osteosarcoma pheroids model and a xenegraft tumor in mice. *J Biol Inorg Chem* 21:1009–1020.

Levina A, Lay PA. 2017. Stabilities and biological activities of vanadium drugs: What is the nature of the active species? *Chem Asian J* 12:1692–1699.

Lima LMA, Murakami H, Gaebler DJ, et al. 2021. Acute toxicity evaluation of non-innocent oxidovanadium(V) Schiff base complexes. *Inorganics* 9(6):42.

Lyonnet B, Martz X, Martin E. 1899. Des derivés du vanadium. *La Presse Médicale* 32:191–192.

Majunder M, Rajak KK. 2020. Oxidovanadium(IV and V) complexes incorporating cumarin based ONO ligand: Synthesis, structure and catalytic activities. *Polyhedron* 176:114241.

Manniko D, Stoll S. 2019. Vanadyl porphyrin speciation based on submegahertz ligand proton hyperfine coupling. *Energy Fuels* 33:4237–4243.

Marberger A, Ferri D, Elsener E, Kröcher O. 2016. The significance of Lewis acid sites in the selective catalytic reduction of nitric oxide on vanadium-based catalysts. *Angew Chem Int Ed* 55:11989–11994.

Maurya MR. 2018. Vanadium complexes based polymer supported catalysts: A brief account of research from our group. *Topics Cat* 61:1500–1513.

Missina JM, Gavinho B, Postal K, et al. 2018. Effects of decavanadate salts with organic and inorganic cations on Escherichia coli, Gardia intestinaleis and vero cells. *Inorg Chem* 57:11930–11941.

Mukhopadhyay S, Parker DS, Sales BC, Puretzky AA, McGuire MA, Lindsay L. 2018. Two-channel model for ultralow thermal conductivity of crystalline Tl_3VSe_4. *Science* 360:1455–1458.

Naglah AM, Al-Omar MA, Almehizia AA, et al. 2018. A novel oxidovanadium(IV) orotate complex as an alternative antidiabetic agent: Synthesis, characterization and biological assessments. *Biomed Res Int* 2018:8108713.

Naglah AM, Al-Omar MA, Bhat MA, et al. 2019. Synthesis and biological evaluations of a novel oxidovanadium(IV) adenosine monophosphate complex as anti-diabetic agent. *Crystals* 9:208.

Naso LG, Lezama L, Valcarcel M, et al. 2016. Bovine serum albumin binding, antioxidant and anticancer properties of an oxidovanadium(IV) complex with luteolin. *J Inorg Biochem* 157:80–93.

Natalio F, André R, Hartog AF, et al. 2012. Vanadium pentoxide nanoparticles mimic vanadium haloperoxidase and thwart biofilm formation. *Nat Nanotech* 7: 530–535.

Owolabi IA, Mandiwana KL, Panichev N. 2016. Speciation of chromium and vanadium in medicinal plants. *S Afr J Chem* 69:57–71.

Pisano M, Arru C, Serra M, et al. 2019. Antiproliferative activity of vanadium compounds: Effects on the major malignant melanoma molecular pathways. *Metallomics* 11:1687–1699.

Popovich A, Sufiiarov V, Borisov E, Polozov I. 2015. Microstructure and mechanical properties of Ti-6Al-4V manufactured by SLM. *Key Engin Mat* 651–653:677–682.

Pranczk J, Jacewicz D, Wyrzykowski D, Wojtczak A, Tesmar A, Chmurzński L. 2015. Crystal structure, antioxidant properties and characteristics in aqueous solutions of the oxidovanadium(IV) complex [VO(IDA)phen]·$2H_2O$. *Eur J Inorg Chem* 2015:3343–3349.

Pranczk J, Tesmar A, Wyrzykowski D, Inkielewicz-Stepniak I, Jacewicz D, Chmurzynski L. 2016. Influence of primary ligands (ODA, TDA) on physicochemical and biological properties of oxidovanadium(IV) complexes with bipy and phen as auxiliary ligands. *Biol Trace Elem Res* 174:251–258.

Rehder D. 2021. Vanadium in catalytically proceeding natural processes. Catalysis series no. 41, Vanadium catalysis (Sutradhar et al., eds.). *Royal Soc. Chem* 24: 535–547.

Reytman L, Braitbard O, Hochmann J, Tshuva EY. 2016. Highly effective hydrolytically stable vanadium(V) amino phenolato antitumor agents. *Inorg Chem* 55:610–618.

Rodriguez-Mercado, JJ. 2011. DNA damage induction in human cells exposed to vanadium oxide in vitro. *Toxicology in Vitro* 25:1996–2002.

Ścibior A, Kurus J. 2019. Vanadium and oxidative stress markers—In vivo model: A review. *Curr Med Chem* 26:5456–5500.

Ścibior A, Llopis J, Holder AA, Altamirano-Lozano M. 2016. Vanadium toxicological potential versus its pharmacological activity: New developments and research. *Oxid Med Cell Longev* 2016:7612347.

Scior T, Guevara-Garcia JA, Do Q-T, Bernard P, Laufer S. 2016. Why antidiabetic vanadium complexes are not in the pipeline of "Big Pharma" drug research? A critical review. *Curr Med Chem* 23:2874–2891.

Shao M, Deng J, Zhong F, et al. 2018. An all-vanadium aqueous ion battery with high energy density and long lifespan. *Energ Stor Mat* 18:92–99.

Sutradhar M, Martins LMDRS, Guedes da Silva, MFC, Pombeiro AJL. 2015. Vanadium complexes: Recent progress and oxidation catalysis. *Coord Chem Rev* 301–302:200–239.

Thomson KH, Orvig C. 2007. Vanadium in diabetes: 100 years from phase 0 to phase I. *J Inorg Biochem* 100:1925–1935.

Tsave O, Petanidis S, Kiosaeoglou S, et al. 2016. Role of vanadium in cellular and molecular immunology: Association with immune-related inflammation and pharmacotoxicology mechanisms. *Oxid Med Cell Longev* 2016:4013639.

Ulbricht C, Chao W, Costa D, et al. 2012. An evidence-based systematic review of vanadium by the Natural Standard Research Collaboration. *Diet Suppl* 9: 223–251.

Wang D, Wilhelmy SAS. 2009. Vanadium speciation and recycling in coastal waters. *Mar Chem* 117:52–58.

Wang Y, Feng Y, Li H, Yang C, Shi J. 2018. Reduction of vanadium(V) in a microbial fuel cell: V(IV) migration and electron transfer mechanism. 2018. *Int J Electrochem Sci* 13:11024–11037.

Willsky GR, Chi L-H, Godzala II M, et al. 2011. Anti-diabetic effects of a series of vanadium dipicolinate complexes in rats with streptozotozin-induced diabetes. *Coord Chem Rev* 255:2259–2269.

Wong C. 2019. The health benefits of vanadium. *Verywell Health* 1–4.

Wu J-X, Hong Y-H, Yang X-G. 2016. Bis(acetylacetonato)-oxidovanadium(IV) and sodium metavanadate inhibit cell proliferation via ROS-induced sustained MAPK/ERK activation but with elevated AKT activity in human pancreatic cancer AsPC-1 cells. *J Biol Inorg Chem* 21:919–929.

Yu X, Zhao C-C, Gu J-X, et al. 2021. Transition-metal modified vanadoborate clusters as stable and efficient photocatalysts for CO_2 reduction. *Inorg Chem* 60(19): 7364–7371.

Zhang L, Huang Y, Liu F, Zhang F, Ding W. 2016. Vanadium(IV)chlorodipicolinate inhibits 3T3-L1 preadipocyte adipogenesis by activating LKB1/AMPK signaling pathway. *J Inorg Biochem* 162:1–8.

Index

Note: Page locators in **bold** indicate a table. Page locators in *italics* indicate a figure.

A

acid
- acid-soluble, 50–51, 77–78
- citric, 139
- humic, 19, 22, 58, 60, 139–141
- lactic, 245
- resistant, 2
- sulfuric, 4

Actinobacteria, 224, 226, *230*, 233, 236, **244**
aluminum oxide, 58, 60, 83, 189
aluminum (Al), 18, 23, 34, 207, 216, 276
anthropogenic
- activities (mining, smelting, spent catalysts, combustion of fossil fuel and waste disposal), 2, 6, 8–9, 73–74
- pollution, 216
- sources, 6, 12, 23, 74, 115, 207

aquifers, 8, 248–249
Arabian Gulf, 12
ash, derived from coal, 8, 12, 141, 159, 168, 177
athletes, 2
Australia, 8, 35
availability in soils, 14–15, 18, 53, 83, 87, 129, 141, 149, 160

B

bacteria
- fermentative, 245, 260
- V-reducing, 137, 233

Bacteroidetes, 146, 224, 226, **244**, **255**
batteries, 2, 270, 276, 278–279
bio-surfactants, 139
bioaugmentation, **255**, 258
bioavailability (of V), 15, 21, 142, 146, 148–149, 168, 207
biochar
- alkaline effect of, 21, 140–142, 172
- modified, 173–174, *175*
- properties, 168–169, *170–171*, 172
- research outlook, 177–179
- risk management, 175–176
- soil remediation, 159–161, *162–163*, 165, 168

bioconversion, 249242
biodetoxification, 249
bioelectrical reactor (BER), 258, 260
biogeochemical cycle, 6, *7*, 101, 136, 208, 224, 239, 247
bioinorganic implications, 22
biological permeable reaction barrier (bio-PRB), 257
bioreactors, 245, *246–247*, 248–249, *252–253*, 256
bioremediation
- environmental bioremediation strategies, **255**
- of vanadium, 146

biotic ligand model (BLM), 190, 197–198, *199*, 200

C

calcium carbonate, 12, 20, 23
Canada, 12, 35, 129
cancer
- bone, 274
- cell lines, 273
- incidents of, 12
- ovarian, *274*, 275

chemical properties, 1, 3, 187, 189, **191–192**
Chile, 2, 14
China, 12, 14–16, 136–137, 206, 212, 224, 234
Chloroflexi, 233
chlorosis, 140, 206, 211–212
chromium, 139, 206, 224, 226–227, 252–253, **255**

clay
content, 12, 19–20, 56
minerals, 18–20, 23, 51–52, 56, 60, 75, 80, 87, 124, 140, 158
Community Bureau of Reference (BCR), 16, **17**, 77–79
compounds
anionic, 49, 80, 83, 87, 106–107, 139, 142, 169, 172
catatonic, 20–21, 49, 80, 87, 141, 190
concentration
high, 9, 21, 50, 117, 120, 213
low, 9, 117, 120, 122, 128–129, 169
in soils, 19, 74, **75**, 137
contamination, 2
geogenic and/or anthropogenic discharges, 12, 23
soil, 2, 34, 37, 140, 159–160
content and mobilization in soils, 8, *9–10*, 142, 207
conversion, 146, 159, 172, 196, 277–279
crude oil, 2, *270*, 278
Czech Republic, 35, 39

D

diethylenetriaminepentaacetic acid (DTPA), 15–16, **17**, 20
discharges, 12, 14, 149
DNA damage, 3, 5–6, 270, 273, 277

E

ecotoxicity
current evidence, 190, **191–193**, 195
in soils, 188–190, 197, 200
electron donors, 99, 242, 245–246, 254, 257
electron probe micro-analysis (EPMA), 76, 77
environmental sources, 6
equilibrium, 52, 59–60, 62–64
Ethylenediaminetetraacetic acid (EDTA), 15–16, **17**, 51, 81, **82**, **87**, 139
Europe, 9, 35
evolution (microbial), 229–230, 233
exposure
human, 2, 5–6, 34, 37, **38**, **41**, 43, **44**, 270
setting limits, 34
extracellular polymeric substances (EPS), 148, 239
extraction methods, 15, **17**, 81–82

F

fertilizer, 2, 53, 73–74, 158, **193**, 207
Firmicutes, 146, 224, 233, **243**
Fluvisol, 9, 12
fly ash (coal), 8, 12, 141
fractions
geochemical, 14–15, 77, *166*
residual, 15, 51, 77, 80
soluble, 193

G

glucose, 2, 5, **244**, 245, 260, *270*, 272
groundwater
contamination, 53, 139, 158, 160, 226, 257
groundwater
limitations, 37

H

health risk assessment (HRA)
dermal contact, 34, **38**, 43
ingestion, 5, 34, 37, **38**, 43, **44**, 270
inhalation, 5–6, 34, 37, **38**, 43, **44**
maximum allowable limits, 34, **38**, *40*, **41**, **44**
risks to humans, 5, 34, 37, 73, 83, 176, 207, 216
heavy metal
contamination, 140, 146, 216
reducers, 138, 226
high-performance liquid chromatography (HPLC), 50
humic acid, 19, 22, 58, 60, 139–141
hydrogen, 52, 206, 213, 248, 256, 278

I

industrial uses, 3, 5–6, 34–35, 50, 74, 80, 252, 270
ionic strength, 22–23
iron, 18, 23, 172, 189, 207, 216, 254

iron oxides, 55
Italy, 9, 14, 35

K

kinetics, 52, 57–59, 62

M

mackinawite, 241, 249
manganese, 18, 23, 207, 213
metabolic function, 223
metallurgy, 2, 12 *see also* smelting
methane, 176, 248
methylomonas, 248
Mexico, 12, 74, 206
microbes
  functional, 242, 246, 260
  reducing, 223, 229–230
  soil, 139, 216
microbial
  remediation, 239, 254
  responses, 260
microbial fuel cell (MFC), 258, 260
mobility
  bioavailability and, 14–15, 87, 107, 114, 125, 163
  in soils, 33, 51, 53, 129, 148, 161
multi-reaction model (MRM), 62–63, *63–66*

N

New Caledonia, 12

O

organic matter (OM)
  content, 41, 140, 142, 207
  dissolved, 19, 114, *116*, 172
  soluble, 163, 245, 254
organic/inorganic substances, 19, 52, 56, 160, 256
orthovanadate, 20, 208, 269, 274

P

peat, 18, 96, 140–141
pentachlorophenol (PCP), 253
periodic table, 1–2
pharmaceutical issues, 271
phenanthrene (PHE), 249, *251*
phosphate, 4, 52–53, *54*, 59, 140, 176, **191**, 195
phytoremediation, 143, **144–145**, 149, 214
phytotoxicity, 14, 21, 88, 165, 168
plant
  absorption, 50
  bioremediation, 216
  development and growth, 210–213
  phytoremediation, 214, **215**, *215*
polyacrylamide (PAM), 140–141
polysaccharides, 196, 210, 239, 271
potentially mobile fractions (PMF), 77–80, 83, 87
potentially toxic element (PTE), 77
protein-tyrosine-phosphatase (PTP), 272
Proteobacteria, 146, 224, 226, 233, 236, **243–244**

R

regulation categories, 34–37, 41, 43, 211
remediation, 149
  microbial, 239, 254
  soil, 135, 137–140
  strategies, 9, 146, 148, *148*
Russia, 8, 137, 212

S

seawater, 2–3, 12
silicates, 6, 74
single extraction (SE), 80–82, 87, 224
sludge, 140, 158, 169, 207, 242, 245, 248
smelting, 12, 14–15, 73–74, **79**, 80, 127, 138, 224
soil
  amendments, 74, 108, 140
  conditioner, 141
  contamination, 34, 74, 159
  interaction, 52
  matrix, 18–19, 22, 138, 178
  pH (potential of hydrogen), 21, 52, 101
  redox, 20, 23, 97, 103–104
  soil biota, 146–147, 168, 176
  sorption/desorption, 56, *61*, 75
  washing, 138–139, 148–149, 159

soil organic matter (SOM), 15, 18–21, 102
sorption/desorption, 52, 56, 58–60, *61*, 62, 65
sources, 33, 49, 74, 80, 207, 212, 223, 269
South Africa, 2, 8, 14, 22, 35, 37, 74, 136, 137, 212
synthesized in 1925, 1

T

toxic metals, 138–139, 209
toxicity
- effects of, 210, *211*, 211–212, 227, 233
- environmental, 65, 83, 124, 137, 158, **192**
- human health risks, 5, 50
- pH-dependent, 198

trace element, 33, 35, **36**, 57, 135
transition metal, 206, 223

U

United States, 12, 22, 97, 137, 212
uses, industrial, 3, **4** *see also* industrial uses

V

vanadate
- reduction, 147, 242, 248
- sorption, 3, 18, 54, 128, 199–200
- uptake, 194–195

vanadium (V)
- bioaccumulation, 3, 208–210
- bioavailability, 15, 18, 21, 146, 148–149, 168, 207
- chemistry of, 187–190
- compounds, 22, 270–273, *278*
- ecotoxicity, 197–199
- extractability, 15, 73, 77, 80–81, 171
- kinetics, 49, 52, 57–59, 62
- microbial community, 224, *225*, 226–227, *227*, 228–229, *230–231*, 239
- risk assessments, 200
- sources of, 33, 49, 74, 80, 207, 212, 223, 269
- spatiotemporal dynamics (space and time), 231–232, *232*, 233–234, *234–235*, 236, *237–238*
- as trace element, 33, 35, 57, 135
- translocation of, 190, 196, 207–208, *208*, 210
- uptake, 190, **191–193**, 193–196
- water-soluble, 80, **81**, *98*, **102**, 108, 161, 163, *164*, 165, *166–167*, **191**

vanadium pentoxide, 208, 278
vanadium reducing bacteria (VRB), 233, *234*
volatile fatty acids (VFAs), 139, 216

X

X-ray absorption fine structure (EXAFS), 55, 107–108, *121*, 122
X-ray absorption spectroscopy (XAS), 107–108, 114, 118–121, 129

For Product Safety Concerns and Information please contact our EU representative GPSR@taylorandfrancis.com
Taylor & Francis Verlag GmbH, Kaufingerstraße 24, 80331 München, Germany

www.ingramcontent.com/pod-product-compliance
Lightning Source LLC
Chambersburg PA
CBHW060340220326
41598CB00023B/2769

* 9 7 8 1 0 3 2 0 0 2 3 8 5 *